Evolutionary Biology

VOLUME 15

Evolutionary Biology

VOLUME 15

Edited by

MAX K. HECHT

Queens College of the
City University of New York
Flushing, New York

BRUCE WALLACE

Virginia Polytechnic Institute
and State University
Blacksburg, Virginia

and

GHILLEAN T. PRANCE

New York Botanical Garden
Bronx, New York

PLENUM PRESS • NEW YORK AND LONDON

Library of Congress cataloged the first volume of this title as follows:

Evolutionary biology. v. 1-   1967–
   New York, Appleton-Century-Crofts.

          v. illus. 24 cm. annual.

   Editors: 1967–   T. Dobzhansky and others.

   1. Evolution — Period. 2. Biology — Period.   I. Dobzhansky, Theodosius
Grigorievich, 1900–
QH366.A1E9   575′.005   67-11961

Library of Congress Catalog Card Number 67-11961
ISBN 0-306-41042-7

©1982 Plenum Press, New York
A Division of Plenum Publishing Corporation
233 Spring Street, New York, N.Y. 10013

Printed in the United States of America

Contributors

CHARLES F. AQUADRO • *Department of Molecular and Population Genetics, University of Georgia, Athens, Georgia 30602*

JOHN C. AVISE • *Department of Molecular and Population Genetics, University of Georgia, Athens, Georgia 30602*

ALWYN H. GENTRY • *Missouri Botanical Garden, St. Louis, Missouri 63166*

BARRY G. HALL • *Microbiology Section, University of Connecticut, Storrs, Connecticut 06268*

KARIN HOFF • *Department of Anatomy, Dalhousie University, Halifax, Nova Scotia, Canada B3H 4H7*

ANTONI HOFFMAN • *Geological Institute, University of Tübingen, Tübingen, West Germany; permanent address: Wiejska 14 m. 8, PL-00-490, Warszawa, Poland*

ROSS J. MACINTYRE • *Section of Genetics and Development, Cornell University, Ithaca, New York 14853*

M. F. MICKEVICH • *Department of Ichthyology, American Museum of Natural History, New York, New York 10024*

JOAN M. MIYAZAKI • *Department of Biology, Queens College, City University of New York, Flushing, New York 11367*

WOLF-ERNST REIF • *Department of Geology and Paleontology, University of Tübingen, Tübingen, West Germany*

W. VAN DELDEN • *Department of Genetics, University of Groningen, 9751 NN Haren, The Netherlands*

RICHARD J. WASSERSUG • *Department of Anatomy, Dalhousie University, Halifax, Nova Scotia, Canada B3H 4H7*

Preface

Fifteen volumes and one supplement have now appeared in the series known as *Evolutionary Biology*. The editors continue to seek critical reviews, original papers, and commentaries on controversial topics. It is our aim to publish papers primarily of greater length and depth than those normally published by society journals and quarterlies. The editors make every attempt to solicit manuscripts on an international scale and to see that no facet of evolutionary biology—classical or modern—is slighted.

Manuscripts should be sent to any one of the following: Max K. Hecht, Department of Biology, Queens College of the City University of New York, Flushing, New York 11367; Bruce Wallace, Department of Biology, Virginia Polytechnic Institute and State University, Blacksburg, Virginia 24061; Ghillean T. Prance, New York Botanical Garden, Bronx, New York 10458.

The Editors

Contents

5. Developmental Changes in the Orientation of the Anuran Jaw Suspension: A Preliminary Exploration into the Evolution of Anuran Metamorphosis **223**

Richard J. Wassersug and Karin Hoff

6. Regulatory Genes and Adaptation: Past, Present, and Future ... **247**

Ross J. MacIntyre

7. **Evolution of Dermal Skeleton and Dentition in Vertebrates: The Odontode Regulation Theory** **287**

   Wolf-Ernst Reif

1

Patterns of Neotropical Plant Species Diversity

ALWYN H. GENTRY

Missouri Botanical Garden
St. Louis, Missouri 63166

INTRODUCTION

Diversity has been given a central role in attempts to develop a general theory of ecology (Johnson and Raven, 1970). Much attention has been focused on large-scale trends, such as the increase in species diversity with decreasing latitude (Pianka, 1966; MacArthur, 1965; Fischer, 1960; Tramer, 1974), but there is surprisingly little documentation of the equally striking changes in diversity within the tropics. Moreover, additional information on tropical plant species diversity patterns is critically needed because of their potential importance in resolving some of the fundamental and theoretically significant differences of opinion as to the nature and manner of regulation of species diversity.

There is some controversy as to how diversity should be measured. Some authors have advocated that the best measure of diversity is simply the number of species, or species richness, in communities or samples of communities, while others have argued that an indication of the evenness of the importance values of different species, or equitability, should be included (Whittaker, 1977). Whittaker (1977) argues that species richness is the most biologically appropriate measure of diversity, and this chapter uses that measure.

One limitation of species richness as a measure of diversity is its dependence on a standard sample area. The way in which diversity S increases curvilinearly with sample area is also of fundamental importance. The relationship between species number and sample area is known

1

as the species–area curve. This somewhat controversial relationship (Connor and McCoy, 1979) is now generally accepted as becoming linear when log–log transformed, following a power function equation formulated by Arrhenius (1921) and popularized by Preston (1948, 1962) as

$$S = kA^z \quad (\text{or} \quad \log S = \log k + z \log A)$$

where A is area, z is the slope of the species–area curve, and k, a measure of the biotic richness of a habitat, is the y intercept of the species–area regression. According to Preston, z is constant when S varies with A through any homogeneous set of areas.

Most small-scale studies of the species–area relationship in plants have been based on islands or samples of temperate zone vegetation (Williams, 1943, 1964; Connor and McCoy, 1979). A number of species–area relationships have been described for trees greater than 10 cm dbh (or 20 or even 30 cm dbh) of individual tropical areas, but only Knight (1975) has included plants less than 10 cm dbh in such studies for a noninsular tropical forest. A few investigations have compared species–area patterns of several edaphically different tropical communities. Ashton (1964, 1977) compared dipterocarp forest species richness with that of "heath" forest on white sand as well as differences between ridge and slope forests. Davis and Richards's (1933, 1934) classical studies of Guyana forest structure and diversity compared different edaphically determined forest types in the same area. On a more limited scale Pires and his co-workers (Pires *et al.*, 1953; Black *et al.*, 1950) compared forests on upland and flooded soils near Belém, and Maas (1971) has constructed species–area curves for a series of forests on a sandy to loamy soil gradient in Suriname. Two general conclusions have emerged: tropical plant communities have more species (and a larger k) than their temperate zone counterparts, and tropical plant communities on poorer soils, like their temperate zone counterparts (Monk, 1967), usually have fewer species (and smaller k) than do core habitat plant communities on good soils (Ashton, 1964, 1967, 1977; Davis and Richards, 1933, 1934; Richards, 1952; Whitmore, 1975).

The positive correlation between plant species diversity and soil fertility has recently been challenged by Huston (1979, 1980), who proposed that rich soils correlate with higher growth rates, which give rise to diversity-lowering competitive displacement by the strongest competitors. He found a significant negative correlation between tree species diversity and certain soil nutrients, using data from Holdridge *et al.* (1971) on Costa Rican forests. However, the significance of these results seems open to question, since the Costa Rican sites are all on relatively rich soils and intrinsically rich-soil, low-diversity montane sites were included in the analysis. Ashton (1977) suggested that in southeast Asia this relationship is more complex; soils of intermediate fertility generally have the richest plant communities. Apparently, if a general increase of diversity with soil richness does exist, it is manifested only over a large range of soil richness.

In another approach to explaining tropical diversity, Connell (1978) has suggested that degree of habitat disturbance may be the critical factor in determining tropical forest species diversity. High numbers of species should occur in forests where disturbances are intermediate in frequency and intensity; relatively low species diversities are found not only in recently disturbed sites, but also in undisturbed climax communities where those species with superior competitive ability would exclude the rest were not a strong element of chance replacement maintained by unpredictable habitat disturbances. Recently Anderson and Benson (1981) have cited Connell's "intermediate disturbance" hypothesis in trying to explain the strikingly low plant species richness of an Amazonian caatinga forest on nutrient-poor white sand near Manaus compared with the high diversity of a nearby forest on lateritic soil. They suggest that the observed difference in diversity is due more to differential rates of regeneration after prior disturbance than to intrinsic diversity differences controlled by edaphic conditions.

An important facet of tropical diversity patterns which has received little attention is the strong positive correlation between precipitation and community level plant species richness. This trend is well known (Hall and Swaine, 1976; Holdridge *et al.*, 1971; Webb *et al.*, 1967) but its documentation has been largely incidental (Gentry, 1978a, 1978b, 1980b), except for the comparisons by Holdridge *et al.* (1971) of diversities of trees over 10 cm dbh on different Costa Rican sites. It is not known whether the positive correlation of increased species richness of individual plant communities with precipitation holds for very wet tropical regions or only for tropical rain forest (*sensu lato*) as compared to tropical dry forest. The Holdridge data that address this point (for large trees only) are equivocal (see below) and no pluvial forest regions were sampled. The differential in species richness between wet and dry tropical forest has largely been neglected; this is critical, given that generalities about neotropical forest ecology are commonly based on studies of dry forest plants and communities (Bawa, 1974; Bawa and Opler, 1975; Heithaus, 1974; Heithaus *et al.*, 1975; Hubbell, 1979). Conceivably these might be no more relevant to the structure of rain forest communities than would be generalities derived from extrapolations from temperate zone communities.

Like the correlation between plant species richness and soil fertility, the correlation between precipitation and plant community diversity is beset by complicating factors. The biggest problem is seasonality, the differential distribution of precipitation through the year. Thus one of the chief criticisms of the Holdridge system is its failure to take this seasonality into account in its definition of life zones (Myers, 1969). Intuitively a site with a relatively low annual rainfall evenly distributed through the year might be expected to be more mesic than a monsoonal one with greater total rainfall concentrated in part of the year. For example, Ashton (personal communication) notes that the tree floras of the wettest places in Asia—Mt. Cherrapunji, Assam, and Baguio, Philippines, both of which have several dry months despite annual precipitations approaching 12,000 mm—are only about one-tenth as rich as that of the Andulau Hills, Brunei, which has 1500 mm of evenly distributed rainfall. However, this problem is much less acute in the neotropics, where reduced dry season and high total annual precipitation are tightly linked.

At a different level, it is widely recognized that the tropics, as a whole, have many more species than temperate regions (Johnson and Raven, 1970; MacArthur, 1972; Fischer, 1960; Raven, 1976). It is also becoming apparent that the neotropics, as a whole, may have many more plant species than both paleotropical regions combined (Raven, 1976), although individual southeastern Asian plant communities may have more species of large trees (10 cm diameter) than do their neotropical equivalents (Ashton, 1964; Whitmore, 1975; but see below). Nonequilibrium advocates (Whittaker, 1972, 1977; Connell, 1978) would suggest that the greater species richness of the tropics is due largely to open-ended accumulation of more species in individual tropical plant communities, perhaps as a result of greater evolutionary time (Fischer, 1960; Baker, 1970: 104; Ashton, 1969) or less extinction of archaic types in the equable rain forest environment (Baker, 1970, p. 109). A logical extension of these arguments is that individual neotropical plant communities should have more plant species than their paleotropical equivalents, in order to account for the greater species richness of the neotropical region as a whole. Available data (for trees greater than 10 cm dbh only) does not support this a-diversity explanation of neotropical versus paleotropical diversity, suggesting instead that similar plant communities in both regions have similar diversities (Anderson and Benson, 1981). However, all available data do show notably higher a-diversity in tropical than in temperate plant communities.

The intent of this chapter is to document the relationship between plant species diversities and precipitation for a series of eleven neotropical plant communities, including lianas and all trees and large shrubs over

2.5 cm dbh. Some implications of these data for species–area analyses and for the community equilibrium/nonequilibrium debate will also be discussed.

SITES AND METHODS

During the past 8 years, I have accumulated vegetation samples of various neotropical plant communities. Selection of areas to be sampled has depended on a combination of logistics and an adequate familiarity with a given flora to identify its species in sterile condition. The sites include plant communities from five countries and include five with dry forest vegetation, three with moist vegetation, and two with (marginally) wet forest vegetation (*sensu* Holdridge, 1967; Holdridge *et al.*, 1971; Table I). Except for two gallery forest samples that are excluded from some calculations, all sites were selected to reflect core-habitat (cf. "climatic climax") forest vegetation under a given climatic regime. Some of the areas had been previously subjected to an undetermined amount of disturbance by humans or livestock, although all stands were judged mature or in an advanced state of regeneration.

At the species level there is little floristic overlap between the different geographically separated areas, although many genera and nearly all families are shared between different climatically similar areas. Three samples, all of dry forest vegetations, are of less than 1000 m², as indicated in Table II. The species–area regression lines established by the available samples in these three cases have correlation coefficients exceeding 0.95, only marginally lower than the 0.97–0.99 values for the 1000-m² samples. They are thus probably adequate representatives of their respective vegetations. An incomplete and presumably inadequate eleventh sample of 300 m² from the tropical pluvial forest of the Colombian Chocó, near Tutunendó, the rainiest place in the world (Gentry, 1980b; Sota, 1972), is also included because of its extreme interest, despite the quantitative limitations on its interpretation.* Three 1000-m² samples of the rich temperate zone vegetation of the Missouri Ozarks are included for comparative purposes. Some information on stand structure from five additional 1000-m² samples from Amazonian Peru and two from coastal Ecuador is also included.

---

* Sampling of an additional 700 m² at this site was completed during 1981 while this chapter was in press.

TABLE I. Numbers of Species and Individuals per Sample Site Arranged According to Precipitation

Site	Number of species/1000 m²				Number of individuals/1000 m²					
	Total	Vines	All trees	Large trees	Total	Vines	All trees	Large trees	Palmettos	Precipitation
Temperate zone										
Babler State Park, Missouri	21	2	19	8	149	8	141	45	—	—
Tyson Reserve (woods), Missouri	23	3	20	14	183	9	174	54	—	—
Tyson (chert glade), Missouri	25	2	23	9	167	2	165	71	—	—
Temperate zone average	23	2	21	10	166	6	160	57	—	—
Dry tropical forest										
Boca de Uchire, Venezuela	69	16	53	22	259	75	184	31	—	1200
Estación Biológico de los Llanos (500 m²), Venezuela	59[b]	10[b]	49[b]	24[d]	326[b]	56[b]	270[b]	44[b]	—	1312
Blohm Ranch, Venezuela	68	17	51	27	306	71	235	86	—	1400
Guanacaste Upland (700 m²), Costa Rica	53[b]	6[b]	47[b]	18[b]	437[b]	81[b]	356[b]	34[b]	—	1533
Guanacaste Gallery (800 m²), Costa Rica	68[b]	8[b]	60[b]	27[b]	195[b]	24[b]	171[b]	33[b]	—	1533
Dry forest average[c]	63	12	52	24	332 (305)	71 (61)	260 (242)	49 (46)	—	—

Moist tropical forest									
Curundú, Panamá	90	24	64	30	286	59	225	52	2	1830
INPA, Manaus, Brazil	110	20	90	21	331	30	301	34	—	1995
Madden Forest Panamá	126	31	93	34	324	76	242	38	6	2433
Moist forest average	109	25	82	28	314	55	256	41	3	—
Wet tropical forest									
Río Palenque, Ecuador	119	27	89	25	305	63	221	42	21	2650
(Río Palenque, repeat)	(119)	(22)	(95)	(32)	(324)	(45)	(269)	(52)	(10)	2650
Pipeline Road, Panamá	167	38	129	35	393	67	325	60	1	3000
Wet forest average	143	32	109	30	349	65	273	51	11	—
Pluvial tropical forest									
Tutunendó Colombia	~258	49	~208	55	523	72	367	82	3	9000

[a] Precipitation (mm) estimated from nearest weather station or precipitation map. See text.
[b] Estimated from species/area regressions or average number of individuals per transect.
[c] Average number of individuals without (including gallery forest site.
[d] Calculated from 1.5 species added per transect after 200 m^2 (29 species from regression).

TABLE II. Lowland Neotropical Diversity Summary

Source	Locality[a]	Plot size, ha	Species/ha	No. species	dbh lower limit, cm	No. individuals	Individuals per ha	Precipitation, mm
Prance and Schaller (1982)	Acurizal (Mato Grosso, BR)	1	36	36	15	274	274	1110
Prance and Schaller (1982)	Acurizal (cerrado) (BR)	1	22	22	15	189	189	1110
Gibbs and Leitão (1978)	Mogi-Guaçu (Sao Paulo, BR) (600 m)	0.72	—	47	10	343	476	1280
Holdridge et al. (1971)	Taboga gallery (CR)	0.4	—	18	10	—	167	1525
Holdridge et al. (1971)	Taboga upland (CR)	0.4	—	19	10	—	265	1525
Hubbell (1979)	COMELCO (CR)	13	—	135	2	—	—	1600?
Hubbell (1979)	COMELCO (CR)	13	—	87	"Trees"	—	—	1600?
Hartshorn (1981)	COMELCO (CR)	4	—	68	10	—	219	1600?
Hartshorn (1981)	COMELCO (CR)	4	—	44	10	—	207	1600?
Holdridge et al. (1971)	Bagaces uplands (CR)	0.3	—	23	10	—	583	1600
Holdridge et al. (1971)	Bagaces gallery (CR)	0.3	—	32	10	—	350	1550
Holdridge et al. (1971)	Alajuela (CR)	0.5	—	44	10	—	490	1850
Holdridge et al. (1971)	Los Inocentes (Rio Ayatal)	0.3	—	39	10	—	566	1850
Klinge and Rodrigues (1968)	Manaus (BR)	27	—	470	25	—	102	1995
Klinge and Rodrigues (1968)	Manaus (BR)	1 (av)	65	65	25	—	102	1995
Prance et al. (1976)	Manaus (BR)	1	179	179	15	350	350	1995
Prance et al. (1976)	Manaus (BR)	1[b]	—	235	5	—	—	1995
Holdridge et al. (1971)	Los Inocentes (Río Cafetal)	0.3	—	42	10	—	546	2000
Tschirley et al. (1970)	Maricao (PR) (700 m)	0.9	105	105	10	—	2112	2030
Tschirley et al. (1970)	Maricao (PR) (700 m)	0.9	~61	60	2.5	—	—	2030
Holdridge et al. (1971)	Barranca (CR)	0.6	—	49	10	—	506	2150
Maas (1971)	Winana Creek (Sur)	0.68	—	~70	10	347	510	2200
Maas (1971)	Winana Creek II (Sur)	0.9	—	~100	10	543	603	2200
Maas (1971)	Snake Creek I (Sur)	1.25	125	~135	10	—	474	2200
Maas (1971)	Snake Creek II (Sur)	0.19	—	38	10	—	633	2200
Maas (1971)	Paris Jacob Creek I (Sur)	0.85	—	~100	10	—	504	2200

Reference	Locality							
Maas (1971)	Paris Jacob Creek II (Sur)	0.53	—	88	10	—	489	2200
Maas (1971)	Kamisa Falls (Sur)	0.91	—	~110	10	—	608	2200
Maas (1971)	Blanche Mari Falls (Sur)	1.21	108	~110	10	—	540	2200
Holdridge et al. (1971)	Potrero Grande Valley (CR)	0.6	—	38	10	—	353	2250
Holdridge et al. (1971)	Potrero Grande Ridge (CR)	0.1	—	21	10	—	500	2250
Schulz (1960)	Mapane, plot 1 (Sur)	3 (5.6)	120	168	(5)	—	(990)[c]	2300
Schulz (1960)	Mapane, plot 1 (Sur)	3 (5.6)	—	87	25 (2)	—	120[c] / (2000)[c]	2300
Schulz (1960)	Mapane, Voucapoa (Sur)[d]	3	112	199	5	—	—	2300
Schulz (1960)	Coesewijne (Sur)	1	116	116	10 (5)	—	—	2300
Schulz (1960)	Coesewijne (Sur)	1	—	43	25 (2)	—	(880)[c] / 140[c] / (4100)[c]	2300
Holdridge et al. (1971)	Los Inocentes (CR)	1.2	~106	108	10	—	539	2400
Holdridge et al. (1971)	Los Inocentes (CR)	0.1	—	37	10	—	590	2400
Davis and Richards (1934)	Moraballi Cr. (Guy)	1.5	—	91	10	—	432	2500+
Holdridge et al. (1971)	Turrialba (CR)	0.4	—	71	10	—	590	2500
Pires et al. (1953)	Castanhal (BR)	3.5	108	179	10	1482	423	2600?
Knight (1975)	Barro Colorado Island (Pan)	1.5	118	130	2.5	—	1889 (av)	2670
Knight (personal commun.)	Barro Colorado Island (Pan)	0.1 (av)	—	22	10	38 (av)	380 (av)	2670
Knight (1975)	Barro Colorado Island (Pan)	—	—	63 (tot)	18	—	199 (av)	2670
Briscoe and Wadsworth (1970)	Luquillo (PR)	2	8	80	4	3142	1571	2690
Briscoe and Wadsworth (1970)	Luquillo (PR)	2	—	59	10	1283	642	2690
Tschirley et al. (1970)	Luquillo (PR)	1.4	—	85	2.5	3265	2332	2690
Tschirley et al. (1970)	Luquillo (PR)	1.4	—	61	10	—	—	2690
Cain et al. (1956)	IPEAN (BR)	2	144	153	10	897	449	2700?
Black et al. (1950)	Belem (terra firme) (BR)	1	87	87	10	423	423	2700+
Vega (1968)	Carare-Opón valley (Col)	1.9	60	89 (90)	10	589	310	2750
Vega (1968)	Carare-Opón hills (Col)	3.5	50	70 (76)	10	910	260	2750
Vega (1968)	Carare-Opón terrace (Col)	1.5	40	50 (70)	10	309	206	2750
Vega (1968)	Carare-Opón rolling hills (Col)	5	65	98[e]	10	1479	296	2750

Continued

TABLE II. (Continued)

Source	Locality[a]	Plot size, ha	Species/ha	No. species	dbh lower limit, cm	No. individuals	Individuals per ha	Precipitation, mm
Holdridge et al. (1971)	Volcan Orosi (CR)	0.3	—	54	10	131	436	2900
Holdridge et al. (1971)	Helechales (CR)	0.3	—	36	10	172	573	2950
Holdridge et al. (1971)	Helechales (CR)	0.3	—	37	10	190	633	3100
Holdridge et al. (1971)	Helechales (CR)	0.3	—	39	10	175	583	3300
Holdridge et al. (1971)	Volcan (San Isidro) (CR)	1.0	67	67	10	503	503	3600
Hartshorn (1982)	La Selva swamp (CR)	2	—	115	10	353	706	3650
Hartshorn (1982)	La Selva upland (CR)	2	—	118	10	423	845	3650
Hartshorn (1982)	La Selva "residual soil" (CR)	4	—	112	10	487	1947	3650
Hartshorn (1982)	La Selva alluvial (CR)	"4"	—	88	10	—	330	3650
Holdridge et al. (1971)	La Selva (Sarapiqui) (CR)	0.8	—	65	10	328	410	3650
Holdridge et al. (1971)	Osa (alluvial) (CR)	0.1	—	24	10	54	540	4200
Holdridge et al. (1971)	Osa (alluvial) (CR)	0.1	—	36	10	44	440	4200
Holdridge et al. (1971)	Osa (hill) (CR)	0.7	—	103	10	365	521	4300
Holdridge et al. (1971)	Osa (Río Riyito) (CR)	0.2	—	61	10	138	690	4350
Holdridge et al. (1971)	Siquirres (CR)	0.4	—	78	10	239	597	4600
Holdridge et al. (1971)	Valle Escondido (CR)	0.8	—	101	10	509	636	5500
Holdridge et al. (1971)	Valle Escondido (CR)	0.2	—	51	10	128	640	5500
Holdridge et al. (1971)	Valle Escondido (CR)	0.3	—	61	10	161	537	5500
Holdridge et al. (1971)	Rio Colorado slope (CR)	0.2	—	48	10	103	513	5700
Holdridge et al. (1971)	Rio Colorado ridge (CR)	0.2	—	39	10	104	520	5700
Holdridge et al. (1971)	Rio Colorado slope (CR)	0.3	—	44	10	131	436	5700

[a] BR, Brazil. CR, Costa Rica. PR, Puerto Rico. Sur, Suriname. Guy, Guyana. Pan, Panamá. Col, Colombia.
[b] Mixed.
[c] Excluding palms.
[d] Determinations not complete.
[e] Graph looks like 4 ha.

Sample Sites

All of the sites sampled are in tropical lowland forest below 300 m altitude and all, except two gallery forest sites, have soils and drainage conditions which seem representative of regional norms.

Dry Forest Sites

Boca de Uchire, Venezuela (10°09' N, 65°25' W). The forest is at 150 m altitude in a range of hills overlooking the Caribbean about 16 km SE from the town of Boca de Uchire, Anzoátegui State. Precipitation at the town of Boca de Uchire, immediately adjacent to the Caribbean on a flat plain, averages only 603.2 mm/year with a very strong dry season from January (11.1 mm average precipitation) to April (17.1 mm average). The sample area has a much more mesic vegetation than the immediate vicinity of the town and most of the surrounding coastal area. The precipitation map of South America (World Meterological Organization, 1975) shows the 1200-mm isohyet passing very close to the sample site and that value has been taken as the best approximation to sample site precipitation. The sampled area is mapped as premontane dry forest according to the Holdridge system (Ewel *et al.*, 1976). It consists predominantly of well-developed, mostly deciduous open forest along the valley of a small stream and on the adjacent slopes. Two of the transects crossed a hilltop with a thorn scrub vegetation dominated by *Acacia* and *Cassia emarginata*.

Estación Biológico de Los Llanos, Venezuela (8°56' N, 67°25' W). This 300-ha field station lies in the center of the Venezuelan Llanos, 12 km SE of the town of Calabozo, at an altitude of about 100 m. The annual precipitation at Calabozo is 1193 mm (1969–1977) and that at the field station is reported by Monasterio (1971) and Monasterio and Sarmiento (1976) as 1312 mm. There is a pronounced dry season and almost 90% of the annual rainfall occurs during the rainy season from May to October. The mean annual temperature is 27.5°C, with a variation in monthly means of no more than 2.3°C. The area is classified as Tropical Dry Forest under the Holdridge system. The vegetation of the area consists of open grass savannah interrupted by gallery forest along streams and dotted with small patches of deciduous to semideciduous forest ranging from a few trees to a couple of hectares in size. The area has been extensively studied (Monasterio and Sarmiento, 1976; Monasterio, 1971) and a florula of its plant species has been compiled (Aristeguieta, 1966). My sample transects were located within the largest of the forest patches on the field station. The commonest species in the wooded areas is *Copaifera officinalis* (12 plants/500 m²), followed by *Lonchocarpus ernestii* and *Cochlospermum vitifolium* (11 plants/500 m²). The commonest liana species are *Arrabidaea oxycarpa* (9 plants/500 m²) and *Pleonotoma clematis* (6 plants/500 m²). Only 500 m² was sampled, in part because that used up most of the available forested areas with the desired diameter of 50 m or more.

Blohm Ranch, Venezuela (8°34′ N, 67°35′ W). The Blohm Ranch lies approximately 45 km S of Calabozo in southern Guárico State. It is predominately a wetter and more poorly drained savanna than the Estación Biológico. Rainfall 8 km south at Corozo Pondo is variable, but averages 1462 mm/year (1953–1977) (Troth, 1979). The ranch's biota have been studied and described by various scientists working under the auspices of the Smithsonian Institution (Eisenberg, 1979); Troth (1979) describes the Blohm ranch vegetation in some detail. The vegetation of the region is primarily open, seasonally waterlogged grass savanna with scattered trees and small patches of scrubby forest. Extensive tracts of well-developed gallery forest up to several kilometers wide occur along the rivers. Most of the gallery forest area is covered by an almost completely deciduous open forest usually about 15 m tall dominated by *Pterocarpus rohrii* and *Copernicia tectorum*, although there is a narrow, patchy strip of semievergreen forest along the Río Calabozo. My transects were in the more upland gallery forest dominated by large *Pterocarpus* trees.

Guanacaste, Costa Rica. My Guanacaste data are conglomerated from two neighboring sites, the Hacienda La Pacífica between Cañas and Bagaces and the COMELCO Area E site of what has since become the Rafael Lucas Rodríguez National Wildlife Refuge near Bagaces. Both regions have been extensively studied (Heithaus, 1974; Heithaus *et al.*, 1975; Fleming *et al.*, 1972; Frankie *et al.*, 1975) and their vegetation described in some detail. Annual precipitation at La Pacifica is 1533 mm (Frankie *et al.*, 1975), with a strong dry season from December through April. Holdridge *et al.* (1971) estimate 1550–1600 mm of rainfall at COMELCO and classify the area as Tropical Dry Forest. I have lumped my transects from the two areas, which are floristically similar, together. In both areas deciduous forest, now mostly cleared, occurs on upland sites and evergreen or semievergreen gallery forest in narrow strips along the streams. A total area of 700 m² of upland deciduous forest and 800 m² of semievergreen gallery forest was sampled. A total of 600 m² of the gallery forest was located along the Río Corobicí at La Pacífica and 200 m² at COMELCO. The upland forest sample included 500 m² at La Pacífica and 200 m² at COMELCO.

Moist Forest Sites

Curundú Forest, Panamá (8°59′ N, 79°33′ W). The Curundú Forest is on the Pacific side of the former Canal Zone immediately adjacent to

Panamá City but separated from the city by a chain link fence. Much of this forest has been disturbed, although it is probably one of the most mature forest tracts on the whole Pacific coast of central or western Panamá. As interpreted by Blum (1968), the bulk of this forest is about 60-year-old mature second growth, thus comparable in vegetational history to the younger part of the Barro Colorado Island forest (Knight, 1975). It has a high but broken canopy of about 25 m, but the heavy covering of climbing bamboo in many areas is probably indicative of past disturbances. Annual precipitation at Balboa Heights, Canal Zone, is 1830 mm and is very similar to that at Curundú. There is a strong dry season from January through April and a mean annual temperature of 26.9°C. The semideciduous Curundú forest is classified as Premontane Moist Forest, a drier vegetation type than the Tropical Moist Forest which originally covered most of the Canal Zone. Two studies of the ecology of Curundú forest have been made: an unpublished study of its succession and vegetational history (Blum, 1968) and a study of its small mammals (Fleming, 1970, 1972). My transects were located in the area of the Curundú forest near the Army Tropical Survival School.

INPA Forest, Brazil. My site in Brazilian Amazonia, also in formerly highly disturbed moist forest, is described in greater detail by Prance (1975) and Gentry (1978c). It is located on lateritic soil on the campus of the Instituto Nacional de Pesquisas da Amazônia (INPA) near Manaus. The forest is of the regenerative stage locally called capoeira and has been allowed to regenerate undisturbed for at least 40 years (Prance, 1975; Rodrigues, personal communication). This forest is floristically very similar to remaining tracts of undisturbed forest in the Manaus region, unlike most capoeira in the vicinity of Manaus (Prance, 1975; Gentry, 1978c), probably because it was never burned and only partially cut for charcoal. Many of the original climax species have regenerated from old stump sprouts, restoring a reasonable floristic facsimile of the site's original vegetation. The 1995 mm of rain in the Manaus vicinity suggest Holdridge system classification as Premontane (rather than Tropical) Moist Forest. There is a pronounced dry season with 3 months below 60 mm of rainfall and 4 with 100 mm or less. The altitude of Manaus is 100 m and the mean annual temperature is 27°C.

Madden Forest, Panamá (9°06′ N, 79°36′ W). The Madden Forest has been a Forest Reserve since the 1920s and most of it is physiognomically one of the most mature forests in central Panamá, with a closed canopy about 30 m high. Some areas have been subjected to more recent disturbance, and storage bunkers and remnants of buildings dating from World War II are scattered in part of the area. The physiognomically mature part of the forest is traversed by the old Las Cruces trail, the main transisthmian route prior to construction of the Panamá Canal, which suggests a long history of at least some human disturbance. The vegetation is very similar to that of Barro Colorado Island described by Knight (1975) and Croat (1979). Rainfall at Madden Dam, about 9 km N of Madden Forest, averages 2433 mm/year, compared with 2750 mm on Barro Colorado Island. The 4-month dry season lasts from January through April. Both Madden Forest and Barro Colorado Island have predominantly lateritic soils; both also have small areas of limestone-derived soils characterized by a distinctive floristic element (Gentry, 1979), but this unusual flora is better represented in Madden Forest, which is nearer the chief central Panamanian limestone outcrop surrounding Madden Lake. The floras of Madden Forest and Barro Colorado Island are otherwise virtually identical. Only seven species in my sample area do not also occur on Barro Colorado Island and at least five of these are commoner in the limestone outcrop area around Madden Lake. Madden Forest straddles the continental divide, which here is about 300 m above sea level.

Wet Forest Sites

Río Palenque, Ecuador (0°34' S, 79°20' W). This 1.7-km² field station (elevation 150–220 m) is located in Los Rios Province halfway between Santo Domingo de los Colorados and Quevedo in the narrow strip of wet forest along the base of the Andes in coastal Ecuador (Dodson and Gentry, 1978). Only 0.8 km² of the field station is in natural forest, apparently the last remnant of the rich forests which once covered this region (Gentry, 1977, 1979). The sampled area is entirely in the mature forest. The nearest rain gauge, 8 km north at Puerto Ila, reports an annual rainfall of about 2650 mm, but the records are very incomplete. At Puerto Ila the dry season lasts from July to November, with 2 months averaging less than 50 mm of precipitation. Santo Domingo, farther north and at 500 m elevation, has 3000 mm of annual rainfall and a less marked dry season. Río Palenque, nearer the base of the Andes than Puerto Ila, probably gets about 50 mm of rain in even the drier months. The site is 130 km inland from the northern terminus of the Humboldt Current and enjoys relatively cool temperatures, with an annual average of 23°C, despite its low latitude. The canopy is closed and 30–40 m tall. Physiognomically the vegetation is clearly a wet forest, one characterized by the extreme prevalence of epiphytes, and it is mapped as Premontane Wet Forest under the Holdridge system (Vivanco de la Torre *et al.*, 1963). The flora, which includes an unusually strong endemic element, has been thoroughly described (Dodson and Gentry, 1978).

Pipeline Road, Panamá (9°10' N, 79°45' W). The Pipeline Road extends north from Gamboa paralleling the Panamá Canal. About 5 miles north of Gamboa the road climbs a steep hill, reaching an altitude of about 300 m and entering an area strikingly different floristically than most of the rest of the former Canal Zone. Many of the plant species common in this area, like *Welffia georgii, Eschweilera pittieri, Laxoplumeria tessmannii, Pouteria neglecta, Pourouma scobina, Psychotria calophylla, Leandra consimilis, Inga heterophylla, Malouetia isthmica,* and *Tabebuia chrysantha,* occur elsewhere in Panamá only in wet forest vegetations. Of the 167 species encountered on the 1000-m^2 sample, 36 (22%) do not occur on nearby Barro Colorado Island, which has a moist forest vegetation. Although the Pipeline Road sample area is indicated on the Holdridge life zone map of Panamá as Tropical Moist Forest, its flora is clearly a wet forest one. There are no rainfall or temperature records from this area, but its annual precipitation is probably at least 3000 mm, extrapolating from the 3280 mm recorded at Colón to the north and 2750 mm on Barro Colorado Island to the west. To the south at Gamboa, which probably has a slight rain shadow effect, annual precipitation is only 2210 mm. It is possible that the precipitation on the low hills of the Pipeline Road area is significantly greater than 3000 mm, perhaps even approaching 4000 mm to judge from the wet forest floristic elements, but the more conservative 3000 mm estimate has been used here. The average annual temperature at Barro Colorado Island is 27°C with a 2.2°C variation in monthly averages (Croat, 1979); presumably the Pipeline Road wet forest area has a similar or slightly lower temperature. The forest is physiognomically mature, with a closed canopy about 30 m tall.

Pluvial Forest Site

Quibdó-Tutunendó Road, Colombia (5°46' N, 76°35' W). This site is in Holdridge system tropical pluvial (i.e., "rain") forest between Quibdó, with 8558 mm of annual precipitation, and Tutunendó, the rainiest place in the world, with almost 12,000 mm of rain a year (Sota, 1972; Gentry, 1980b). Probably the precipitation at the actual sample area is between 9000 and 10,000 mm a year; there is no dry season. The sample area is along a trail leading to Taudó through apparently undisturbed forest (selective logging was beginning as the area was being studied) at about 90 m altitude. The closed canopy of the study area is 30–40 m tall with prevalent epiphytes, especially hemiepiphytic vines. Only a 300-m^2 sample was available when this chapter was submitted for publication. The full 1000-m^2 sample was completed in 1981, but identifications for the last 700 m^2 are incomplete as this chapter goes to press. Except as otherwise indicated, only the original 300-m^2 sample is included in these analyses.

Identifications

Although the study reported here addresses ecologic questions, it is dependent on the taxonomic data base. Identification of sterile representatives of poorly known tropical floras is as difficult as it is critical to diversity studies. In the absence of comprehensive floras, identification depends on a thorough herbarium representation of an area's flora. As an example of the problems faced in identifying the species encountered in tropical vegetation samples, the Pipeline Road, Panamá, sample is illuminating. Although the former Canal Zone is perhaps the best known part of the neotropics floristically, and this sample was taken from an area only 8 km from especially intensively studied Barro Colorado Island, 11 species new to Panamá, at least four of them apparently undescribed, were encountered in the 1000-m^2 Pipeline Road sample. Similarly, the sample from Río Palenque, Ecuador, included nine undescribed species. The lack of any relevant flora or herbarium representation of fertile collections from the area virtually mandated preparation of a local florula (Dodson and Gentry, 1978) as a prerequisite to analysis of the transect data. Voucher specimens for these vegetation samples (excluding many of the species positively identified in the field) are deposited in the Missouri Botanical Garden Herbarium.

It is noteworthy that all previous studies of neotropical rain forest composition except that of Prance *et al.* (1976) relied heavily or exclusively on "materos" or skilled local woodsmen to supply common-name identifications. In my experience this convenient technique has severe limitations, since the taxonomic discrimination of materos with whom I have worked has invariably been different from a botanist's. For example, a single common name is often applied to many species and even genera of Myristicaceae or Lauraceae. Native materos are incompetent at differentiating lianas, which may be one reason for the general neglect of lianas in earlier studies. Leaves are usually needed for accurate identification by botanists, whereas materos can supply their version of identifications based entirely on characters of the trunk and slash. Thus an important obstacle to verification of the identifications of the sampled species is the need to climb many of the large trees and lianas to obtain specimens of leaves. In Peru a professional tree climber has done most of the climbing. In other sites the author did most of the necessary climbing himself, sometimes aided by telephone lineman's spikes. Use of an expanded set of tree-trimming poles made possible collection without climbing of leafy branches up to 15 m from the ground. During a single ascent, the poles also were used frequently to obtain material from several adjacent trees or to collect branches of a large tree from a perch in the top of a smaller one.

That many, perhaps most, of the diversity studies of neotropical trees published to date are seriously compromised by the lack of independent and critical taxonomic identifications has been noted previously (Richards, 1963; Schultz, 1960, p. 171), but often overlooked. Indeed, I strongly suspect that much of the data responsible for the contention by Whitmore (1975), Ashton (1964, 1977; Gan *et al.*, 1977), and others that paleotropical forests have more tree species than do neotropical ones is due to lack of adequate taxonomic resolution in the available neotropical data. Thus, the importance of the taxonomic expertise and supporting herbarium analyses on which this study is based is worth stressing.

Sampling Techniques

Each 1000-m^2 sample represents the sum of ten 50 m × 2 m subplots. The vegetation of each subplot was sampled within 1 m on either side of a 50-m-long transect line oriented in a predetermined direction (usually perpendicular to or parallel to a trail or stream) from a randomly selected starting point. All plants of 2.54 cm dbh and rooted within the transect area are included in the sample. If a plant overlapped the plot boundary, the location of the midpoint of its diameter measured perpendicular to the transect line determined whether or not it was included. The samples thus include lianas, trees, larger shrubs, many hemiepiphytes, and even a few coarse herbs and low-growing epiphytic vines. Some large stemless palms were included on account of petiole diameters reaching 2.54 cm. Each plant was identified or recorded as a unique "morphospecies" and its diameter recorded in the field. Diameters of erect plants were measured at approximate breast height (1.37 m), except that buttressed trees were measured above the buttresses. If an erect plant branched below breast height, the diameter of each stem 2.54 cm dbh or over was recorded separately. For lianas rooted within the plot, greatest stem diameter was measured, even though this diameter frequently falls below breast height. Any hemiepiphytes with descending roots less than 1 cm dbh (cf. many Araceae) would be excluded, whereas an ascending terrestrial liana of equivalent dimensions would be included in the sample.

Determination of what constitutes an "individual" is frequently problematic, especially for lianas. For density tabulations, above-ground connections are taken to define an individual plant, even though this sometimes results in treating as separate individuals several lianas which were obviously derived as vegetative shoots from a no longer evident common stem. In the case of colonial palms a cluster of stems from a common base is taken as a single individual.

Sample Size

Although the sample area used is relatively small in absolute dimensions, it is unique except for a few insular studies (Tschirley *et al.*, 1970; Tanner, 1977) and Knight's (1975) Barro Colorado Island study in including individual trees as small as 2.5 cm dbh. Schultz (1960, p. 171) included treelets down to 2 cm dbh, but specific identifications of individuals below 10 cm dbh was tentative and incomplete. Since this study places as much emphasis on small trees and lianas as on large trees, logistics dictated the small sample area.

Even for large trees, a sample area of 1000 m² is the basic unit employed in Holdridge's studies (although three such plots were recommended when possible), although it is only half that suggested by Lang *et al.* (1971) and Knight (1975) as appropriate for analysis of complex tropical forest plant communities. It is much smaller than the hectare or several hectare samples measured by most authors (Black *et al.*, 1950; Pires *et al.*, 1953; Prance *et al.*, 1976). Thus it is appropriate to ask whether a 1000-m² sample adequately reflects the composition of the plant community in which it was taken. We may approach this question from several directions.

Circumstantial evidence that 1000 m² is an adequate sample size to indicate community plant species diversity comes from comparing the numbers of individuals and species encountered with the results of other investigators. My samples actually include more species and individuals than most of the other samples which have been previously reported for the neotropics. The number of individuals per plot (260–437; to 514 in Amazonian Peru) is in exactly the same range as the 230–564 individuals included by previous neotropical investigators using much larger 1-ha plots and a 10-cm (or 20-cm) dbh cutoff (Black *et al.*, 1950; Pires *et al.*, 1953; Table II). While these samples would certainly not be appropriate for studies of the population structures of highly dispersed canopy tree species, they do provide a good indication of floristic composition of the forest community as a whole. In this respect it is useful to remember, for example, that the Ecuadorian samples on which Grubb and Whitmore (1966) and Grubb *et al.* (1963) based their important studies of neotropical forest structure and composition were only about half as large as mine.

One direct test of the replicability of these 1000-m² samples was made at Río Palenque, Ecuador, where a second 1000-m² sample of the same vegetation was made subsequent to that reported here. Data from this second sample [reported parenthetically in Table I and to be developed more fully elsewhere (Gentry and Dodson, in preparation)] are extremely similar to those generated by the first sample. The first 1000-m^2 sample included 119 species, 25 of which were vines. The second included 119 species, 22 of which were vines. Moreover, these virtually identical species diversity values were obtained from samples of strikingly dissimilar taxonomic composition. Indeed, only half of the species of either sample are shared with the other one; 58 of the 119 species included in the second sample were not included in the first one. That both these floristically very different samples had the same richness strongly suggests that diversity as measured by such 1000-m^2 samples is largely independent of which taxonomic subset of a community's species happens to be included in the sample.

Another indication of the adequacy of these 1000-m^2 samples comes from the internal consistency of the species–area curves derived from their subplots (see Table III). Moreover, a complete census of plant species is available for the sampled plant community of the Río Palenque Science Center in Ecuador (Dodson and Gentry, 1978). The 245 (counting only mature forest species) to 355 (total with appropriate habit in Flora)

TABLE III. Parameters for Log–Log species–Area Curves

Site	a'	b	a' (X_{10})	S (predicted for 1000 m^2)	r
Temperate zone forest					
Babler	0.798	0.503	1.301	20	0.988
Tyson (glade)	0.862	0.457	1.319	21	0.972
Tyson (oak woods)	0.885	0.497	1.382	24	0.966
Dry forest					
Guanacaste (upland)	1.127	0.660	1.787	61	0.992
Guanacaste (gallery)	1.224	0.612	1.835	68	0.981
Boca de Uchire	1.118	0.691	1.879	76	0.977
Llanos (500 m^2)	1.334	0.436	1.771	59	0.961
Blohm Ranch	1.262	0.598	1.860	72	0.994
Moist forest					
Curundú	1.144	0.780	1.924	84	0.991
Manaus	1.296	0.775	2.070	118	0.995
Madden forest	1.199	0.928	2.127	134	0.994
Wet forest					
Río Palenque	1.285	0.809	2.093	124	0.997
Río Palenque (No. 2)	1.3799	0.7105	2.090	123	—
Pipeline Road	1.505	0.716	2.222	167	0.997
Pluvial forest					
Tutunendó (300 m^2)	1.722	0.770	2.491	309	0.999

species with habits that should be detected by the sample method (trees, larger shrubs, lianas, low-growing thick-stemmed epiphytic vines) that occur in the 0.8-km^2 forested area of Río Palenque is in fair agreement with the 286 species projected for such an area by the regression line calculated from the 1000-m^2 sample and suggests only a slight leveling of the species–area curve for an area much larger than any which has ever been actually sampled in tropical forest vegetation. If Barro Colorado Island (BCI) is taken to represent the same plant community as the Madden forest sample reported here, a similar projection can be made by extrapolation to an area of 15.6 km^2 from the regression line. This gives a ridiculously large overestimate of species diversity for a BCI-sized area, which has 579 potentially includable species by my count from the Flora (Croat, 1979), suggesting that the species–area curve of neotropical forests asymptotes somewhere between 1 and 15 km^2. Although the statistical problems and extremely large confidence limits associated with extrapolation of species–area curves make it unclear how meaningful such suggestions are (Simberloff, 1978), the trend seems clear.

Use of the formula proposed by Evans *et al.* (1955),

$$S = \frac{s}{\log(n + 1)} \log(N + 1)$$

where S is the number of species expected in an area of N units and s is the actual number of species in n units selected at random, gives a more reasonable approximation to the number of plant species to be expected in larger areas based on a given sample. Calculating from the Madden Forest sample, an area the size of Barro Colorado Island should have 633 species of plants, which is reasonably close to the 579 potentially includable species which are known there. A similar calculation for Río Palenque suggests 550 species for an area of 0.8 km^2, a much greater overestimation, about double the number of potentially sampleable species. However, for an area of the order of magnitude of 1–10 ha this formula should give reasonably accurate results. Use of this formula to extrapolate from my 1000-m^2 samples to an area of 1 ha makes possible comparisons of the large tree data subset with data from other 1-ha neotropical sites.

Most previous studies of species–area relationships for plants have either included all habit classes of plants based on complete floristic lists, sometimes supplemented by various importance value calculations [most temperate zone studies, e.g., Johnson and Raven (1970)] or have included only trees [most tropical studies, e.g., Ashton (1964), Schultz (1960), Black *et al.* (1950), Prance *et al.* (1976)]. Nevertheless, lianas are an extremely important component of tropical forests, and their inclusion in ecologic studies is highly desirable. Use of a 2.5-cm-diameter cutoff was specifically aimed at including most lianas. Small tree and large shrub species, which would be eliminated by the forestry-oriented standard tropical sampling techniques, are also included in these samples. One important result of including the smaller-dbh plants is that many more individuals and many more species are contained in a given sample area than might be anticipated, so that a relatively small, and thus logistically more manageable, sample area gives a more complete floristic representation, even for large tree species, which are often represented by juveniles.

While inclusion of herbs and epiphytes would also have been desirable, sampling of epiphytes was not feasible and identification of the many seedlings encountered in samples of the herb layer was impossible. Another drawback to sampling herbs is the marked seasonal changes in representation of seedlings of the different tree species. Complete samples of all plant species in 1000-m^2 samples at two Ecuadorian sites have been completed but will be reported separately (Gentry and Dodson, in preparation).

RESULTS

1. Diversity of core habitat tropical forest plant communities is strongly correlated with precipitation (Fig. 1). Plant species diversities for 1000-m^2 samples (using predicted 1000-m^2 S values for calculated regression lines for three communities with sample size less than 1000 m^2) give a highly significant linear correlation of 0.93 when regressed on precipitation values for the different communities. Species diversity as measured by these samples approximately doubles from dry to moist forest (*sensu* Holdridge, 1967) and triples from dry to wet forest. Dry forest plant community diversity, in turn, is more than double that for several similar samples of temperate zone forest (Table I). Thus the difference in diversity between tropical dry and wet forests is even greater than that between temperate and tropical forests. Moreover, the strong positive correlation between species diversity and precipitation appears to be retained even for very wet perhumid plant communities, since the species–area regression line calculated from the 124 species in the incomplete 300-m^2 sample of perhumid Chocó pluvial forest predicts 309 species in a 1000-m^2 sample of pluvial forest.

Clearly, and not unexpectedly, precipitation is a major diversity-

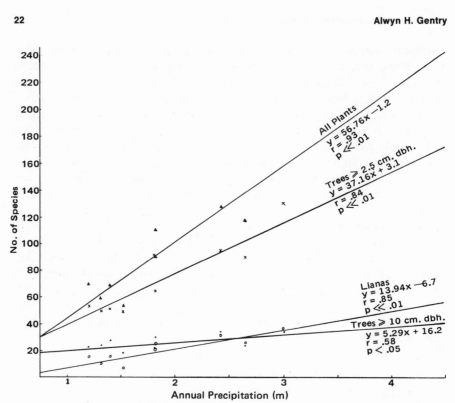

FIG. 1. Change in species richness with precipitation for plants ≥2.5 cm dbh in 1000-m² samples.

determining factor for tropical plant communities and a-diversity is much greater for wetter plant communities. Far less evident, though equally significant to evolutionary theory, the floristic diversity of wet forest communities is as much greater than that of moist forests as the latter are richer than dry forests; the wettest pluvial forests are apparently by far the most species-rich of all.

These sites do not provide a good test of whether seasonality of precipitation or total annual rainfall is the better predictor of plant species richness, since length of dry season is strongly correlated with annual precipitation. All of the dry forest sites have 4–5 months with ≤60 mm of rain and 6 months with less than 100 mm. All of the moist forest sites have 3 months with less than 60 mm and 4 months with less than 100 mm. Both wet forest sites probably have only 2 months with rainfall below 60 mm, though this figure is largely conjecture; one has (probably) 4 below 100 mm, while the other might have as many as 5. The pluvial site has no month with less than 100 mm of rain.

2. Diversity of large trees (≥10 cm dbh) increases barely significantly

with precipitation as measured by 1000-m^2 samples. Although the differences between average number of large tree species in dry (24), moist (28), and wet (30) forest samples or even between dry versus moist-plus-wet forest samples is not significant, the overall trend is ($r = 0.583$, $p < 0.05$; Fig. 1). The weaker correlation between diversity and precipitation for larger trees than for vines or total diversity could either reflect intrinsically less sensitivity of tree diversity to changes in precipitation or be a spurious result of inadequate sample size for this group.

Since previous workers (Ashton, 1964; Davies and Richards, 1933, 1934; Holdridge et al., 1971; Jones, 1955/1956; Paijmans, 1969; Schultz, 1960; Black et al., 1950; Pires et al., 1953) have used a 10-cm-diameter cutoff almost exclusively, the response of this large-tree subset of my samples is critical to any effort to compare these data with the extant data from other areas. Table II records species diversities for the various larger samples of neotropical vegetation that have been previously compiled and which provide a much more representative sample for this habit class than do the 1000-m^2 samples reported here.

The correlation between large-tree diversity and precipitation has already been suggested by Holdridge et al. (1971) using 1000-m^2 sample plots in different Costa Rican forests, and the Complexity Index developed by Holdridge (1967) as a means of objective differentiation of tropical life zones largely depends on this correlation. Averages of species–area curves for wet, moist, and dry forest sites gave curves indicating 83, 53, and 30 species, respectively, for 5000-m^2 samples. The Holdridge analysis calculated these relationships using an asymptotic equation $y = N(1 - 3^{-Kx})$ that approaches a horizontal limit of species diversity; the respective asymptotic limits for wet, moist, and dry forest average curves for trees 10 cm dbh are 98, 55, and 30. It seems evident, however, that the small sample areas used in the Costa Rican analysis are inadequate to suggest such fixed limits on intracommunity species diversity, especially since in both cases where a sample was extended to 1 ha or more the counted number of species in the sample was well in excess of the calculated asymptote. As an example of the inappropriateness of the equation used, similar curves for adjacent 1000-m^2 and 12,000-m^2 samples at Los Inocentes gave predicted asymptotes of 40 species from the smaller sample and 79 species from the larger sample even though the 37 species counted for the smaller area was well *over* the 29.9 average number of species per 1000 m^2 for subsets of the larger sample. Another problem with the analysis of the Costa Rican species–area curves is that they are summarized by the Holdridge Life Zone so that intermediate precipitation values for lowland tropical sites are excluded (as premontane forest) from the tropical dry, moist, and wet forest averages, artificially accentuating the differences between the respective curves. However, a different analysis of the Holdridge data comparing species diversity values for a given sample area with precipitation also gives a significant correlation (Huston, 1980).

We may conclude from the Costa Rican data set that species–area curves for trees ≥10 cm dbh do not approach horizontal asymptotes for tropical areas of less than 1 ha except possibly in the driest forest types. However, since diversity of the larger 1-ha or more samples from this series of sites also shows a marked increase correlated with greater precipitation (when edaphically limited extreme sites are eliminated), the generality of the proposed linear relationship between a-diversity and precipitation in the neotropics is supported.

Although my data and the Costa Rican data set support the hypothesis that species diversity of trees ≥10 cm dbh increases with precipitation, the data summarized in Table II do not. Unfortunately, different workers have used very different analyses, making comparison of the data difficult. The most commonly used samples are of trees ≥10 cm dbh in 1-ha plots. Thirteen such samples of mainland neotropical forests are included in Table II, and these data sets show a weak, but apparently negative correlation ($r = -0.61$, $p < 0.05$) of species richness with precipitation. However, the range of annual precipitation for these sites is only from 2200 to 3600 mm, and addition of data for dry forest and wetter forest sites might give a very different picture. Moreover, four of the wetter sites are those of Vega from the Magdalena Valley of Colombia; these sites seem highly atypical in having tree densities, as well as diversities, only about half as great as for other sites. Elimination of these sites eliminates any correlation between precipitation and tree species richness for the remaining data set. My data extrapolated to anticipated 1-ha diversities using the formula of Evans *et al.* (1955) give no significant correlation between diversity and precipitation ($r = 0.44$, n.s.). It is difficult to determine to what extent these results reflect sampling artifact. My large-tree data are mostly based on fewer than 50 individuals and the 1000-m^2 samples include only a few hundred individual trees. In Southeast Asia, Ashton (1977, personal communication) has suggested that species per thousand individuals is an appropriate measure of a site's species richness and that species/individual (or species–area) curves based on less than 100 individuals are very similar despite the massive differences in species richness indicated by larger samples. Unfortunately, no comparable data are available from the neotropics, although one might expect similar patterns. Thus it remains unclear to what extent, or even whether, species diversity of large trees in plots of 1 ha or more increases with precipitation in the neotropics. Although my data suggest a weak but significant correlation of large-tree diversity with precipitation for plots as small as 1000 m², this may be largely due to sampling artifact.

3. Diversity of trees ≥2.5 cm dbh increases directly with precipitation (Fig. 1). For small sample plots, at least, this increase is much more pronounced than is that for trees of ≥10 cm dbh. Part of this differential is due to more individuals per sample. Part of it is no doubt a real increase in species diversity of understory species. In either case, like similar values for lianas, diversity of all trees ≥2.5 cm dbh is a stronger predictor for complexity indices than is diversity of trees ≥10 cm dbh, at least for relatively small plots.

Only Knight (1975) has previously included all trees ≥2.5 cm dbh in a neotropical forest vegetation analysis. [Studies on Puerto Rico (Tschirley et al., 1970) and Jamaica (Tanner, 1977) have also included lower diameter trees, but peculiarities associated with the island status of these vegetations make them generally unsuitable for comparative purposes.] Knight's data for Barro Colorado (48–83 species per 1000 m²) are very comparable to my more limited data for similar moist forest sites (64–94 species), although my density values for this size class of trees are much higher than his (225 individuals per 1000 m² at Curundú and 242 in Madden Forest versus averages of 140 and 136 individuals per 1000 m² for young and old forest, respectively, on Barro Colorado Island). Whether this reflects actual differences in the vegetations or differences in sample techniques is not clear. Knight's data are significant in showing that species–area curves for trees ≥2.5 cm dbh continue to rise with increasing sample area. There is only a slight plateau in 1.5-ha samples, which may have 130 species of trees ≥2.5 cm dbh.

4. Liana diversity, considered separately, increases directly ($r = 0.852$, $p \ll 0.01$) with precipitation at the same rate as does total diversity (Fig. 1). Thus lianas make up an average of about 20% of the species sampled in dry forest, moist forest, and wet forest vegetations. There are an average of 12 liana species in 1000-m² samples of dry forest, 25 in moist forest, and 32 in wet forest (Table I). This increase in species diversity is in marked contrast to liana density, which remains relatively constant across the tropical communities sampled (see below). Liana species diversity, at least in small sample areas, seems much more responsive to increased precipitation than diversity of trees ≥10 cm dbh. Number of liana species should thus be a more powerful predictor than number of tree species in calculating such measures of tropical forest structure as complexity indices.

Webb et al. (1967), working in Queensland, and Hall and Swaine (1976), working in Ghana, also found liana species to be extremely well correlated with environmental factors. In Webb's studies lianas and large trees provided the most information for site classification; to arrive at the conclusion that large trees contributed a better floristic classification than other synusiae, it was necessary to exclude one common liana species from the analysis! Hall and Swaine concluded that "the indicator value of climbers, herbs, and understorey trees is at least as high as that of canopy trees."

5. The correlation of diversity trends between trees and shrubs or trees and herbs has not been examined thoroughly, even in temperate zone forests (Auclair and Goff, 1971). The expectation that plant communities richest in tree species would also be richest in shrubs and herbs is contradicted by many of the data comparing different temperate zone forests (Whittaker, 1956, 1960, 1977; Whittaker and Niering, 1965; Glenn-Lewin, 1977) and a distinctly nonlinear, even generally inverse, relation between canopy and understory diversity has been found across large gradients in the Siskiyou Mountains (Whittaker, 1960), the Cascades (Zobel et al., 1976), and the Wenatchee Mountains (Moral, 1972). On the other hand, Auclair and Goff find strong positive correlations between shrub and herb diversities and significant but less marked correlation between tree diversities and shrub or herb diversities in the western Great Lakes region.

Given this background, it is not at all obvious that diversity trends for tropical trees, shrubs, and lianas would be correlated, especially in view of the subjectively more conspicuous presence of lianas in many dry tropical forests than in wetter ones. That diversity trends, though not densities, of both lianas and of small trees and shrubs are strongly correlated with those of large trees in the neotropics is thus of more than trivial interest.

6. Other synusiae. The only tropical diversity studies to include all species are those of Webb et al. (1967) in Australia and Hall and Swaine (1976) in Ghana. Using a rather different data set based on presence or absence of "established" species (including epiphytes and lianas, which were "identified when accessible") in 0.1-ha plots, Webb et al. (1967) found a generally similar correlation of species per site with precipitation, with a mean of 139.8 species per mesic site but only 72.1 species per site for a second group of generally less mesic sites (but also including two climatically very wet but also swampy sites on gley soil) in northern Queensland. Webb's analyses indicated that the primary difference between sets of sites reflected a largely climatic moisture difference, with species of all synusiae showing parallel responses to environmental parameters. Hall and Swaine (1976), using presence or absence of all species in small 0.06-ha plots, also found rainfall to be the most important factor affecting forest composition and diversity.

It is well known that epiphyte diversities in the neotropics are greater under wetter conditions (Madison, 1977). I know of no data specifically comparing understory herb diversities of different tropical communities, but my impression is that the diversity of terrestrial forest herbs also increases with precipitation.

Several authors (Poore, 1963) have pointed out that members of different synusiae of tropical forests are exposed to different microenvironments and questioned whether their distributional patterns are correlated. Poore (1963) singled out this and the nature of the species/area relationship as especially critical problems needing investigation in rain forest communities. My data only establish that lianas, trees, and shrubs show similar patterns in the correlation of diversity with rainfall.

Recently data have become available which make possible a tentative examination of these trends for other synusiae as well. Reasonably complete floristic analyses have been completed for three comparable neotropical sites, one in dry forest (Santa Rosa National Park, Costa Rica; Janzen and Liesner, 1980), one in moist forest (Barro Colorado Island, Panamá; Croat, 1979), and one in wet forest (Río Palenque, Ecuador; Dodson and Gentry, 1978).

Direct comparison of the species diversity of the floras of the three areas is complicated by species–area effects resulting from their different sizes (Río Palenque is 1.7 km^2, Barro Colorado Island 15 km^2). Nevertheless, it appears that species diversity of each synusia (perhaps excepting parasites) increases with precipitation. Despite the differences in absolute numbers of species, there is very little difference between these areas in percent of the flora that is made up of tree species, shrub species, lianas, or parasites (Table IV).

However, close to 50% of the species of the dry forest area are herbs and subshrubs (inclusion of grasses and sedges would probably raise the figure in Table IV at least 5%), whereas only about one-third of the floras of the moist and wet forest sites belong to this category. At the other extreme, epiphytes comprise a minuscule 2% of the Santa Rosa species but 13% of the Barro Colorado Island species and 22% of the Río Palenque species. Thus the rate of change in diversity with precipitation, but not its direction, is lower for herbs and subshrubs and much greater for epiphytes than for other synusiae. Obviously, many more complete censuses are needed to quantify these suggested trends, but their general outline seems clear.

7. The species–area curves for these samples become linear for a log-species versus log-area plot but concave for a species versus log-area plot, contrary to the expectations of the logarithmic series model (Williams, 1964) but in accordance with Preston's (1962) power function model.

The slope of the log–log species–area curve represents the rate at

TABLE IV. Composition of Local Floras by Habit[a]

	Santa Rosa (dry forest)		Barro Colorado (moist forest)			Río Palenque (wet forest)		
	No. species	%	No. species		%	No. species		%
Trees ≥10 cm potential dbh	(142)	(20)	291	(291)	22 (26)	154 (153)		15 (17)
Small trees and large shrubs (2.5–10 cm potential dbh)	(64)	(11)	134	(128)	10 (12)	99	(94)	10 (10)
Herbs and subshrubs	(250)	(43)	439	(281)	33 (25)	376 (286)		36 (32)
Epiphytes	(12)	(2)	180	(140)	13 (12)	227 (201)		22 (22)
Parasites	(6)	(1)	8	(8)	1 (1)	6	(6)	1 (1)
Lianas	(52)	(9)	149	(147)	11 (13)	87	(85)	8 (9)
Small vines	(62)	(11)	117	(110)	9 (10)	84	(81)	8 (9)
Total species	(588)		1318 (1105)			1033 (906)		

[a] Values in parentheses exclude ferns, grasses, and sedges.

which species accumulate with increments in area. The values of this slope for the samples reported here (Table III and Fig. 2) are generally above the range of slope values found by most previous workers [see Connor and McCoy (1979) for a summary], which generally range between 0.2 and 0.4 (Connor and McCoy, 1979) or 0.2 and 0.35 (MacArthur and Wilson, 1967) and were supposed by Preston (1962) to reflect uniformly the canonical slope of 0.262. Several researchers have recently suggested that concentration of slope values in these ranges is a mathematical artifact (May, 1975; Schoener, 1976). Thus the higher slope values noted here are especially interesting from a theoretical viewpoint.

The slopes of the log–log species–area curves for moist and wet tropical forest communities are not parallel to those for temperate forest communities ($t = 6.6702$, $p \ll 0.01$, 6 df). The log-species/log-area regression lines for dry forest communities parallel those for temperate zone forests rather than those for moist and wet forests ($t = 3.6071$, $p \ll 0.01$, 8 df versus $t = 1.910$, 6 df, n.s.). That log-species/log-area curves for these different sites are not parallel (Table III and Fig. 2) is contrary to Preston's (1962) supposition but hardly surprising in view of Johnson and Raven's (1970) interpretation of the series of species–area relationships they analyzed (see also Johnson and Simberloff, 1974). One outcome of these nonparallel species–area curves is that the smallest plot sizes (where number of individuals in a sample becomes as much as one-half or one-third the total species in 1000 m^2) of low-diversity dry forest communities tend to have as many species as, or more species than, similar samples

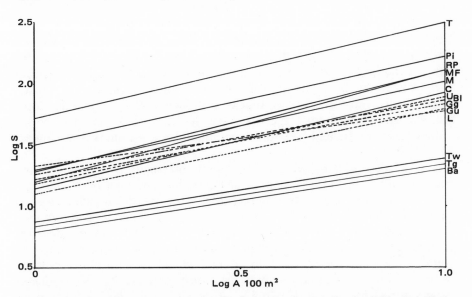

FIG. 2. Log–log species–area curves for 1000-m² samples of vegetation. Ba, Babler State Park; Tw, Tyson oak woods; Tg, Tyson chert glade; L, Estación Biológico de los Llanos; Gu, Guanacaste upland; Gg, Guanacaste gallery forest; Bl, Blohm Ranch; U, Boca de Uchire; C, Curundú; M, Manaus; MF, Madden Forest; RP, Río Palenque; Pi, Pipeline Road; T, Tutunendó. Dotted lines denote dry forest. From bottom to top: temperate zone forest, dry forest, moist forest, wet forest, pluvial forest.

of higher diversity plant communities. For larger plot sizes the steeper slope of the species–area regression of moist and wet forest communities results in samples containing more species, as expected. That Preston's z is not a constant means that k in Preston's formula $S = ka^z$ is not a simple measure of relative diversity but is dependent on the relationship of plot size to density. Another way of expressing this is that forest structure varies relatively little with climate as compared to species diversity.

Structural Trends

Several major trends in forest structure, as well as in species composition, are indicated by the accumulated data. One pronounced structural difference between the tropical forests examined and their temperate zone counterparts is their much greater density of individuals. Hartshorn (1978) found a similar difference between a single tropical–temperate site pair. The tropical forests generally have about twice as many plants 1 in.

or over in stem diameter as do the temperate zone forests. That the tropical forests studied are significantly denser than the temperate zone ones ($p < 0.001$, t test, 9 df) is hardly surprising, although it does not support the idea of an open cathedral-like rain forest. What is surprising is the constancy in density between the different types of tropical forest. Superficially, most dry forests appear much denser than do mature tropical wet and moist forests, yet the stem densities of the sampled core-habitat dry, moist, and wet forest communities are remarkably similar ($\bar{X} = 330 \pm 50$).

The Costa Rican gallery forest sample has many fewer individuals per 1000 m^2 than the core-habitat samples, but this may be in part an artifact caused by concentration of cattle in the evergreen gallery forest during the dry season, with resultant destruction of much of the undergrowth and saplings, which normally contribute greatly to the density figures. This interpretation is supported by a normal density of larger trees in the gallery forest sample (33 individuals ≥ 10 cm dbh versus $\bar{X} = 36 \pm 6$ for the three core-habitat dry forest samples). The Venezuelan gallery forest has the anticipated total density (306 individuals), but is unique in an unusually high density of large trees (86 individuals ≥ 10 cm dbh).

Lianas, considered separately, show a similar lack of obvious changes in density between the different tropical habitat types, although site-to-site variation is proportionally greater. Dry forest core-habitat sites have slightly but not significantly higher densities than moist forest and wet forest ones ($\bar{X} = 74 \pm 15$ versus $\bar{X} = 59 \pm 17$, $t = 1.2295$ with 6 df, n.s.). As was the case for total density, the Costa Rican gallery forest site has a lower liana density than other sites. There is also an apparent negative correlation between liana density and stand maturity. An intuitive (but patently unquantifiable) ranking of the sites according to degree of disturbance would suggest that two of the three sites with lowest liana densities (Manaus and Curundú) are the most disturbed. Lianas also parallel the overall density pattern in a very marked differential between the liana-rich tropical forests and liana-poor temperate zone ones. While the prevalence of lianas in tropical forests has been cited as perhaps the single most important differentiator between tropical and temperate forests (Croat, 1979), the extent of this differential has apparently never been quantified. The data of Table I suggest an average tenfold difference in liana density between temperate and lowland tropical forests, with a threefold difference between the most liana-rich temperate forests and the most liana-poor lowland tropical ones.

Although overall liana density, as measured here to include all climbers 2.5 cm in diameter or larger, does not decline significantly in wet forest samples, the composition of the liana flora does change. Many more of the climbers are epiphytic or hemiepiphytic in wet forest communities. Thus ten of the 27 species sampled as "lianas" at Río Palenque and 13 of the 49 at Tutunendó actually are epiphytic or hemiepiphytic, while all of the lianas in the combined dry forest samples (except one hemiepiphytic *Ficus* at Blohm Ranch) are free-climbing. Relative density of hemiepiphytic versus free-climbing lianas should provide a good index of a site's mesicity.

Density patterns of trees considered separately are also very similar to the overall density pattern, with a significant ($t = 3.01$, $p < 0.02$, 9 df) difference between tropical and temperate sites but no clear trends distinguishing the different tropical sites. Both the site with the most trees over 2.54 cm dbh (Guanacaste upland forest) and that with the fewest (Boca de Uchire) are dry forest ones. Consideration only of trees ≥ 10 cm dbh gives a slightly different result. For trees 10 cm dbh or greater, there is at least a weak correlation between precipitation and tree density. Beard (1946, p. 83) reported a doubling of number of trees per acre from drier to wetter districts in Trinidad. However, the correlation of precipitation with tree density based on my data is not significant ($r = 0.587$, 6 df) unless the Tutunendó pluvial forest is included. Core-habitat dry forest sites average 35 trees ≥ 10 cm dbh, moist forest ones average 41 trees ≥ 10 cm dbh, and wet forest ones 51 trees ≥ 10 cm dbh. Also in contrast to the other density patterns, the temperate zone sites tend to have slightly (but not significantly) greater densities of larger trees than do the tropical sites (excluding the Tutunendo pluvial forest) ($t = 2.0268$, 9 df, $0.05 < p < 0.1$).

Interestingly, the subjectively most open site, the temperate zone Tyson chert glade, had the second highest density of larger trees, while the most mature and physiognomically most cathedral-like tropical forest, the Pipeline Road wet forest, had the third highest density of larger trees. Apparently there is a strong correlation between the subjective impression of "openness" and density of larger trees which seems relatively independent of overall stem density (the Tyson chert glade had the second lowest overall density, while the Pipeline Road wet forest had the second highest).

To summarize, the density of all individuals, all lianas, and all trees differs significantly only between temperate and tropical sites and not from one tropical site to another. Density of larger trees varies slightly with precipitation between different tropical sites but not between the tropics and the temperate zone. Structurally, the greater densities of tropical as compared to temperate forests is accounted for entirely by lianas and small trees.

Community Organization

Given that the a-diversity of neotropical plant communities shows a consistent pattern of greater species diversity with increased precipitation, one can ask what properties, if any, are shared by the additional species of the wetter communities. Taxonomic composition, gross dispersal strategies, and generalized pollination syndromes are three important aspects of community organization for which even a rather superficial analysis of replicating patterns might prove of interest in this connection. Each of these will be examined in turn.

Floristic Composition

The floristic composition of such a diverse series of sites is predictably varied. Altogether 500 species belonging to 309 genera and 83 families (excluding the incompletely identified Manaus sample) are represented in these samples. I have not attempted a formal analysis of intersite floristic similarities at the generic and specific levels, because of the restricted sample size, which fails to include many fairly common species and genera, especially of large trees, at each site. At the species level we can say little more than that no individual species is shared by more than a few sites. *Cochlospermum vitifolium* and *Allophyllus occidentalis* are the most widespread species, occurring in six sample areas. Four other species occur in five sample areas—*Chomelia spinosa, Genipa americana, Spondias mombin,* and *Guazuma ulmifolia.* Four species are represented at four different sites—*Phryganocydia corymbosa, Bursera simaruba, Tetracera volubilis,* and *Lacistema aggregatum*—and many species occur in sets of three different samples. The species shared between samples are invariably shared between ecologically similar sites, occurring in dry forest sites or moist and dry sites or moist and wet sites but never only in dry and wet sites. Moreover, no one species occurs in both the driest and the wettest sites. Each site is basically an independent assemblage of different species.

By contrast, some striking patterns become apparent at the familial and generic level. Some genera are fairly consistent in their occurrence. Only one genus—*Arrabidaea*—occurs in all the samples. Another—*Machaerium*—was represented in all but two of the samples and is present in both areas where it was not sampled. Four genera—*Allophyllus, Cordia, Casearia,* and *Lonchocarpus*—were present in seven samples, and nine more—*Spondias, Tabebuia, Cochlospermum, Acacia, Trichilia, Coccoloba, Erythroxylon, Inga,* and *Paullinia*—were represented at six sites. All of these genera (except *Cochlospermum,* which was represented by a single widespread species) are virtually omnipresent in most lowland neotropical forests and all (except *Cochlospermum*) were probably present even in the sites where they were not sampled.

Some of the prevalent genera owe their frequency of occurrence to one or two widespread species. Besides *Cochlospermum vitifolium, Allophyllus occidentalis* accounted for six of the seven records of the genus, *Spondias mombin* for five of the six records of *Spondias,* and *Chomelia spinosa, Genipa americana,* and *Guazuma ulmifolia* for all five of the records of their genera. Five other genera had a single species sampled at four different sites and four of these—*Lacistema aggregatum, Bursera simarouba, Xylophragma seemannianum,* and *Phryganocydia corymbosa*—provided the only records of their respective genera.

A more common pattern than that of a single widespread representative of a genus is a series of congeneric species which replace each other ecologically. Thus, three yellow-flowered species of *Tabebuia* were included in the total sample. Of these, *Tabebuia billbergii* occurs only at the driest sites, *T. ochracea* at dry to somewhat moist sites, and *T. chrysantha* only at the wettest sites. Another species of yellow-flowered *Tabebuia, T. guayacan* (or its close relative *T. serratifolia,* which replaces it through most of South America) occurs at all the moist forest sites but was not included in any of the samples. Similarly, *Luehea* has a series of three species which replace each other on a moisture gradient—from driest to wettest, *L. candida, L. speciosa,* and *L. seemannii*—although two of the three species overlap in many intermediate areas.

These samples are inadequate to demonstrate a third, and complementary, trend of geographic replacement, although many of the sampled species, and even genera, are exclusively South American and many others exclusively found in Central America and northwestern South America.

We can safely conclude that each site's flora, though specifically distinct, is largely drawn from the same generic pool.

Another point of interest is that the species/genus ratio is remarkably constant for all the sites (Table V). All of the sites except the driest one have between 1.2 and 1.3 species per genus; the Boca de Uchire site, which has several genera represented by three species and one, *Capparis,* with five, is comparable to the temperate zone sites where species/genus ratios range from 1.5 to 1.8. As diversity increases in more mesic sites, additional genera of woody plants are added to the plant community at the same rate as additional species. Although the biologic significance of this ratio has recently been challenged by Simberloff (1970, 1978), who notes that the ratio is strongly correlated with the number of species in

TABLE V. Ratios of Species per Genus, Species per Family, and Individuals per Species for Each Site

Site	Species/genus	Species/family	Individuals/species
Boca de Uchire	1.53	3.29	3.77
Llanos (500 m²)	1.17	2.05	—
Blohm Ranch	1.21	2.27	4.5
Guanacaste Upland (700 m²)	1.23	2.45	—
Guanacaste Gallery (800 m²)	1.11	1.91	—
Curundú	1.23	2.14	3.18
Manaus	—	—	3.01
Madden Forest	1.27	2.82	2.55
Río Palenque	1.27	2.34	2.62
Pipeline Road	1.27	2.88	2.35
Babler State Park	1.50	1.62	7.10
Tyson oak woods	1.53	1.92	7.96
Tyson chert glade	1.79	2.27	6.68

a sample, it is still a widely used statistic, especially among botanists (Carlquist, 1974; Burger, 1981), and thus is of interest at least for comparative purposes. It is especially noteworthy that the sites with fewer species have more species per genus, contrary to the mathematical expectation.

Fundamentally different patterns are shown if the total floras of Santa Rosa National Park, Barro Colorado Island, and Río Palenque are compared (Table VI). For these complete floras (ferns, grasses, and sedges are omitted from the Santa Rosa data, and so are also excluded in calculating statistics for the other sites) the species/genus ratio generally increases with precipitation, from 1.5 at Santa Rosa to 1.9 at BCI and Río Palenque. This is exactly the pattern that would be expected purely on the basis of the larger number of species in the latter floras (Simberloff, 1970). However, the interesting point is that this greater diversity of sympatric congeners in wetter habitats is not evident among trees and lianas. Such intrageneric diversification occurs almost exclusively among epiphytes and a few genera of shrubs.

At Río Palenque 21 genera have seven or more species; 11 of these are entirely or predominantly epiphytic, and three others, including the two largest genera in the flora (*Piper,* with 22 species, and *Solanum* with 18 species) are mostly shrubs and subshrubs. At Barro Colorado Island 12 genera have ten or more species; five of these are almost entirely epiphytic, and four more, including the two largest genera in the flora (*Piper* and *Psychotria,* with 20 species each) are mostly shrubs and subshrubs. In general the same genera tend to have many sympatric species

TABLE VI. Species/Genus Ratios for Complete Floras[a]

	Number of Species	Number of Genera	Species/Genus
Santa Rosa (99 km²)	(588)	(369)	(1.5)
Barro Colorado Island (15.6 km²)	1318 (1105)	674 (581)	2.0 (1.9)
Río Palenque[b] (1.7 km²)	1033 (907)	545 (479)	1.9 (1.9)

[a] Numbers in parentheses exclude grasses, sedges, and ferns.
[b] Includes 56 species discovered after publication of the flora.

in different moist and wet areas (Table VII). The explosion of epiphyte diversity in wetter sites results mostly from an increase in the diversity of congeners and is largely responsible for the very different trend in species/genus ratios when complete floras rather than the essentially woody plants measured in my transect samples are compared.

Returning to the transect results, patterns at the familial level are similar to those at the generic level. The species/family ratio is between 1.9 and 2.9 at all sites except the driest one (Boca de Uchire), where it is almost 3.3. However, there is no obvious correlation between number of species per family and precipitation for these samples (Spearman rank correlation, $r_s = 0.179$, n.s.): the second highest ratio (2.88) is for the

TABLE VII. Genera with Ten or More Species in the Río Palenque (Wet Forest) and Barro Colorado Island (Moist Forest) Floras[a]

Río Palenque		Barro Colorado Island	
Piper	22	Piper	20
Solanum	18	Solanum	11
Ficus[b]	18	Ficus[b]	16
Philodendron[c]	15	Philodendron[c]	13
Anthurium[c]	14	Anthurium[c]	12
Peperomia[c]	14	Peperomia[c]	10
Epidendrum[c]	11	Epidendrum[c]	13
Pleurothallis[c]	11	Psychotria	20
		Inga	18
		Miconia	14
		Cyperus	11
		Panicum	10

[a] Note that all but one of the largest Río Palenque genera are also among the largest on Barro Colorado Island. Dry forest Santa Rosa National park has only two genera with ten or more species.
[b] Hemiepiphytic.
[c] Predominantly epiphytic.

wettest Pipeline Road sample. Though not statistically correlated with number of species (Spearman rank correlation, $r_s = 0.479$, n.s.), this difference may be largely due to sampling artifact. Nevertheless, the additional diversity of mesic sites generally seems to result more from the presence of additional families than from additional species of the families represented in drier areas.

One striking pattern at the familial level is the prevalence of Leguminosae at all the sites. There are more species of Leguminosae than of any other family in all of the dry forest and moist forest samples and Leguminosae rank second in number of species in the two wet forest sites. Moreover, the number of species of Leguminosae in each community sampled is rather constant at nine of the ten sites, ranging from eight to 16 species. The Manaus sample is distinctly richer in legume species with 21. (The three sites in Amazonian Peru for which tentative analysis has been completed also have between 19 and 25 legume species.)

Bignoniaceae, the predominant vine family in all samples, and probably in all lowland neotropical forests, show a similar diversity pattern. At all five dry forest sites Bignoniaceae rank second to Leguminosae in number of species in the samples. In moist forest sites Bignoniaceae rank third to fourth among families in number of species in a 1000-m² sample, while in the two wet forest samples Bignoniaceae rank seventh and eighth among families in number of species sampled. These samples might suggest a slight absolute decrease in number of species of Bignoniaceae from an average of 8.2 in dry forest sites to an average of 6 in moist forest sites to an average of 5 in wet forest sites, but this difference is hardly significant. Moreover, a thorough analysis of all the Bignoniaceae species present was made at each site. The numbers of Bignoniaceae species per site turns out to be remarkably constant, with about 20 species occurring at each site (Gentry, 1976, 1980a). The lower number of sampled species in wetter sites reflects a greater distance between individuals of a given species. This is automatically correlated with the additional species of other families which are added to the more mesic plant communities. About half of the total Bignoniaceae flora of dry sites is represented in a 1000-m² sample.

The other families best represented in dry areas include Rubiaceae (average 3.8 species/sample), Flacourtiaceae (average 2.2 species/sample), Sapindaceae (average 2.2 species/sample), and Myrtaceae (average 1.4 species/sample). All are noticeably better represented in moist forest sites and all except Flacourtiaceae are at least equally well represented in wet forest sites. We may conclude that most families occurring in dry forest are at least as diverse in more mesic sites as in dry forest ones, even though they may not be as obvious.

Two families, Capparaceae and Cactaceae, show a strikingly different pattern. Capparaceae is the third largest family in number of species at Boca de Uchire, the driest of all the sites. Cactaceae are represented by two arborescent genera in the Boca de Uchire sample, and by single species at each of the other dry Venezuelan sites. Capparaceae and Cactaceae seem to be specialized for, and most diversified in, dry areas, whereas the other families present in dry forest vegetations might better be interpreted as components of more mesic vegetations which have a few species capable of surviving dry forest conditions.

Dispersal Ecology

Another way in which these samples differ is in the relative importance of different dispersal strategies (Table VIII). It is generally supposed that wind-dispersed seeds are more prevalent in dry forest than in wet forest communities. It has also been suggested that mammal-dispersal is more important in wet forest communities than in dry or moist forest ones (Gentry, 1980b). To test whether these trends are apparent for this set of sites, each species was characterized as to probable chief dispersal vector based on gross fruit morphology. In general, wings on indehiscent fruits or winged seeds in dehiscent fruits were taken as evidence of wind dispersal. Differentiation between primarily bird-dispersed and primarily mammal-dispersed fruits or seeds is on more tenuous ground. I based my division in part on the data and observations summarized in Croat (1979), supplemented (and in several cases, reclassified—e.g., *Rinorea,* which is autochorous) according to my own observations or suggestions of R. Foster. I assume that congeners of BCI species which share similar fruit morphologies are similarly dispersed. When data on dispersal of a genus were not available I arbitrarily assigned fleshy diaspores 2 cm or more long (which are usually greenish, yellowish, or orangish at maturity) to the mammal-dispersed category and those less than 2 cm (which are usually blue-black or red at maturity) to the bird-dispersed category. While this is not an entirely satisfactory procedure, it makes possible indications of general dispersal trends which would be impossible if based only on the meagre hard data available from most tropical forest communities. Trees and vines were tabulated separately (Fig. 3). A few species were classified as primarily autochorous or water-dispersed. I did not attempt to classify the tree species of the Manaus sample or the Boca de Uchire sample as to dispersal vector, since several of the generic identifications are too tentative to be sure that some wind-dispersed plants, especially legumes, might not be misidentified as mammal-dispersed genera or vice versa.

TABLE VIII. Dispersal Agents for Species Represented in Transects and in Local Florulas[a]

Site	Habit	Wind No. species	Wind %	Mammal No. species	Mammal %	Bird No. species	Bird %	Autochorous No. species	Autochorous %	Water No. species	Water %
Transects												
Dry forest average (per 1000 m²)	Trees	15	30	13	27	21	42	1	1	>1	—
	Lianas	8	81	>1	6	1	11	>1	3	>1	3
Moist forest average (per 100 m²)	Trees	12	15	23	29	41	51	3	3	—	—
	Lianas	15	60	>1	2	9	35	>1	2	—	—
Wet forest average (per 1000 m²)	Trees	6	6	37	34	63	58	3	2	—	—
	Lianas	15	50	4	12	11	36	>1	2	—	—
Local florulas												
Santa Rosa National Park (dry forest)	Large trees	40	29	44	32	44	32	8	6	2	1
	Lianas	35	71	3	6	8	16	3	6	—	—
	Small trees	5	8	9	15	37	62	9	15	—	—
Barro Colorado Island (moist forest)	Large trees	48	16	99	34	129	44	13	4	2	1
	Lianas	85	56	17	11	43	28	4	3	2	1
	Small trees	6	5	23	18	91	69	7	5	4	3
Rio Palenque (wet forest)	Large trees	6	4	66	46	69	48	3	2	1	1
	Lianas	27	33	14	17	38	46	3	4	1	1
	Small trees	5[b]	5	18	19	69	72	2	2	2	2

[a] Large trees ≥10 cm dbh; small trees 2.5–10 cm dbh.
[b] All ferns.

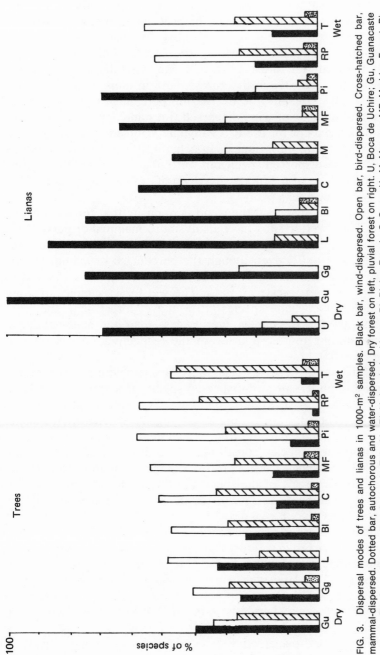

FIG. 3. Dispersal modes of trees and lianas in 1000-m² samples. Black bar, wind-dispersed. Open bar, bird-dispersed. Cross-hatched bar, mammal-dispersed. Dotted bar, autochorous and water-dispersed. Dry forest on left, pluvial forest on right. U, Boca de Uchire; Gu, Guanacaste upland; Gg, Guanacaste gallery forest; L, Estación Biológico de los Llanos; Bl, Blohm Ranch; C, Curundú; M, Manaus; MF, Madden Forest; Pi, Pipeline Road; RP, Río Palenque; T, Tutunendó.

The dispersal data are summarized in Fig. 3. There is a clear and obvious trend toward reduction in the prevalence of wind-dispersed tree species in the wetter areas (χ^2, $p \ll 0.01$). For the dry sites an average of 30% of the tree species are wind-dispersed, for moist forest sites, 15%, and for the two wet forest sites, 6%. For wind-dispersal in lianas the average values are 81% in dry forest, 60% in moist forest, and 48% in wet forest sites. Despite the decreasing percentages of wind-dispersal, the absolute number of wind-dispersed trees and lianas remains remarkably constant except for the Río Palenque sample ($r = -0.585$, n.s. for trees; $r = 0.606$, n.s. for lianas). In general, it is not so much that there are fewer wind-dispersed species in wetter sites, but that they seem less prevalent because of the increasing diversity of bird-dispersed and mammal-dispersed species.

For mammal-dispersal the trend is opposite that for wind-dispersed species. Thus wetter sites have many more mammal-dispersed species (37 tree species/sample in wet forest, 23 in moist forest, 13 in dry forest; $r = 0.956$, $p \ll 0.01$) even though the percentage of mammal-dispersal remains rather constant (27% of tree species in dry forest, 29% in moist forest, 34% in wet forest; χ^2, n.s.). The wet forest samples average 4.5 mammal-dispersed liana species (15% of wet forest liana species), but moist and dry forest samples average less than one mammal-dispersed liana, accounting for 2% and 6% of the respective liana samples ($r = 0.759$, $p < 0.05$).

Bird-dispersal increases even more strongly with precipitation, both in absolute numbers of bird-dispersed species (for trees $r = 0.972$, $p < 0.01$; for lianas $r = 0.781$, $p < 0.05$) and as a percent of the total flora of a sample. Thus dry forest samples average 21 species of bird-dispersed trees constituting 42% of the tree species, moist forest samples average 38 bird-dispersed species constituting 51% of the total sample, and the average 63 tree species of wet forest samples constitute 58% of all tree species. For lianas 6% of the species of the dry forest samples are bird-dispersed as compared to 35% of the moist forest lianas and 33% of the wet forest lianas (χ^2, $p < 0.05$).

Lianas as a group are much more consistently wind-dispersed than are trees. Thus, even in wet forest sites where wind-dispersed tree species are almost nonexistent, wind-dispersal tends to be the predominant strategy among lianas. However, all hemiepiphytic climbers (except ferns) are bird-dispersed. As hemiepiphytic climbers become increasingly prevalent in wetter sites, there is a concomitant shift from wind- to bird-dispersal.

Although bird-dispersed species also become proportionally more important in wetter sites when all trees and large shrubs are considered together, as above, this increase is not uniform in all strata. The increase in bird-dispersal is concentrated in understory species. When only larger tree species (≥ 10 cm dbh) are considered, the increase in diversity from dry to wet forests is accounted for to a much greater extent by additional mammal-dispersed species.

The same conclusions are reached if the three complete florulas are compared (Fig. 4). Percentage of wind-dispersal in trees drops from 30% to 6% (χ^2, $p \ll 0.01$) and in lianas from 70% to 33% (χ^2, $p < 0.01$) in going from the dry to the wet forest site. Percentage of mammal-dispersal is relatively constant in trees (χ^2, n.s.), going from 32% at the dry site to 46% at the wet site, while mammal-dispersal in lianas changes from 6% of the species to 17% of the species. Bird-dispersal in trees increases slightly more than mammal-dispersal from 32% at the dry site to 48% (χ^2, $p < 0.01$) at the wet site and bird-dispersal in lianas goes from 16% to 46% (χ^2, $p < 0.01$). Wind-dispersal is always most prevalent among lianas

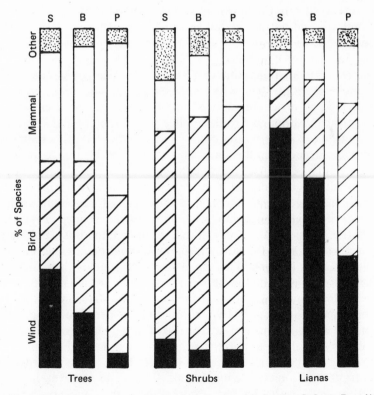

FIG. 4.   Dispersal modes for trees, shrubs, and lianas in complete florulas. S, Santa Rosa National Park (dry forest); B, Barro Colorado Island (moist forest); P, Río Palenque (wet forest).

and negligible among small trees and shrubs. Small trees and shrubs are always overwhelmingly bird-dispersed.

Since the samples (which might be faulted for being inadequate representations of their sites) and the complete local floras give such similar results, we may conclude both that the species represented in 1000-m² samples are sufficient to give a good idea of the dispersal ecology of a community and that the observed trends are real ones.

Pollination Strategies

The organization and floristic composition of tropical plant communities is strongly and intricately influenced by interactions with pollinators (Frankie, 1975; Stiles, 1975, 1978). Accordingly, it should be of interest to analyze trends in the prevalence of different pollination systems in my series of communities. Frankie (1975) compared a dry forest and a wet forest site in Costa Rica, and suggested that in the wet forest hummingbird- and bat-pollinated species are mostly restricted to the understory, while most canopy species are probably pollinated by a diverse array of opportunistic pollinators, including small bees, butterflies, wasps, and beetles. Frankie noted that these opportunistically pollinated species, which include about 70% of the species of the wet forest canopy, are characterized by producing many small flowers, each containing a small nectar reward. Although Frankie called these species "massively flowering" because of their production of so many flowers, this seems a very different syndrome from the "mass-flowering" of dry forest species. The flowers of most of the wet forest canopy species are inconspicuous, not only individually but also in mass, while mass-flowering species (*sensu* Heinrich and Raven 1972; Gentry, 1978d) of dry forest communities are usually exceedingly conspicuous when in flower. Indeed, Frankie categorized the species of his dry forest sample into those pollinated by medium to large bees (mostly anthophorids and euglossines) and those opportunistically pollinated by small bees and an assortment of other visitors. Moreover, Frankie showed a distinct seasonal differentiation between these two pollination strategies in the dry forest community, with most species pollinated by medium to large bees blooming in the dry season and most of the wet-season flowering species opportunistically pollinated by small bees, butterflies, moths, wasps, and beetles.

Frankie did not further analyze his wet forest data, but the relative concentration of small, inconspicuous flowers in the wet forest plant community and, in the dry forest community, the concentration of such flowers in the rainy season suggests a general correlation of species with small, inconspicuous flowers and wetter conditions. Unfortunately, little is known about the pollen vectors of most tropical plant species, especially those of wet forests. While it is often possible to predict pollen vectors from floral morphology (e.g., tubular red flowers are probably humming-bird-pollinated), this procedure is not entirely satisfactory. Moreover, the concentration of species with such specialized pollination systems as hummingbird pollination and bat pollination in the understory and her-baceous layers eliminates most such species from my samples. Never-theless, I have tried to address the question of whether the proportion of different pollination strategies in a community is predictably correlated with precipitation by categorizing each species occurring in my samples as either conspicuous-flowered (C in the Appendix) or inconspicuous-flowered (I in the Appendix). In order to focus on the dominant oppor-tunistic pollination syndrome of the wet forest, I have used somewhat subjective criteria based on appearance as well as flower size. Thus, included under "conspicuous" are mass-flowering species with small, brightly colored flowers which are mostly pollinated by the same medium and large bees that also pollinate larger flowers (e.g., many papilionate legumes). Large but solitary and greenish and thus not very conspicuous flowers such as those of most Annonaceae (probably beetle-pollinated) are also included under "conspicuous." Generally small flowers which are greenish cream, tan, or whitish are judged inconspicuous, while sim-ilar-sized massively produced flowers which are purple or yellow are not.

TABLE IX. Pollination Syndromes: Species with "Conspicuous" and "Inconspicuous" Flowers[a]

Site	Number of species[b]		Percent of species[b]	
	C	I	C	I
Llanos	28	11	71	28
Blohm Ranch	34	30	53	47
Guanacaste (Upland)	34	15	69	31
Guanacaste (Gallery forest)	29	31	49	51
Curundú	41	44	48	52
Madden Forest	61	61	50	50
Río Palenque	41	74	36	64
Pipeline Road	52	68	32	68
Tutunendó[c]	43	69+	38	62+

[a] The former are mostly pollinated by large and medium-size bees (a few also by hum-mingbirds, bats, or sphingids), the latter by small bees and a diverse array of butterflies, wasps, moths, and other relatively generalized insects.
[b] C, Conspicuous flowers; I, inconspicuous flowers.
[c] Identifications incomplete.

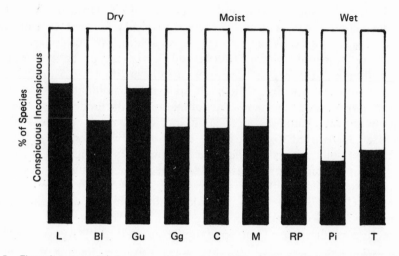

FIG. 5.   Flowering strategies of species ≥2.5 cm dbh in 1000-m² samples. Driest site on left, wettest on right. L, Estación Biológico de los Llanos; Bl, Blohm Ranch; Gu, Guanacaste Upland; Gg, Guanacaste gallery forest; C, Curundú; M, Madden Forest; RP, Río Palenque; Pi, Pipeline Road; T, Tutunendó.

When I am not familiar with a species's flowering in the field, I have guessed from herbarium material whether it should be classified as conspicuous or inconspicuous.

The results of this classification are presented in Table IX and Fig. 5. Almost three-fourths of the species of the driest sites, but only about one-third of the species of the wettest sites, have conspicuous flowers. The pattern observed by Frankie at his Costa Rican sites seems to be a general and predictable one: the wetter the forest, the greater the percentage of species with small, inconspicuous flowers. The absolute number of conspicuous-flowered species actually increases slightly in 1000-m² samples of wetter forests ($r = 0.68$, n.s.) but not nearly so fast as the number of inconspicuous-flowered species ($r = 0.93$, $p < 0.01$).

CONCLUSION

Although the plant communities of neotropical forests are extremely diverse, they are very far from random assemblages of species. Precipitation is strongly correlated with neotropical plant community diversity and organization. In general, wetter sites have more diverse plant communities. This trend holds for all synusia (habit classes) taken separately, but the sensitivity of the different synusia to changes in precipitation differs. Epiphyte diversity increases most rapidly with precipitation, diversity of large trees least rapidly, lianas and herbs at intermediate rates. For any given neotropical plant community, diversity, at least of 1000-m^2 samples, can be accurately predicted from rainfall data.

Rate of accumulation of species with added sample area (i.e., slope of the species–area curve) is also predictable from precipitation. The log–log species–area curves for these sites are linear; dry forest species–area curves ($b = 0.44–0.66$; $\bar{X}_b = 0.60$) parallel temperate zone ones ($b = 0.46–0.50$; $\bar{X}_b = 0.49$) rather than those for tropical moist and wet forest ($b = 0.71–0.93$; $\bar{X}_b = 0.79$).

Structurally, tropical forests are significantly denser for stems ≥ 2.5 cm in diameter than are temperate zone ones, but core-habitat tropical dry, moist, and wet forest communities are remarkably uniform in stem density. Much of the difference in density between tropical and temperate forests comes from an average tenfold increase in liana density. Density of small (2.5–10 cm dbh) trees, but not of larger trees (≥ 10 cm), is also greater in tropical forests. Within the tropics there is a tendency for density of larger trees to increase from dry to wet forest communities.

Repeating patterns are also apparent in the floristic composition of these different communities. The increase of species diversity of trees and lianas in wetter tropical sites is due mostly to additional families rather than to adding more species or genera of families also represented in dry forest communities. Species per genus of woody plants in 100-m^2 samples is constant at 1.2–1.3 species/genus for most tropical communities. Only the driest tropical site has the higher species/genus ratio characteristic of temperate zone samples. In contrast, the increase of diversity of epiphytes and small shrubs in wetter communities comes mostly from great proliferation of species in relatively few genera.

Several important aspects of community dynamics can also be predicted from precipitation. Prevalence of wind-dispersal decreases regularly with precipitation, while prevalence of mammal-dispersal and bird-dispersal increases. Lianas, as a whole, are much more dependent on wind-dispersal than are trees, but show a similar decrease in prevalence of wind-dispersal in wetter sites. Pollination strategies are also correlated with precipitation; most woody species (about two-thirds) of dry forest communities are conspicuous-flowered and pollinated by large and medium-sized bees; in wet forest communities, although there are more conspicuous-flowered species in absolute numbers, this prevalence is reversed, so that about two-thirds of the woody species have small, greenish, whitish, or tannish inconspicuous flowers and are pollinated by a generalized spectrum of opportunistic pollinators.

It is striking that the representation of primarily conspicuous-flow-ered families like Leguminosae and Bignoniaceae remains rather constant from dry to wet forest communities, although such groups do change in the relative importance in them of wind-dispersal and animal-dispersal in wet versus dry sites. These two families dominate dry forest plant communities, where they are represented almost entirely by wind-dis-persed species. Bombacaceae is another conspicuous-flowered family which shows an exactly similar pattern. The community-wide contrast in pollination and dispersal strategies between wet and dry forest com-munities results largely from the increasing prevalence in the canopy of additional families which are uniformly animal-dispersed and have small inconspicuous flowers, such as Chrysobalanaceae, Humiriaceae, Laur-aceae, Menispermaceae, Moraceae, Palmae, and Sapotaceae.

At the same time there is additional stratification of the forest cor-related with increasing prevalence of epiphytes, mostly representing greater species diversity in very few genera (e.g., *Anthurium, Epiden-drum, Peperomia, Philodendron, Pleurothallis*) and increased differen-tiation of a specialized shrub layer, again dominated by numerous species of very few genera (*Piper, Solanum, Miconia, Psychotria, Palicourea, Clidemia, Conostegia*). All of these have bird-dispersed berries except *Piper*, which is apparently dispersed by bats, especially the genus *Car-ollia*. It is noteworthy that this pattern of explosive adaptive radiation in a few genera is similar to that on islands, where relatively few genera are very successful at radiating into many "empty" niches. This might be taken to suggest that these genera represent fairly recent radiation into previously empty niches. Perhaps this has happened only in the neotrop-ics, which would account for the apparent relative paucity of understory and epiphytic species in paleotropic plant communities (Janzen, 1977). One might predict that the dispersal agents for these groups will be re-cently (co)evolved, relatively specialized taxa which are absent or poorly represented in the Old World and that these groups should account for a disproportionate share of the differences in diversity between Old World and New World bird and bat communities. The recent evolution of these groups would also imply a necessity for modification of Regal's (1977) theory that the tripartite coevolution between angiosperms and their pol-linators and dispersal agents which resulted in the former's sudden ev-olutionary success probably occurred in the subcanopy plant community.

The predictability of the compositions of different neotropical plant communities composed of very different sets of species is the kind of evidence that zoologists have taken to suggest niche saturation and equi-librium communities (MacArthur, 1965; MacArthur and Wilson, 1967).

It is tempting to attribute the apparently constant species diversity of conspicuous-flowered lianas and canopy trees in different neotropical communities to saturation of the available pollination niches with number of sympatric species determined by the potential for niche subdivision of the community's pollinator resource. The nearly constant occurrence of about 20 species of Bignoniaceae, each with a different pollination niche (Gentry, 1976), in each neotropical plant community is a case in point. Diversity of the shrub stratum similarly reflects partitioning of a biotic resource, in this case the birds, which are the seed dispersal agents for this stratum. Sympatric shrub species show a striking displacement of fruiting seasons, with each species fruiting in turn, the shrub community as a whole providing a year-long fruit source for their dispersal agents. Diversity of the genera which show this seasonal displacement of fruiting seasons is strikingly constant from site to site. Thus, Snow (1965) reported 19 species of *Miconia* in his Trinidad study area, 18 with staggered fruiting seasons. Hilty (1981) reported a different set of 19 species of *Miconia* in his study area in western Colombia, again with staggered fruiting seasons, and the 14 Barro Colorado species of *Miconia* show similar patterns. *Psychotria* (or *Psychotria* plus closely related *Palicourea*) is another shrub genus which seems to reach a local limit of about 20 species with staggered fruiting seasons. Since each species fruits for over a month in these systems there is considerable overlap. It is quite possible that shrub species numbers are generally controlled by the degree to which these bird-dispersed species can subdivide the year. The only non-bird-dispersed shrub genus is *Piper*. In a dry forest community with only three species of *Piper* in the depauperate shrub stratum, the three species fruit cyclically but out of phase, providing a continuous food source for the *Carollia* bats which disperse them (Fleming *et al.*, 1977; personal communication); it seems likely that the more numerous *Piper* species of wetter forests will be found to show a similar strategy. Is it only coincidental that *Piper* diversity in many moist and wet forest communities is also about 20 species?

It is hard to imagine theoretical maxima on species diversity of epiphytes, although the greater climatic constancy of wetter sites is undoubtedly correlated with greater niche specificity with respect to light. Thus, even though the Río Palenque epiphytes are subdivided into rather clear-cut height groupings [17 understory species (below 2 m), 120 midcanopy species (2–20 m), 111 canopy species (C. Dodson, personal communication)], there is no hint of any kind of limitation on diversity within any of these groups. Similarly, it is difficult to imagine any kind of determinant of species diversity of the inconspicuous-flowered species which make up the bulk of wet forest canopy species. Why the diversity patterns of these families show such striking similarities at so many climatically similar sites remains unknown.

The gradient of tropical plant community diversity with precipitation is germane to several of the hypotheses which have been advanced to account for the greater diversity of tropical communities. Thus, from the available time theory—that tropical forests have more species largely because they are older (Whittaker, 1977)—it would follow from this precipitation/diversity gradient that, in general, wetter neotropical forests are older than drier ones. Since very rich tropical forests have developed in very wet but geologically young areas [e.g., the Chocó; see Gentry (1980b)], this conclusion is hardly tenable and the alternative of dynamic ecologic determinants of diversity is favored. The diversity patterns of groups like Bignoniaceae and the understory shrubs strongly suggest that competition for pollinators or seed dispersal agents plays a prominent role in determining community diversity; that taxa like Bignoniaceae, *Psychotria, Miconia,* and *Piper* seem limited to around 20 strictly sympatric species differentiated in a predictable way with respect to their vectors suggests further that there are limits to the diversity which results from competitive niche specialization.

Another potential biotic determinant of diversity is predation. Hubbell (1979) has recently shown that individuals of dry forest species are clumped rather than evenly dispersed as would be predicted by the predation theory. However, the fact that similar numbers of individuals are distributed among more and more species in progressively wetter habitats agrees with the predictions of the predation theory. Perhaps only in wetter, less seasonal (i.e., more "tropical") communities does this factor become important. Perhaps the clumping of individuals of a species with respect to statistical randomness becomes largely irrelevant biologically as the statistical "clumps" become progressively more diffuse in wetter habitats.

Since wetter tropical forests are generally more productive, the idea that diversity generally increases with productivity is supported by this gradient. The almost diametrically opposite theory recently proposed by Huston (1979), that low-fertility sites have higher species diversity, is not supported by these data except incidentally in that higher rainfall tends to leach the soil to a greater extent. Moreover, preliminary analysis of the data from sites in Peruvian Amazonia shows directly that on adjacent sites plant species diversity is greatest in core-habitat (climatic climax) communities and decreases on nutrient-poor upland white sand soils; Ashton (1977) has shown a similar pattern of low diversity on nutrient-poor soils for southeast Asian trees, although he also finds a secondary decrease in tree diversity on the sites with the richest soils.

The idea that lack of extinction in "permissive environments" is critical to high tropical diversity would seem negated by the fact that the very taxa with numerous sympatric congeners that have been cited as evidence for lack of extinction are those that show the strongest evidence for recent adaptive radiation, generally correlated with coevolutionary interaction with such demonstrably recent organisms as bats (*Piper*), passerine birds (*Miconia, Psychotria*), or euglossine bees (orchids, Bignoniaceae).

Diversity in these samples is correlated with climatic stability in that the wettest communities, which, at least in the neotropics, generally have the least seasonality, are the most diverse. Presumably the biotic fine-tuning which makes possible so many sympatric species of *Miconia* or *Psychotria* is only possible in less seasonal sites.

One final theory of tropical diversity—that it may be due to greater spatial heterogeneity—seems irrelevent in view of the surprising physiognomic and structural constancy of the range of sites reported here.

That exactly those theories of tropical diversity that are based on dynamic interactions and suggest potential ecologic equilibria are those supported by these data, while those that suggest open-ended accumulation of species are not, is not likely to be entirely coincidental.

Whittaker (1977) has recently emphasized that "diversities (of land plant communities) show relatively few consistent and predictable relationships," and suggested that the intuitively and theoretically appealing idea that community diversities reach equilibria limited by saturation of the available niches may no longer be tenable. Whittaker does grant that there might be exceptions to the general rule of nonsaturated plant community diversities, a potential exception being rich tropical rain forests. That this first extensive comparison of plant community diversity of neotropical forests shows such high predictability of diversity suggests that, given the constraints posed by the familial and generic composition of the neotropical flora, individual core-habitat neotropical plant communities may in fact approximate equilibrium species diversities in evolutionary as well as ecologic time.

ACKNOWLEDGMENTS

This work has been supported both directly and indirectly by grants from the National Geographic Society and National Science Foundation

(GB-40103, INT-7920783, DEB-8006253) and by an NSF graduate fellowship. I thank Drs. P. Ashton, W. Burger, C. Dodson, L. Emmons, R. Foster, P. Harcombe, G. Hartshorn, M. Johnson, P. Raven, D. Simberloff, A. Templeton, and J. Terborgh for review comments and S. McCaslin for computational assistance.

APPENDIX. SITES AND COMMUNITIES STUDIED[a]

	Frequency	Density	Mode of dispersal	Flowering strategy
Estación Biológico de Los Llanos (500 m²)				
Anacardiaceae				
Spondias mombin	0.4	2	M	I
Annonaceae				
Annona jahnii	1.0	9	M	C
Xylopia aromatica	0.2	1	B	C
Bignoniaceae				
*Arrabidaea oxycarpa	0.6	9	W	C
*Arrabidaea mollissima	0.4	2	W	C
Godmania aesculifolia	0.2	1	W	C
Jacaranda obtusifolia	0.2	1	W	C
Tabebuia ochracea	0.2	2	W	C
*Pleonotoma clematis	0.6	6	W	C
*Xylophragma seemannianum	0.4	3	W	C
Cactaceae				
Pereskia guamacho	0.2	1	B	C
Cochlospermaceae				
Cochlospermum vitifolium	0.8	11	W	C
Connaraceae				
Connarus venezuelensis	0.2	3	B	I
Dilleniaceae				
Curatella americana	0.2	1	B	I
Erythroxylaceae				
Erythroxylon orinocensis	0.4	7	B	I
Flacourtiaceae				
Casearia 10299	0.2	1	B	I
Flacourt?	0.2	1	B	(I)
Leaves dried and doubly saw-toothed	0.2	2	—	—
Leguminosae				
Acacia? (long, paired spines, 10298)	0.2	1	—	—
Bowdichia virgiloides	0.2	1	W	C
aff. Bowdichia 10283	0.2	1	W	C
Cassia moschata	0.8	4	M	C
Copaifera officinalis	0.4	12	B	C

[a] Two sites are omitted from the Appendix because of incomplete identifications as indicated in the text. Key to symbols: M, mammal; B, bird; W, wind; I, inconspicuous; C, conspicuous. Asterisk denotes liana.

	Frequency	Density	Mode of dispersal	Flowering strategy
Leguminosae (*continued*)				
Lonchocarpus ernestii	0.6	11	W	C
Machaerium cf. *aculeatum*	0.2	1	W	C
Pterocarpus cf. *rohrii*	0.2	2	W	C
Malpighiaceae				
Byrsonima crassifolia	0.6	4	B	C
Myrtaceae				
Eugenia biflora	0.4	3	B	I
Nyctaginaceae				
Guapira pacurero	0.2	5	B	I
Ochnaceae				
Ouratea guildingii	0.2	2	B	C
Passifloraceae				
Passiflora serrulata	0.2	2	M	C
Polygalaceae				
Securidaca diversifolia	0.2	1	W	C
Rubiaceae				
Chomelia spinosa	0.8	8	B	C
Genipa americana	0.2	2	M	C
Guettarda "elliptica"	0.8	10	B	C
Randia aculeata	0.2	5	M?	C
Sapindaceae				
Allophyllus occidentalis	0.8	7	B	I
Sterculiaceae				
Guazuma ulmifolia	0.2	1	M	I
Tiliaceae				
Luehea candida	0.6	8	W	C
Vochysiaceae				
Vochysia venezuelensis	0.2	1	W	C
Blohm Ranch				
Anacardiaceae				
Spondias mombin .	0.1	1	M	I
Asclepiadaceae				
Marsdenia margaritaria	0.2	3	W	I
Bignoniaceae				
Arrabidaea corallina	0.4	5	W	C
Arrabidaea oxycarpa	0.5	22	W	C
Cydista aequinoctialis	0.1	1	W	C
Macfadyena unguiscati	0.1	1	W	C
Tabebuia billbergii	0.1	2	W	C
Xylophragma seemannianum	0.1	5	W	C
Bombacaceae				
Bombacopsis quinata	0.1	2	W	C
Boraginaceae				
Cordia colococca	0.3	4	B	I

(*continued*)

	Frequency	Density	Mode of dispersal	Flowering strategy
Cactaceae				
Cereus jamacaru	0.2	2	B	C
Capparaceae				
Capparis hastata	0.7	10	M	C
Capparis odoratissima	0.3	3	M	C
Cochlospermaceae				
Cochlospermum vitifolium	0.1	2	W	C
Combretaceae				
Combretum fruticosum	0.1	1	W	C
Dilleniaceae				
Tetracera volubilis	0.3	4	B	I
Ebenaceae				
Diospyros	0.6	11	M	I
Erythroxylaceae				
Erythroxylon orinocensis	0.1	1	B	I
Euphorbiaceae				
Adelia? 24792	0.1	1	Auto	I
Margaritaria nobilis	0.5	9	B	I
Flacourtiaceae				
Hecatostemon completus	0.8	12	B	I
Casearia 24783	0.2	3	B	I
Hippocrateaceae				
Hippocratea volubilis	0.5	9	W	I
Leguminosae				
Acacia glomerosa	0.2	2	W	C
cf. *Acacia*	0.1	2	—	—
Acacia (spines)	0.1	1	—	—
Albizzia (big leaflets)	0.2	2	W	C
Caesalpinia coriacea	0.4	5	M	C
Entada gigas	0.1	1	Water	I?
Caesalpinia granadilio	0.1	1	M	C
Erythrina	0.1	1	B	C
Lonchocarpus (little leaflets)	0.1	2	W	C
Lonchocarpus (velutinous leaflets)	0.1	1	W	C
Lonchocarpus (large puberulent leaflets)	0.1	1	W	C
Machaerium? (little leaves 24780)	0.1	1	W	C
Machaerium moritzianum	0.9	10	W	C
Leucaena leucocephala	0.2	2	W	C
Pterocarpus rohrii	0.6	11	W	C
Malpighiaceae				
Malpighia glabra	0.3	5	B	C
Meliaceae				
Trichilia trifolia	0.2	3	B	I
Trichilia aff. acuminata	0.1	2	B	I
Moraceae				
Ficus pertusa	0.2	2	B	I

	Frequency	Density	Mode of dispersal	Flowering strategy
Moraceae (*continued*)				
Ficus trigonata	0.1	1	M	I
Sorocea sprucei	0.5	8	B	I
Myrtaceae				
cf. Nyctag.	0.6	17	B	I
Nyctaginaceae				
Guapira olfersoniana	0.1	1	B	I
Palmae				
Copernicia tectorum	0.7	18	M	I
Passifloraceae				
*Passiflora serrulata	0.3	5	M	C
Polygonaceae				
Ruprechtia ramiflora	0.3	3	W	I
Coccoloba caracasana	0.3	9	B	I
Rhamnaceae				
*Gouania lupulina	0.1	1	W	I
Zizyphus saeri	0.2	3	M?	I
Rubiaceae				
Chomelia spinosa	0.2	3	B?	C
Genipa americana	0.2	2	M	C
Guettarda elliptica	0.6	11	B	C
Randia	0.2	3	M	C
Psychotria microdon	0.1	1	B	I
Rutaceae				
Zanthoxylum culantrillo	0.6	10	B	I
Sapindaceae				
Allophyllus occidentalis	0.1	1	B	I
*Paullinia (three leaflets)	0.1	1	B	I
Sapindus saponaria	0.1	1	M	I
*Serjania	0.1	1	W	I
Sterculiaceae				
Guazuma ulmifolia	0.5	22	M	I
Verbenaceae				
Clerodendron ternifolium	0.3	3	B	C
Vitex compressa	0.3	4	B	C
Vitex orinocensis	0.3	6	B	C
Violaceae				
*Corynostylis arborea	0.4	4	W (Water?)	C
Undetermined				
Close veins (flacourt?) 24746	0.1	1	—	—
Guanacaste Upland (700 m²)				
Anacardiaceae				
Spondias purpurea	0.7	11	M	I
Annonaceae				
Sapranthus palanga	0.3	2	M	C

(*continued*)

	Frequency	Density	Mode of dispersal	Flowering strategy
Apocynaceae				
Stemmadenia obovata	0.3	3	B	C
Bignoniaceae				
Arrabidaea mollissima	0.4	19	W	C
Arrabidaea patellifera	0.7	23	W	C
Cydista diversifolia	0.4	2	W	C
Godmania aesculifolia	0.4	15	W	C
Pithecoctenium crucigerum	0.1	1	W	C
Tabebuia impetiginosa	0.4	10	W	C
Tabebuia ochracea	1.0	24	W	C
Xylophragma seemannianum	0.7	10	W	C
Bixaceae				
Bixa orellana	0.1	1	B	C
Boraginaceae				
Cordia alliodora	0.4	10	W	C
Cordia panamensis	0.3	2	B	I
Burseraceae				
Bursera simarouba	0.1	1	B	I
Cochlospermaceae				
Cochlospermum vitifolium	0.7	29	W	C
Erythroxylaceae				
Erythroxylon	0.1	1	B	I
Euphorbiaceae				
Margaritaria nobilis	0.1	1	B	I
Flacourtiaceae				
Casearia sp.	0.1	3	B	I
Casearia sp. 2	0.1	2	B	I
Casearia sp. 3 (three-carpeled)	0.1	2	B	I
Hippocrateaceae				
Hemiangium excelsum	0.3	8	W	I
Leguminosae				
Acacia cornigera	0.1	1	M	I
Dalbergia retusa	0.1	1	W	C
Enterolobium cyclocarpum	0.1	1	M	C
Hymenaea courbaril	0.3	2	M	C
Lonchocarpus costaricensis (siete cueros)	0.3	3	W	C
Lonchocarpus cf. *atropurpureus* (chaperno with rough bark)	0.1	1	W	C
Lonchocarpus minimiflorus (chaperno)	0.6	41	W	C
Lonchocarpus cf. *hondurensis* (large leaves)	0.1	1	W	C
Lonchocarpus sp. (pavo)	0.3	3	W	C
Albizzia caribaea	0.4	5	W	C
Machaerium biovulatum	0.1	1	W	C
Myrospermum frutescens (arco)	0.1	1	W	C?
Pithecellobium saman	0.1	1	M	C
Pterocarpus rohrii	0.1	1	W	C

	Frequency	Density	Mode of dispersal	Flowering strategy
Malpighiaceae				
Bunchosia cornifolia	0.3	5	B	C
Meliaceae				
Trichilia americana	0.6	7	B	I
Olacaceae				
Ximenia americana	0.1	1	M	I
Rubiaceae				
Calycophyllum candidissimum	0.4	4	W	C
Chomelia spinosa	0.3	2	B	C
Genipa americana	0.4	4	M	C
Randia lasiantha	0.3	3	M	C
Sapindaceae				
Allophyllus occidentalis	0.7	7	B	I
*Serjania	0.1	2	W	I
Sterculiaceae				
Guazuma ulmifolia	0.1	1	M	I
Theophrastaceae				
Jacquinia	0.4	8	B	C
Tiliaceae				
Apeiba tibourbou	0.4	4	M	C
Luehea candida	0.1	2	W	C
Guanacaste Gallery Forest (800 m²)				
Acanthaceae				
Aphelandra deppeana	0.1	1	Auto	C
Anacardiaceae				
Anacardium excelsum	0.3	2	M	I
Astronium graveolens	0.1	3	W	I
Spondias mombin	0.1	1	M	I
Annonaceae				
Annona holosericea	0.4	6	M	C
Sapranthus palanga	0.4	6	M	C
Apocynaceae				
Stemmadenia obovata	0.1	1	B	C
Bignoniaceae				
*Arrabidaea patellifera	0.2	3	W	C
Crescentia alata	0.1	1	M	C
*Cydista diversifolia	0.2	9	W	C
Godmania aesculifolia	0.1	1	W	C
*Macfadyena unguiscati	0.1	1	W	C
Tabebuia ochracea	0.2	2	W	C
Tabebuia rosea	0.1	1	W	C
*Xylophragma seemannianum	0.1	1	W	C
Bixaceae				
Bixa orellana	0.2	2	B	C
Boraginaceae				
Cordia alliodora	0.1	1	W	C
Cordia panamensis	0.1	2	B	I

(continued)

	Frequency	Density	Mode of dispersal	Flowering strategy
Burseraceae				
Bursera simarouba	0.2	2	B	I
Capparaceae				
Capparis (?) (typical petiole)	0.1	1	?	?
Caricaceae				
Carica papaya	0.1	1	M	C
Chrysobalanaceae				
Licania arborea	0.5	5	M	I
Cochlospermaceae				
Cochlospermum vitifolium	0.1	1	W	C
Dilleniaceae				
Tetracera volubilis	0.1	1	B	I
Elaeocarpaceae				
Sloanea terniflora	0.3	2	B	I
Erythroxylaceae				
Erythroxylon	0.1	1	B	I
Flacourtiaceae				
Casearia aculeata	0.1	1	B	I
Casearia sp. 2	0.1	1	B	I
Casearia sp. 3 (three-carpeled fruit)	0.1	3	B	I
Hippocrateaceae				
Hemiangium excelsum	0.3	4	W	I
Hippocratea volubilis	0.1	1	W	I
Leguminosae				
Andira inermis	0.1	1	M	C
Acacia cornigera	0.1	1	M	I
Bauhinia ungulata	0.1	5	Auto	C
Enterolobium cyclocarpum	0.1	2	M	C
Hymenaea courbaril	0.4	4	M	C
Piscidia carthaginensis	0.1	1	W	C
Pithecellobium longifolium	0.3	2	Water	C
Pterocarpus rohrii	0.1	3	W	C
Once pinnate legume, soft semismooth bark	0.1	1	?	?
Malpighiaceae				
Bunchosia cornifolia	0.1	2	B	C?
*Banisteriopsis?	0.1	1	W	C
Meliaceae				
Swietenia humilis	0.1	1	W	I
Trichilia americana	0.4	3	B	I
Myrsinaceae				
Ardisia revoluta	0.5	7	B	I
Myrtaceae				
Eugenia salamensis	0.1	4	B	I
Psidium salutarum	0.1	1	B	I
Psidium sp.	0.1	1	B	I

	Frequency	Density	Mode of dispersal	Flowering strategy
Olacaceae				
Ximenia americana	0.1	1	M	I
Palmae				
Bactris guianensis	0.1	3	M	I
Piperaceae				
Piper aboreum	0.1	10	M	I
Polygonaceae				
Coccoloba padifolia	0.1	1	B	I
Rubiaceae				
Calycophyllum candidissimum	0.3	2	W	C
Chomelia spinosa	0.1	1	B	C
Rutaceae				
Zanthoxylum procerum	0.1	1	B	I
Alternating simple fragrant leaves with punctations	0.1	1	?	?
Sapindaceae				
Allophyllus occidentalis	0.1	1	B	I
*Paullinia	0.1	2	B	I
Thouinidium decandrum	0.6	9	W	I
Simaroubaceae				
Quassia amara	0.4	11	B	C
Sterculiaceae				
Guazuma ulmifolia	0.3	2	M	I
Theophrastaceae				
Jacquinia	0.3	2	B	?
Tiliaceae				
Luehea candida	0.1	1	W	C
Curundú Forest				
Ancardiaceae				
Anacardium excelsum	0.3	6	M	I
Spondias mombin	0.5	6	M	I
Annonaceae				
Annona hayesii	0.6	12	M	C
Annona purpurea	0.3	6	M	C
Annona spraguei	0.2	2	M	C
Xylopia aromatica	0.1	1	B	C
Xylopia frutescens	0.1	1	B	C
Apocynaceae				
Stemmadenia grandiflora	0.1	1	B	C
Araceae				
*Philodendron sp.	0.1	1	B	C?
Araliaceae				
Dendropanax arboreus	0.1	1	B	I
Didymopanax morototonii	0.2	2	B	I
Aristolochiaceae				
*Aristolochia chapmanniana	0.1	1	W	C

(continued)

	Frequency	Density	Mode of dispersal	Flowering strategy
Bignoniaceae				
Anemopaegma orbiculatum	0.1	1	W	C
Arrabidaea candicans	0.1	2	W	C
Arrabidaea patellifera	0.2	2	W	C
Phryganocydia corymbosa	0.2	2	W	C
Stizophyllum riparium	0.2	3	W	C
Bombacaceae				
Bombacopsis quinata	0.1	1	W	C
Boraginaceae				
Cordia alliodora	0.3	5	W	C
Cordia panamensis	0.1	1	B	I
Tournefortia bicolor	0.1	1	B	I
Burseraceae				
Bursera simarouba	0.1	1	B	I
Chrysobalanaceae				
Hirtella racemosa	0.5	16	B	C
Licania arborea	0.1	1	M	I
Cochlospermaceae				
Cochlosperma vitifolium	0.1	1	W	C
Connaraceae				
Cnestidium rufescens	0.2	5	B	I
Rourea glabra	0.2	2	B	I
Combretaceae				
Combretum decandrum	0.1	2	W	C
Terminalia amazonica	0.2	5	W	I
Compositae				
Wulffia baccata	0.1	1	B	C
Dilleniaceae				
Doliocarpus major	0.1	1	B	I
Dilleniac (smooth red bark)	0.1	1	B	I
Tetracera volubilis	0.4	6	B	I
Elaeocarpaceae				
Sloanea terniflora	0.1	1	B	I
Euphorbiaceae				
Croton panamensis	0.1	1	Auto	I
Flacourtiaceae				
Banara guianensis	0.2	3	B	?
Casearia guianensis	0.3	6	B	I
Lacistemaceae				
Lacistema aggregatum	0.7	13	B	I
Lauraceae				
Nectandra gentlei	0.3	4	B	I
Nectandra latifolia	0.2	2	B	I
Phoebe mexicana	0.4	4	B	I
Leguminosae				
Acacia melanoceras	0.2	2	?	I

	Frequency	Density	Mode of dispersal	Flowering strategy
Leguminosae (*continued*)				
Albizzia adenocephala	0.1	1	W	C
Andira inermis	0.1	1	M	C
Inga cf. *hayesii*	0.1	1	M	C
Inga cf. *pezizifera*	0.1	1	M	C
Inga cf. *sapindoides*	0.1	2	M	C
Inga vera	0.1	2	M	C
Lonchocarpus sp.	0.1	3	W	C
*Machaerium	0.1	1	W	C
Ormosia panamensis	0.1	1	B	C
Pithecellobium rufescens	0.3	3	B	C
Malpighiaceae				
*Mascagnia nervosa	0.5	18	W	C
Melastomataceae				
Conostegia speciosa	0.1	1	B	I
Miconia impetiolaris	0.5	7	B	C?
Monimiaceae				
Siparuna guianensis	0.3	9	B	I
Moraceae				
Ficus sp. (flying buttresses)	0.1	1	M	I
Meliaceae				
Carapa guianensis	0.1	1	M?	I
Trichilia tuberculata	0.6	13	B	I
Musaceae				
Heliconia latispatha	0.1	1	B	C
Myrsinaceae				
Rapanea panamensis	0.1	1	B	I
Myrtaceae				
Eugenia origanoides	0.1	3	B	I
Nyctaginaceae				
Guapira standleyana	0.2	2	B	I
Neea delicatula	0.5	8	B	I
Palmae				
Astrocaryum standleyanum	0.2	3	M	I
Piperaceae				
Piper aboreum	0.3	6	M	I
Polygonaceae				
*Coccoloba parimensis	0.1	1	B	I
Coccoloba caracasana	0.2	2	B	I
Rubiaceae				
Alibertia edulis	0.6	19	M	C
Alseis blackeana	0.1	1	W	C
Amaioua corymbosa	0.1	1	M	C
Genipa americana	0.4	4	M	C
Bertiera guianensis	0.1	1	B	I
Palicourea guianensis	0.1	1	B	C
Pittionitis trichantha	0.1	1	B	I

(*continued*)

	Frequency	Density	Mode of dispersal	Flowering strategy
Sapindaceae				
Allophyllus occidentalis	0.3	5	B	I
Cupania costaricensis	0.1	1	B	I
Cupania fulvida	0.1	1	B	I
Cupania sylvatica	0.2	2	B	I
Serjania racemosa	0.1	1	W	I
Serjania nessites	0.2	3	W	I
*Serjania (triangular stem, = racemosa?)	0.1	1	W	I
Sapotaceae				
Sapotaceae	0.1	1	M	I
Solanaceae				
Solanum extensum	0.1	1	B	I
Tiliaceae				
Apeiba tibourbou	0.2	2	M	C
Luehea seemannii	0.2	2	W	C
Luehea speciosa	0.1	2	W	C
Verbenaceae				
Petrea volubilis	0.1	1	W	C
Vitaceae				
Vitis tiliifolia	0.1	1	B	I
Zingiberaceae				
Costus guanaiensis	0.1	1	B	C
Madden Forest				
Anacardiaceae				
Anacardium excelsum	0.4	4	M	I
Astronium graveolens	0.2	2	W	I
Spondias mombin	0.3	3	M	I
Annonaceae				
Annona hayesii	0.1	2	M	C
Annona purpurea	0.2	4	M	C
Xylopia frutescens	0.1	1	B	C
Apocynaceae				
Odontadenia puncticulatum	0.3	3	W	C
Prestonia portobellensis	0.1	1	W	C
Araliaceae				
Dendropanax arboreus	0.1	1	B	I
Bignoniaceae				
Arrabidaea candicans	0.1	3	W	C
Arrabidaea patellifera	0.1	1	W	C
Arrabidaea verrucosa	0.1	1	W	C
Cydista heterophylla	0.2	2	W	C
Paragonia pyramidata	0.3	3	W	C
Phryganocydia corymbosa	0.1	2	W	C
Stizophyllum riparium	0.3	3	W	C
Bombacaceae				
Cavanillesia platanifolia	0.2	3	W	C
Pseudobombax septinatum	0.1	1	W	C

	Frequency	Density	Mode of dispersal	Flowering strategy
Boraginaceae				
Cordia alliodora	0.1	1	W	C
Burseraceae				
Bursera simarouba	0.2	2	B	I
Protium panamense	0.4	6	B	I
Capparaceae				
Capparis frondosa	0.1	3	M?	C
Chrysobalanaceae				
Hirtella racemosa	0.1	2	B	C
Cochlospermaceae				
Cochlospermum vitifolium	0.2	2	W	C
Combretaceae				
Terminalia amazonica	0.1	1	W	I
Convolvulaceae				
Maripa panamensis	0.1	1	M	C
Connaraceae				
Connarus panamensis	0.8	15	B	I
Dilleniaceae				
Davilla nitida	0.1	1	B	I
Doliocarpus dentatus	0.2	2	B	I
Doliocarpus olivaceous	0.3	6	B	I
Doliocarpus? (smooth bark, peeling slightly)	0.1	1	B	I
Tetracera volubilis	0.3	3	B	I
Erythroxylaceae				
Erythroxylon panamensis	0.2	2	B	I
Elaeocarpaceae				
Sloanea terniflora	0.1	1	B	I
Euphorbiaceae				
Acalypha diversifolia	0.1	1	Auto	I
Acalypha sp. (pubescent)	0.2	3	Auto?	I
Margaritaria nobilis	0.1	1	B	I
Flacourtiaceae				
Casearia aculeata	0.3	7	B	I
Casearia commersoniana	0.1	1	B	I
Casearia guianensis	0.1	2	B	I
Casearia sylvestris	0.1	1	B	I
Laetia procera	0.1	1	B	I
Hasseltia floribunda	0.3	4	B	I
Prockia crucis	0.1	2	B	I
Zuelania guidonia	0.1	1	B	I
Lacistemaceae				
Lacistema aggregatum	0.1	1	B	I
Lauraceae				
Nectandra purpurascens	0.1	2	B	I
Nectandra martinicensis	0.2	2	B	I
Phoebe mexicana	0.1	2	B	I

(*continued*)

	Frequency	Density	Mode of dispersal	Flowering strategy
Lecythidaceae				
Gustavia superba	0.3	19	M	C
Leguminosae				
Andira inermis	0.3	3	M	C
Bauhinia hymenifolia	0.1	1	Auto	C
Cassia fruticosa	0.1	1	Water?	C
Dalbergia retusa	0.1	1	W	C
Dioclea (red ring)	0.1	1	M	C
Entada gigas	0.1	1	Water	I
Entada polystachya	0.1	1	W	C?
Inga hayesii	0.1	1	M	C
Inga aff. pauciflora	0.2	5	M	C
Inga sapindoides	0.1	1	M	C
Machaerium arboreum	0.2	2	W	C
Machaerium seemannii	0.1	1	W	C
Machaerium sp. (red sap; connarac type bark)	0.1	2	W	C
Lonchocarpus velutina	0.2	4	W	C
Swartzia simplex	0.3	5	B	C
Vatairea erythrocarpa	0.2	5	W	C
Malpighiaceae				
Bunchosia cornifolia	0.2	2	B	C
Byrsonima crassifolia	0.1	1	B	C
Hiraea sp. nov.	0.1	1	W	C
Tetrapteris macrocarpa	0.1	1	W	C
Meliaceae				
Guarea guidonia	0.2	5	B	I
Cedrela fissilis	0.1	1	W	I
Trichilia tuberculata	0.2	2	B	I
Trichilia pallida	0.1	2	B	I
Melastomataceae				
Conostegia speciosa	0.1	1	B	I
Miconia alba	0.3	4	B	I
Moraceae				
Ficus perforata	0.1	1	B	I
Sorocea affinis	0.1	1	B	I
Musaceae				
Heliconia catheta	0.1	1	B	C
Heliconia latifolia	0.2	5	B	C
Myristicaceae				
Virola sebifera	0.1	3	B	I
Myrtaceae				
Eugenia coloradensis	0.1	1	B	I
Eugenia oerstediana	0.1	1	B	I
Myrciaria floribunda	0.1	1	B	I
Olacaceae				
Heisteria concinna	0.4	13	B	I

	Frequency	Density	Mode of dispersal	Flowering strategy
Palmae				
Astrocaryum standleyanum	0.2	2	M	I
Bactris	0.2	2	M	I
Elaeis oleifera	0.1	2	M	I
Oenocarpus panamensis	0.5	7	M	I
Scheelea zonensis	0.1	1	M	I
Piperaceae				
Piper reticulatum	0.1	1	M	I
Polygalaceae				
Securidaca diversifolia	0.1	1	W	C
Polygonaceae				
Coccoloba parimensis	0.1	1	B	I
Rhizophoraceae				
Cassipourea elliptica	0.1	1	B	I
Rubiaceae				
Alibertia edulis	0.5	5	M	C
Calycophyllum candidissimum	0.3	3	W	C
Coussarea curvigemmia	0.4	8	B	C
Faramea luteovirens	0.1	2	B	C
Faramea occidentalis	0.6	18	B	C
Cephaelis tomentosa	0.1	1	B	C
Genipa americana	0.1	1	M	C
Ixora floribunda	0.1	1	B	C
Guettarda foliacea	0.2	2	B	C
Macrocnemum glabrescens	0.2	2	W	C
Pittionitis tricantha	0.1	1	B	I
Randia armata	0.1	1	M	C
Randia formosa	0.1	1	M	C
Thick-leaved Rub.	0.1	1	?	?
Rutaceae				
Zanthoxylum procerum	0.1	2	B	I
Sapindaceae				
Allophyllus occidentalis	0.1	1	B	I
Cupania fulvida	0.1	1	B	I
Cupania latifolia	0.1	1	B	I
Cupania sylvatica	0.1	1	B	I
Paullinia costaricensis	0.1	1	B	I
Sapotaceae				
Pouteria stipitata	0.1	1	M	I
Simaroubaceae				
Simarouba amara	0.1	1	B	I
Sterculiaceae				
Guazuma ulmifolia	0.1	1	M	I
Tiliaceae				
Apeiba tibourbou	0.2	3	M	C
Luehea seemannii	0.4	4	W	C

(continued)

	Frequency	Density	Mode of dispersal	Flowering strategy
Tiliaceae (*continued*)				
Luehea speciosa	0.2	2	W	C
Trichospermum mexicanum	0.1	1	W	C
Verbenaceae				
Petrea aspera	0.4	11	W	C
Violaceae				
Hybanthus prunifolia	0.1	1	Auto	C
Rinorea sylvatica	0.3	5	Auto	I
Undetermined				
*Vitis? typical bark, two-parted stem, Dilleniaceous rays (Paullina?)	0.1	1	—	—
*Tough vine with radial rays and Callichlamys-type lenticels	0.1	1	—	—
Río Palenque				
Acanthaceae				
Mendoncia brenesii	0.1	1	B	C
Mendoncia reticulata	0.1	1	B	C
Annonaceae				
Guatteria sp.	0.1	1	B	C
Raimondea quinduensis	0.1	1	M	C
Araceae				
Anthurium (dolichostuchyum)	0.1	2	B	I
Rhodospatha sp. nov.	0.2	2	—	C
Xanthosoma sagittifolia	0.1	2	—	C
Bignoniaceae				
Arrabidaea chica	0.1	1	W	C
Arrabidaea verrucosa	0.2	4	W	C
Schlegelia darienensis	0.1	1	B	I
Schlegelia fastigiata	0.3	3	B	C
Bombacaceae				
Quararibea asterolepis	0.2	2	M	C
Quararibea coloradorum	0.7	12	M	C
Quararibea grandifolia	0.5	6	M	C
Boraginaceae				
Cordia dwyeri	0.1	1	B	I
Capparaceae				
Capparis ecuadorica	0.2	2	B	C
Caricaceae				
Carica papaya	0.2	6	M	C
Carica microcarpa	0.1	1	B	I
Jacaratia spinosa	0.1	1	M	I
Chrysobalanaceae				
Hirtella mutisii	0.1	1	B	I
Compositae				
Mikania leiostachya	0.4	6	W	I
Wulffia baccata	0.1	1	B	C

	Frequency	Density	Mode of dispersal	Flowering strategy
Cyclanthaceae				
*Asplundia peruviana	0.1	1	M?	C
*Asplundia	0.1	1	M?	C
Euphorbiaceae				
*Acalypha diversifolia	0.6	13	Auto	I
Alchornea iracuana	0.1	1	B	I
*Omphalea diandra	0.1	1	M	I
Phyllanthus anisolobus	0.1	1	Auto	I
Fern				
*Lomariopsis japurensis	0.5	6	W	—
Trichopteris trichiata	0.1	1	W	—
*Campyloneurum magnificum	0.1	1	W	—
Flacourtiaceae				
Carpotroche ramosii	0.1	1	B	C
Gramineae				
Bambusa guadua	0.3	3	B	I
Guttiferae				
*Clusia	0.1	1	B	C
Tovomita weddelliana	0.2	3	B	I
Hernandiaceae				
Hernandia stenura	0.3	3	M	I
Icacinaceae				
Calatola costaricensis	0.1	4	M	I
Lacistemaceae				
Lacistema aggregatum	0.1	1	B	I
Lauraceae				
Nectandra aff. trianae	0.1	1	B	I
Ocotea cf. dendrodaphne	0.1	1	B	I
Persea theobromifolia	0.1	1	M	I
Lecythidaceae				
Grias tessmannii	0.1	2	M	C
Leguminosae				
*Dioclea reticulata	0.1	1	M	C
Dussia lehmannii	0.1	1	M	C
Erythrina megistophylla	0.1	1	M	C
Inga alatocarpa	0.3	2	M	C
Inga corruscans	0.2	2	M	C
Inga riopalenquensis	0.2	2	M	C
Inga ruiziana	0.2	2	M	C
Pithecellobium macradenium	0.1	1	B	C
Marcgraviaceae				
*Norantea sodiroi	0.1	2	B	I
Melastomataceae				
Clidemia caudata	0.1	1	B	I
Ossaea micrantha	0.2	3	B	I
Meliaceae				
Carapa guianensis	0.3	3	M	I

(continued)

	Frequency	Density	Mode of dispersal	Flowering strategy
Meliaceae (*continued*)				
Guarea cartaguenya	0.1	1	B	I
Guarea glabra	0.2	3	B	I
Guarea guidonia	0.1	1	B	I
Menispermaceae				
Anomospermum sp. nov.	0.1	1	B	I
Disciphania aff. inversa	0.1	1	B	I
Disciphania juliflora	0.1	1	B	I
Monimiaceae				
Siparuna	0.1	1	B	I
Moraceae				
Castilla elastica	0.2	3	M?	I
Cecropia aff. peltata	0.1	1	M	I
Clarisia biflora	0.3	3	B	I
Clarisia racemosa	0.4	4	B	I
Coussapoa eggersii	0.2	2	B	I
Ficus schippi	0.2	2	B	I
Ficus cf. turrialbana	0.1	1	B	I
Maquira costaricana	0.2	3	B	I
Naucleopsis chiguila	0.1	1	M	I
Poulsenia armata	0.2	3	M	I
Pourouma aff. guianensis	0.4	6	B	I
Musaceae				
Heliconia curtispatha	0.4	16	B	C
Heliconia riopalenquensis	0.1	4	B	C
Myristicaceae				
Otoba novogranatensis	0.1	1	M?	I
Otoba gordoniaefolia	0.1	1	M?	I
Virola reidii	0.1	1	B	I
Virola sebifera	0.4	6	B	I
Myrsinaceae				
Ardisia longistaminea	0.1	1	B	I
Stylogyne gentryi	0.1	1	B	I
Nyctaginaceae				
Neea amplifolia	0.1	1	B	I
Olacaceae				
Heisteria cyanocarpa	0.1	1	B	I
Palmae				
Catoblastus velutinus	0.9	36	M	I
Chamaedorea polycarpa	0.2	2	B	I
Pholidostachys dactyloides	0.1	1	B?	I
Prestoea sejuncta	0.1	1	B	I
Synechanthus warscewiczianus	0.2	6	B	I
Wettinia quinaria	0.1	1	M	I
Phytolaccaceae				
Trichostigma octandra	0.1	1	B	I

	Frequency	Density	Mode of dispersal	Flowering strategy
Piperaceae				
Piper augustum	0.3	3	M	I
Piper imperiale	0.2	2	M	I
Piper multiplinervium	0.1	1	M	I
Piper sanctifelicis	0.1	1	M	I
Piper sp. 9871	0.1	1	M	I
Polygonaceae				
Triplaris cumingiana	0.1	1	W	I
Rubiaceae				
Cephaelis gentryi	0.1	1	B	C
Hamelia macrantha	0.2	3	B	C
Palicourea 9879 (fruits)	0.1	1	B	?
Palicourea guianensis	0.1	1	B	C
Palicourea (ternate)	0.1	2	B	?
Pentagonia grandiflora	0.1	1	M	C
Pentagonia macrophylla	0.1	1	M	C
Rutaceae				
Zanthoxylum tachuelo	0.1	1	B	I
Sabiaceae				
Meliosma occidentalis	0.1	1	M	I
Sapindaceae				
Allophyllus amazonicus	0.1	1	B	I
Sapotaceae				
Pouteria lucentifolia	0.2	2	M	I
Simaroubaceae				
Simarouba amara	0.3	3	B	I
Solanaceae				
Cestrum megalophyllum	0.2	3	B	I?
Cyphomandra hartwegii	0.1	1	M	C?
Solanum palenquense	0.2	6	B	C
Witheringia riparia	0.2	3	B	I
Staphyleaceae				
Turpinia occidentalis	0.1	2	B	I
Sterculiaceae				
Herrania balaensis	0.2	3	M	C
Thymelaeaceae				
Schoenobiblus panamensis	0.3	5	B	I
Ulmaceae				
Trema integerrima	0.1	1	B	I
Urticaceae				
Urera baccifera	0.1	2	B	I
Verbenaceae				
Aegiphila alba	0.4	5	B	C
Citharexylum gentryi	0.1	1	B	I
Zingiberaceae				
Costus lima	0.1	1	B	C

(continued)

	Frequency	Density	Mode of dispersal	Flowering strategy
Pipeline Road				
Anacardiaceae				
Anacardium excelsum	0.1	1	M	I
Mosquitoxylon jamaicense	0.1	1	B	I
Tapirira guianensis	0.4	4	B	I
Annonaceae				
Desmopsis panamensis	0.9	13	B	C
Desmopsis sp. 2	0.2	2	B	C
Guatteria dumetorum	0.2	2	B	C
Oxandra longipetala	0.6	12	B	C
Unonopsis cf. *floribunda*	0.3	3	B	C
Xylopia frutescens	0.2	2	B	C
Xylopia macrantha	0.1	1	B	C
Apocynaceae				
Aspidospermum cruentum	0.4	4	W	I
Laxoplumeria tessmannii	0.2	2	W	I
Odontadenia puncticulosa	0.1	1	W	C
Tabernaemontana arborea	0.1	1	B	C
Thevetia ahouai	0.1	1	M	C
*Prestonia??	0.1	1	W	C
Araliaceae				
Dendropanax arboreus	0.1	1	B	I
Dendropanax stenodontus	0.1	1	B	I
Bignoniaceae				
Cydista aequinoctialis	0.1	1	W	C
Jacaranda copaia	0.1	2	W	C
Arrabidaea verrucosa	0.1	3	W	C
Phryganocydia corymbosa	0.2	3	W	C
Stizophyllum inaequilaterum	0.1	1	W	C
Tabebuia chrysantha	0.1	1	W	C
Boraginaceae				
Cordia lasiocalyx	0.1	1	B	I
Cordia sericicalyx	0.1	1	B	I
Burseraceae				
Protium costaricensis	0.1	1	B	I
Protium panamensis	0.4	6	B	I
Tetragastris panamensis	0.2	2	B	I
Trattinickia aspera	0.2	2	B	I
Chrysobalanaceae				
Licania cf. *morii*	0.1	1	M	I
Maranthes corymbosa	0.3	6	M	I
Combretaceae				
Combretum laxum	0.1	1	W	C
Terminalia amazonica	0.2	2	W	I
Compositae				
Mikania tonduzii	0.2	2	W	I

	Frequency	Density	Mode of dispersal	Flowering strategy
Convolvulaceae				
*Maripa panamensis	0.1	1	M	C
Connaraceae				
*Connarus panamensis	0.2	2	B	I
*Rourea adenophora	0.1	1	B	I
Dichapetalaceae				
*Dichapetalum cf. bullatum	0.1	1	—	I
Dilleniaceae				
Saurauia laevigata	0.1	1	B	I
*Tetracera hydrophila	0.1	4	B	I
*Dilleniac (wood only)	0.1	1	—	—
Ebenaceae				
Diospyros artanthifolia	0.1	1	M	I
Elaeocarpaceae				
Sloanea zulianensis	0.2	2	B	I
Euphorbiaceae				
Mabea occidentalis	0.2	2	Auto	I
?? (cf. Mortoniodendron)	0.1	1	—	—
Ferns				
*Polybotrya canaliculata	0.2	2	W	—
Tree fern	0.2	2	W	—
Flacourtiaceae				
Ryania speciosa	0.3	7	?	C
Guttiferae				
Calophyllum longifolium	0.3	3	M	I
Marila laxiflora	0.1	1	W	C
Rheedia madruño	0.1	1	M	I
Tovomitopsis nicaraguensis	0.1	1	B	I
Hippocrateaceae				
*Anthodon panamense	0.1	1	W	I
*Hylenaea pracelsa	0.1	1	W	I
Humiriaceae				
Humiriastrum diguense	0.1	1	M	I
Hypericaceae				
Vismia macrophylla	0.1	1	B	I
Lacistemaceae				
Lacistema aggregatum	0.1	1	B	I
Lauraceae				
Beilschmedia pendula	0.1	1	M	I
Nectandra panamensis	0.1	1	B	I
Ocotea dendrodaphne	0.1	1	B	I
Ocotea ira	0.1	1	B	I
Octea aff. pyramidata	0.1	1	B	I
Lecythidaceae				
Couratari panamensis	0.2	2	W	C
Eschweilera pittieri	0.2	2	M	C
Lecythis ampla	0.1	1	M	C

(continued)

	Frequency	Density	Mode of dispersal	Flowering strategy
Leguminosae				
*Bauhinia guianensis	0.2	2	Auto	C
Coumarouna oleifera	0.1	1	M	C
Erythrina	0.1	1	B	C
Inga pezizifera	0.1	1	M	C
Inga cf. sapindoides	0.1	1	M	C
*Machaerium seemannii	0.1	1	W	C
*Machaerium (diamond-shaped fissures)	0.2	2	W	C
*Machaerium (large leaves new to Panamá)	0.3	3	W	C
Swartzia simplex	0.3	3	B	C
Tachigalia versicolor	0.1	1	W	C
Loganiaceae				
*Strychnos darienensis	0.1	1	M	I
Malpighiaceae				
*Banisteriopsis cornifolia	0.1	1	W	C
*Hiraea quapara	0.1	1	W	C
*Hiraea reclinata	0.2	5	W	C
*Hiraea aff. grandifolia	0.1	1	W	C
*Mascagnia hippocrateoides	0.1	1	W	C
*cf. Tetrapteris discolor	0.1	1	W	C
cf. Byrsonima 3328	0.1	1	B	C
Malvaceae				
Pavonia dasypetala	0.3	4	B	C
Melastomataceae				
Conostegia bracteata	0.1	1	B	I
Leandra dichotoma	0.1	2	B	I
Leandra consimilis	0.1	1	B	I
Miconia impetiolaris	0.1	1	B	C?
Miconia "theazans"	0.1	1	B	I
Miconia (strigose pubescence)	0.1	1	B	I
Melastom with cordate leaves	0.1	2	B	I
Meliaceae				
Guarea glabra	0.1	1	B	I
Trichilia tuberculata	0.1	1	B	I
cf. Trichilia 2676	0.1	1	B	I
Menispermaceae				
*Abuta sp. nov.	0.1	1	B	I
Monimiaceae				
Siparuna pauciflora	0.1	1	B	I
Moraceae				
Brosimum alicastrum	0.5	6	M	I
Ficus insipida	0.1	1	M	I
Ficus tonduzii	0.1	1	M	I
Maquira costaricensis	0.2	2	B	I
Perebea xanthochyma	0.6	9	M?	I
Pourouma scobina	0.2	2	B	I

	Frequency	Density	Mode of dispersal	Flowering strategy
Moraceae (*continued*)				
Poulsenia armata	0.6	17	M	I
Sorocea affinis	0.5	6	B	I
Brosimum costaricense	0.1	1	M	I
Myristicaceae				
Virola nobilis	0.1	1	B	I
Virola surinamensis	0.2	3	B	I
Virola sebifera	0.3	4	B	I
Myrsinaceae				
Ardisia bartlettii	0.1	1	B	I
Rapanea	0.1	1	B	I
Stylogyne standleyi	0.1	1	B	I
Myrtaceae				
cf. Psidium 3259	0.1	1	—	—
Eugenia zetekii	0.3	5	B	I
Myrcia fosteri	0.1	1	B	I
Nyctaginaceae				
Guapira	0.1	1	B	I
Ochnaceae				
Cespedezia macrophylla	0.2	4	W	C
Olacaceae				
Heisteria concinna	0.1	1	B	I
Palmae				
Chamaedorea wendlandiana	0.3	3	B	I
Bactris coloniata	0.2	4	M	I
Bactris barronis	0.2	2	M	I
Bactris coloradonis	0.4	8	B	I
Euterpe panamensis	0.2	3	B	I
Geonoma cuneata	0.7	16	B	I
Geonoma interrupta	0.1	1	B	I
Oenocarpus panamensis	0.5	9	M	I
Socratea durissima	0.7	11	M	I
Synechanthus warscewiczianus	0.5	6	B	I
Welfia georgii	0.4	8	M	I
Passifloraceae				
*Passiflora vitifolia	0.1	1	M	C
Piperaceae				
Piper imperiale	0.1	1	M	I
Piper viridicaule	0.1	1	M	I
Polygonaceae				
Coccoloba mazanillensis	0.1	2	B	I
*Coccoloba parimensis	0.1	2	B	I
Rhamnaceae				
*Gouania adenophora	0.1	2	W	I
*Gouania lupuloides	0.2	2	W	I
Rhizophoraceae				
Cassipourea elliptica	0.1	1	B	I

(continued)

	Frequency	Density	Mode of dispersal	Flowering strategy
Rubiaceae				
Amaouia corymbosa	0.2	2	B	C
Faramea luteovirens	0.3	3	B	C
Psychotria calophylla	0.3	3	B	I
Psychotria grandis	0.1	1	B	I
Pentagonia macrophylla	0.1	1	M	I
Large leaves 3288	0.1	1	—	—
Rutaceae				
Zanthoxylum belizense	0.1	1	B	I
Sapindaceae				
Cupania latifolia	0.1	1	B	I
Cupania sylvatica	0.1	1	B	I
Matayba apetala	0.1	1	B	I
Paullinia bracteosa	0.1	1	B	I
Paullinia fibrigera	0.1	1	B	I
Serjania paucidentata	0.1	1	W	I
Talisia nervosa	0.4	6	M	I
Sapotaceae				
Chrysophyllum panamense	0.1	1	M	I
Pouteria neglecta	0.1	1	M	I
Pouteria stipitata	0.1	1	M	I
Simaroubaceae				
Quassia amara	0.2	2	B	C
Simarouba amara	0.1	1	B	I
Solanaceae				
Solanum hayesii	0.1	1	B	I
Theophrastaceae				
Clavija	0.1	1	M	I
Tiliaceae				
Apeiba tibourbou	0.1	1	M	C
Ulmaceae				
Celtis schippii	0.1	1	M	I
Urticaceae				
Myriocarpon yzabalensis	0.2	5	? Water	I
Verbenaceae				
Petrea aspera	0.2	1	W	C
Petrea volubilis	0.6	7	W	C
Violaceae				
Hybanthus prunifolius	0.1	1	Auto	C
Rinorea squamata	0.3	5	Auto	I
Rinorea sylvatica	0.1	1	Auto	I
Indetermined tree with soft bark, oxidizing brown	0.1	1	—	—
Tutunendó				
Annonaceae				
Anaxagorea 24265	0.2	5	B	C
Guatteria pilosula	0.1	1	B	C

	Frequency	Density	Mode of dispersal	Flowering strategy
Annonaceae (*continued*)				
Guatteria aeruginosa	0.1	1	B	C
Guatteria cargadero	0.3	5	B	C
Guatteria aff. cargadero (larger, broader)	0.1	1	B	C
Guatteria (cf. *cargadero*, not obovate)	0.3	3	B	C
Guatteria 30112	0.1	1	B	C
Guatteria (tiny leaves) 24474	0.1	1	B	C
Guatteria 30104	0.1	1	B	C
Guatteria (very large, hairy)	0.2	2	B	C
Unonopsis cf. *pittieri* 30145	0.1	3	B	C
Unonopsis 30200, 30158	0.3	3	B	C
Xylopia (little leaves)	0.1	1	B	C
Annonac 30241	0.1	1	—	—
Annonac 30151	0.1	1	—	—
Apocynaceae				
Apocynac (tiny leaves)	0.3	3	—	—
Couma macrocarpa	0.3	3	M	C
**Macropharynx renteriae*	0.1	1	W	C
Malouetia quadracasorum	0.1	1	W	C
Araceae				
*Anthurium 30109	0.1	1	B	I
*Philodendron aff. *fragrantissimum*	0.1	2	B	C
*Philodendron 30108	0.1	1	B	C
Araliaceae				
Oreopanax anchicayanum	0.1	1	B	I
Schefflera 30134	0.1	1	B	I
Bignoniaceae				
**Martinella obovata*	0.1	1	W	C
**Schlegelia dressleri*	0.2	2	B	C
Bombacaceae				
Phragmotheca fuchsii	0.1	1	M	C
Quararibea hirta	0.5	5	M	C
Quararibea aff. *hirta*	0.1	1	M	C
Quararibea 30325	0.1	1	M	C
Quararibea 30139, 30308	0.2	3	M	C
Boraginaceae				
Cordia dwyeri	0.1	1	B	I
Burseraceae				
Protium 30187, 30133	0.2	2	B	I
Protium 30124	0.1	1	B	I
Protium 30211	0.1	1	B	I
Trattinickia? 30208	0.1	1	B	I
Capparaceae				
Capparis 24473	0.1	1	?	C
Chloranthaceae				
Hedyosmum scaberrimum	0.1	1	B	I

(*continued*)

	Frequency	Density	Mode of dispersal	Flowering strategy
Chrysobalanaceae				
Hirtella 30281, 30201	0.1	1	B	I
Licania calvescens	0.3	4	M	I
Licania caudata	0.1	1	M	I
Licania micrantha	0.1	1	M	I
Licania 30324	0.1	1	M	I
Licania 30304	0.1	1	M	I
Licania 30147	0.1	1	M	I
Combretaceae				
Combretum laxum	0.1	1	W	I
Compositae				
Piptocarpha atratoensis	0.2	2	W	I
Convolvulaceae				
Dicranostyles	0.3	3	M	I
Maripa panamensis	0.1	1	M	C
Cyclanthaceae				
Asplundia (divergent segments)	0.1	1	M?	C
Asplundia (straw-colored base)	0.1	1	M?	C
Dilleniaceae				
Davilla aff. *nitida* (hairy)	0.1	2	B	I
Pinzona coriacea	0.2	2	B	I
Elaeocarpaceae				
Sloanea (opposite leaves, 30306)	0.1	1	B	I
Sloanea grandiflora	0.1	1	B	I
Sloanea (small leaves, 24299)	0.2	2	B	I
Sloanea? (large leaves, 30297)	0.1	1	—	—
Ericaceae				
*Narrow three-veined leaves 30175	0.1	1	B	C
*Sessile, cordate leaves 30120	0.1	1	B	C
Euphorbiaceae				
Plukenetia penninervis	0.1	1	Auto	I
Alchornea grandis	0.2	2	B	I
Mabea (green beneath)	0.1	1	Auto	I
Mabea aff. *occidentalis*	0.1	1	Auto	I
Pausandra trianae	0.3	5	—	I
Tetrorchidium gorgonae	0.5	6	Auto	I
Ferns				
Trichopteris nigripes	0.1	1	W	—
Flacourtiaceae				
Casearia arborea	0.3	4	B	I
Ryania speciosa	0.1	1	?	C
Gesneriaceae				
Drymonia alloplectoides	0.1	1	B	C
Guttiferae				
Chrysochlamys (giant leaves, 30286)	0.2	2	B	I
Clusia mamillata	0.2	3	B	C

	Frequency	Density	Mode of dispersal	Flowering strategy
Guttiferae (*continued*)				
Clusia cf. *polystigma*	0.1	1	B	C
Clusia sp. (long leaves, 30225)	0.1	1	B	C
Clusiella macropetala	0.1	1	B	C
Marila (large leaves)	0.1	1	W	C
Reedia 24324	0.1	1	M	I
Tovomita stylosa	0.2	2	B	I
Tovomita sp. 30183, 30323	0.2	2	B	I
Tovomito weddelliana	0.2	2	B	I
Tomovita cf. *nicaraguensis*	0.1	2	B	I
cf. *Tovomitopsis* (subsessile leaves)	0.1	2	B	I
Hippocrateaceae				
Salacia 30192	0.1	1	M	I
Humiriaceae				
Saccoglottis ovicarpa	0.2	3	M	I
Lauraceae				
Large, fuzzy leaves, pubescent twigs	0.2	3	B	I
Nectandra aff. *membranacea*	0.1	1	B	I
Ocotea auriculata	0.2	3	B	I
Ocotea cooperi	0.3	3	B	I
Ocotea (little leaves, 30235)	0.1	1	B	I
Smooth, coriaceous 30237	0.1	1	—	—
30152, ascending venation	0.2	2	B	I
Long, dark, raised reticulate, 30204	0.1	1	B	I
aff. *Pleurothyrium krukovii* 24300	0.1	1	—	—
Lecythidaceae				
Couratari 24327	0.3	3	W	C
Eschweilera (medium longish leaves)	0.1	1	M	C
Eschweilera integricalyx	0.4	4	M	C
Eschweilera (finely reticulate)	0.2	2	M	C
Eschweilera pittieri	0.3	5	M	C
Gustavia cf. *grandibracteata*	0.4	5	M	C
Gustavia sp. (finely serrulate)	0.1	1	M	C
Leguminosae				
Acacia (spiny stems)	0.1	1	W	I
Bauhinia guianensis	0.2	2	Auto	C
Dussia lehmannii	0.2	2	M	C
Inga (winged petiole, hairy)	0.1	1	M	C
Inga punctata	0.4	4	M	C
Inga (pubescent, 30103, 30343)	0.2	2	M	C
Inga (many secondary veins, subbullate)	0.2	2	M	C
Inga (2-foliolate, strongly winged)	0.1	1	M	C
Inga (4-foliolate, large, reticulate)	0.1	1	M	C
cf. *Ormosia* (or *Lonchocarpus*?)	0.4	4	—	—
Macrolobium archeri	0.2	3	—	C
Macrolobium ischnocalyx	0.4	4	Auto	C

(*continued*)

	Frequency	Density	Mode of dispersal	Flowering strategy
Leguminosae (*continued*)				
Macrolobium (acute base, 30156)	0.1	1	—	—
Ormosia cuatrecassii	0.1	1	B	C
Pentaclethra macrolobum	0.1	1	Auto	C
"Pseudovouapoa" stenosiphon	0.1	1	—	—
Pterocarpus officinalis	0.3	3	Water	C
Pterocarpus rohrii	0.1	1	W	C
Swartzia (simple-leaved, 30218)	0.2	2	—	C
Swartzia brachyrachis	0.1	1	B	C
Swartzia oraria	0.2	3	B	C
Swartzia simplex	0.1	1	B	C
cf. Sclerolobium (pale below, 30280)	0.1	1	—	—
Tachigalia	0.1	1	W	C
*Legum, tiny leaflets, 24485	0.1	1	—	—
Loganiaceae				
*Strychnos (little leaves, 30314)	0.1	1	M	I
?Strychnos (finely reticulate, 30185)	0.1	1	M	I
*Strychons (green intricate venation)	0.1	1	M	I
Malpighiaceae				
*30328	0.1	1	W	C
Marantaceae				
Ischnosiphon arouma	0.1	1	B	C
Marcgraviaceae				
*Marcgravia (medium leaves, 30236)	0.1	1	B	I
*Marcgravia (small, thick leaves)	0.1	1	B	I
*Norantea/Souroubea (sessile)	0.1	1	B	—
Melastomataceae				
*Adelobotrys 30125	0.1	1	W	I
Clidemia aff. acostae	0.1	2	B	I
Conostegia cf. rubiginosa	0.4	5	B	I
Conostegia cf. rubiginosa	0.4	5	B	I
Conostegia montana	0.2	3	B	I
*Graffenriedia anomala	0.3	3	B	I
Miconia punctata	0.4	6	B	I
Miconia (racemose inflor., 30231)	0.2	2	B	I
*Miconia cf. reducens	0.1	1	B	I
*Ossaea rufibarbis	0.2	2	B	I
*Topobaea alternifolia	0.3	4	B	I
Topobaea floribunda	0.1	1	B	C
*Topobaea aff. inflata	0.1	1	B	I
*Topobeae aff. pittieri	0.1	1	B	—
?(thick veins, moniliform twigs)	0.2	2	B	I
Menispermaceae				
*Anomospermum	0.1	1	M	I
Meliaceae				
Carapa guianensis	0.3	3	Water	I
Trichilia 30150	0.1	1	B	I

	Frequency	Density	Mode of dispersal	Flowering strategy
Meliaceae (*continued*)				
Trichilia 30240	0.1	1	B	I
Trichilia (?, lenticellate, 24991)	0.1	1	—	—
Monimiaceae				
Siparuna (big leaves)	0.2	2	B	I
Siparuna (little leaves)	0.1	1	B	I
Moraceae				
Brosimum lactescens	0.2	2	M	I
Brosimum utile ssp. *occidental*	0.4	5	M	I
Brosimum utile ssp. *magdalenense*	0.1	1	M	I
Cecropia cf. *obtusifolia*	0.1	2	M	I
Ficus (?, small asymmetric leaves)	0.1	1	M	I
Ficus (aff. *citrifolia* 30226)	0.1	1	M	I
Maquira costaricensis	0.1	1	B	I
Naucleopsis glabra	0.1	1	M	I
cf. *Naucleopsis* 30335	0.1	1	—	I
Perebea guianensis ssp. *castilloides*	0.1	1	M	I
Pourouma	0.1	1	M	I
Pseudolmedia oxyphyllaria	0.1	2	B	I
Myristicaceae				
Compsoneura atopa	0.3	3	B?	I
Compsoneura cuatrecasii	0.2	2	B	I
Iryanthera megistophylla	0.3	3	B	I
Iryanthera ulei	0.9	11	B	I
Otoba latialata	0.2	2	M?	I
Virola pavonis	0.1	1	B	I
Virola sebifera	0.1	1	B	I
Myrsinaceae				
Ardisia longistaminea	0.1	1	B	I
Ardisia	0.1	1	B	I
Correlliana	0.1	1	B	I
Long punctate leaves	0.1	1	B	I
aff. *Lecythidaceae* 30189, 30292	0.6	11	—	—
Myrtaceae				
Calycorectes 30213	0.1	1	M	I
Eugenia origanoides	0.1	1	B	I
cf. *Mouriri* 24468	0.1	1	—	—
Myrcyia aff. *coumeta*	0.1	1	B	I
Very large, densely punctate 30212	0.1	1	—	—
Large gray leaves 30355	0.1	1	—	—
Nyctaginaceae				
Guapira cf. *costaricana*	0.3	3	B	I
Ochnaceae				
Cespedezia spathulata	0.1	1	W	C
Olacaceae				
Heisteria concinna	0.2	2	B	I

(*continued*)

	Frequency	Density	Mode of dispersal	Flowering strategy
Palmae				
Attalea (stemless)	0.2	2	M	I
Bactris (short, whitish spines)	0.1	1	M	I
Bactris (colonial, large inflor.) 30339	0.2	3	M	I
Bactris cf. *coloniata* 24256, 30059	0.3	3	M	I
Bactris (single stem, few spines)	0.5	10	M	I
Catoblastus? (no stilt roots, 30184)	0.1	1	M	I
Rachis red below 30321	0.1	1	—	—
Euterpe 30188	0.3	4	M	I
Genoma 30188	0.1	1	B	I
Jessenia bataua	0.9	17	M	I
Iriartea/Catoblastus	0.6	8	M	I
Manicaria saccifera	0.3	3	M	I
Oenocarpus	0.1	1	M	I
Prestoea (amargo) 30106	0.3	3	M	I
Philidostachys 30155	0.1	1	M	I
Wettinia quinaria	0.5	9	M	I
Stemless palm; irregular leaflets	0.2	2	—	I
Passifloraceae				
Dilkea 30149	0.1	1	M	C
Passiflora auriculata	0.1	1	M	C
Passiflora ambigua	0.1	1	M	C
Rubiaceae				
Cephaelis 24285, 30144	0.3	6	B	I
Coussarea (terminal cap, 24466)	0.1	1	B	—
Faramea (narrow leaves, gray below)	0.1	1	B	C
Faramea eurycarpa	0.5	8	B	C
Faramea (veins yellow-margined)	0.1	1	B	C
Psychotria (square twigs, 30348)	0.1	1	B	I
Psychotria cooperi	0.2	2	B	I
Psychotria? (clustered leaves)	0.1	1	—	—
Pentagonia	0.2	3	M	C
Sabicea colombiana	0.4	5	B	I
Schradera	0.1	2	M	C
Simira? (large leaves)	0.1	1	—	—
cf. *Siparuna* (pubescent, 24460)	0.1	2	—	—
Narrow leaves 30234	0.1	1	—	—
24334	0.1	1	—	—
*Large stipules, pubescent 30153	0.1	1	—	—
Large, ± hairy leaves, 30157	0.1	1	—	—
Parallel tertiary venation, 30214, 30178	0.3	3	—	—
Corrugated leaf surface, 30203	0.1	2	—	—
Sabiaceae				
Meliosma glabrata	0.2	2	M	I
Meliosma (densely pubescent, 30296)	0.2	2	M	I
Sapindaceae				
Cupania 30220, 24266	0.2	2	B	I

	Frequency	Density	Mode of dispersal	Flowering strategy
Sapindaceae (*continued*)				
Paullinia 24462	0.1	1	B	I
Paullinia 30356	0.1	1	B	I
Talisia (narrow, black leaves, 30218)	0.3	3	M	I
Talisia (large, smooth leaves)	0.1	1	M	I
Talisia? (aff. *Meliac.*, 24283)	0.1	1	—	—
Sapindac? (white below 24493)	0.1	1	—	—
Sapotaceae				
Manilkara 30284	0.1	1	M	I
Pouteria neglecta	0.1	1	M	I
Pouteria 24290, 24321, 30193	0.3	3	M	I
Richardella	0.1	1	M	I
Long, blackish, thin, smooth leaves 30302	0.1	1	M	I
Tan below 30140	0.1	1	M	I
Giant leaves 30138, 30136, 30305	0.2	3	M	I
Large leaves, reticulate below 30127	0.1	1	M	I
Long leaves, plane below 30360	0.1	1	M	I
Long leaves, tertiary venation parallel 30209	0.1	1	M	I
Small leaves drip tip, reddish branches	0.1	1	M	I
Small leaves 30317	0.1	1	M	I
Small leaves 30239	0.1	1	M	I
Simaroubaceae				
Simarouba amara	0.3	4	M	I
Sterculiaceae				
Sterculia aerisperma	0.1	1	M	I
Symplocaceae				
Symplocos? 24317, 30093	0.3	3	B	I
Thymelaeaceae				
Thymelaeac. 24316	0.1	1	—	I
Tiliaceae				
Apeiba membranacea	0.1	1	M	C
Verbenaceae				
Aegiphila 30117, 30359	0.2	2	B	I
Violaceae				
Leonia triandra	0.2	2	M	I
Gloeospermum aequatorialis	0.1	1	M	I
Vochysiaceae				
Qualea lineata	0.3	4	W	C
Undetermined				
cf. *Laurac*, but sweet odor and lenticels	0.1	1	—	—
Aff. *Allophyllus/Meliosma*	0.1	1	—	—

REFERENCES

Anderson, A. B., and Benson, W. W., 1980, On the number of tree species in Amazonian forests, *Biotropica* **12**:235–237.

Aristeguieta, L., 1966, Florula de la estación Biológica de Los Llanos, *Bol. Soc. Venez. Ciencias Naturales* **26**:228–307.

Arrhenius, O., 1921, Species and area, *J. Ecol.* **9**:95–99.

Ashton, P., 1964, Ecological studies in the mixed dipterocarp forests of Brunei State, *Oxford Forestry Memoir* **25**:1–75.

Ashton, P., 1967, Climate versus soils in the classification of Southeast Asian tropical lowland vegetation, *J. Ecol.* **5**:67–68.

Ashton, P., 1969, Speciation among tropical forest trees: Some deductions in the light of recent evidence, *Biol. J. Linn. Soc.* **1**:155–196.

Ashton, P., 1977, A contribution of rain forest research to evolutionary theory, *Ann. Mo. Bot. Gard.* **64**:694–705.

Auclair, A. N., and Goff, F. G., 1971, Diversity relations of upland forests in the western Great Lakes area, *Am. Nat.* **105**:499–528.

Baker, H. G., 1970, Evolution in the tropics, *Biotropica* **2**:101–111.

Bawa, K., 1974, Breeding systems of tree species of a lowland tropical community, *Evolution* **28**:85–92.

Bawa, K., and Opler, P., 1975, Dioecism in tropical forest trees, *Evolution* **29**:167–179.

Beard, J. S., 1946, The natural vegetation of Trinidad, *Oxford Forestry Memoir* **20**:1–152.

Black, G. A., Dobzhansky, T., and Pavan, C., 1950, Some attempts to estimate species diversity and population density of trees in Amazonian forests, *Bot. Gaz.* **111**:413–425.

Blum, K. E., 1968, Contributions toward an understanding of vegetational development in the Pacific lowlands of Panamá, PhD Thesis, Florida State University, Tallahassee.

Briscoe, C. B., and Wadsworth, F. H., 1970, Stand structure and yield in the Tabonaco Forest of Puerto Rico, in: *A Tropical Rain Forest,* (H. Odum, ed.), pp. B79–B89, U.S. Atomic Energy Commission, Oak Ridge, Tennessee.

Burger, W. C., 1981, Why are there so many kinds of flowering plants in Costa Rica? *Brenesia* **17**:371–388.

Cain, S. A., Castro, G. M. O., Murca Pires, J., and da Silva, N. T., 1956, Application of some phytosociological techniques to Brazilian rain forest, *Am. J. Bot.* **43**:911–941.

Carlquist, S., 1974, *Island Biology,* Columbia University Press, New York.

Connell, J. H., 1978, Diversity in tropical rain forests and coral reefs, *Science* **199**:1302–1310.

Connor, E. W., and McCoy, E. D., 1979, The statistics and biology of the species–area relationship, *Am. Nat.* **113**:791–833.

Croat, T., 1979, Flora of Barro Colorado Island, Stanford University Press.

Davis, T. A., and Richards, P. W. 1933, The Vegetation of Moraballi Creek, British Guiana: An ecological study of a limited area of tropical rain forest. Part 1, *J. Ecol.* **21**:350–384.

Davis, T. A., and Richards, P. W., 1934, The vegetation of Moraballi Creek, British Guiana: An ecological study of a limited area of tropical rain forest. Part II, *J. Ecol.* **22**:106–155.

Dodson, C., and Gentry, A. H., 1978, Flora of the Río Palenque Science Center, *Selbyana* **4**:1–628.

Eisenberg, J. (ed.), 1979, *Vertebrate Ecology in the Northern Neotropics,* Smithsonian Institution Press.

Evans, F. C., Clark, P. J., and Brand, R. H. 1955, Estimation of the number of species present on a given area, *Ecology* **36**:342–343.

Ewel, J. J., Madriz, A., and Tosi, Jr., J. A., 1976, *Zonas de Vida de Venezuela,* 2nd ed., Caracas, Venezuela.

Fischer, A. G., 1960, Latitudinal variations in organic diversity, *Evolution* **14**:64–81.

Fleming, T. H., 1970, Notes on the rodent faunas of two Panamanian forests, *J. Mammal.* **51**:473–490.

Fleming, T. H., 1972, Aspects of the population dynamics of three species of opossums in the Panamá Canal Zone, *J. Mammal.* **53**:621–623.

Fleming, T. H., Hooper, E. T., and Wilson, D. E., 1972, Three Central American bat communities: Structure, reproductive cycles, and movement patterns, *Ecology* **53**:555–559.

Fleming, T. H., Heithaus, E. R., and Sawyer, W. B., 1977, An experimental analysis of the food location behavior of frugivorous bats, *Ecology* **58**:619–627.

Frankie, G. W., 1975, Tropical forest phenology and pollinator plant coevolution, in: *Coevolution of Animals and Plants* (L. Gilbert and P. Raven, eds.), pp. 192–209, University of Texas Press, Austin, Texas.

Frankie, G. W., Baker, H. G., and Opler, P. A., 1974, Comparative phenological studies of trees in tropical wet and dry forests in the lowland of Costa Rica, *J. Ecol.* **62**:881–919.

Gan, Y.-Y., Robertson, F. W., Ashton, P. S., Soepadmo, E., and Lee, O. W., 1977, Genetic variation in wild populations of rain-forest trees, *Nature* **269**:323–325.

Gentry, A. H., 1976, Bignoniaceae of southern Central America: Distribution and ecological specificity, *Biotropica* **8**:117–131.

Gentry, A. H., 1977, Endangered plant species and habitats of Ecuador and Amazonian Peru, in: *Extinction is Forever* (G. Prance and T. Elias, eds.), pp. 136–149, New York Botanical Garden.

Gentry, A. H., 1978a, Floristic knowledge and needs in Pacific Tropical America, *Brittonia* **30**:134–153.

Gentry, A. H., 1978b, Extinction and conservation of plant species in tropical America: A phytogeographical perspective, in: *Systematic Botany, Plant Utilization and Biosphere Conservation* (I. Hedberg, ed.), pp. 110–126, Almquist and Wiksell, Stockholm, Sweden.

Gentry, A. H., 1978c, Diversidade e regeneração da capoeira do INPA, com referencia especial as Bignoniaceae, *Acta Amazonica* **8**:67–70.

Gentry, A. H., 1978d, Antipollinators for mass-flowering plants?, *Biotropica* **10**:68–69.

Gentry, A. H., 1979, Buxaceae. In: Flora of Panama, *Ann. Mo. Bot. Gard.* **65**:5–8.

Gentry, A. H., 1980a, Distribution patterns of neotropical Bignoniaceae: Some phytogeographic implications, in: *Tropical Botany* (K. Larsen and L. Holm-Nielsen, eds.), pp. 339–354, Academic Press, New York.

Gentry, A. H., 1980b (1982), Phytogeographic patterns as evidence for a Chocó refuge, in: *Biological Diversification in the Tropics* (G. Prance, ed.), pp. 112–136, Columbia University Press, New York.

Gentry, A. H., and Dodson, C., in preparation.

Gibbs, P. E., and Leitão Filho, H. F., 1978, Floristic composition of an area of gallery forest near Mogi Guaçu, state of São Paulo, S.E. Brazil, *Rev. Bras. Bot.* **1**:151–156.

Glenn-Lewin, C., 1977, Species diversity in North American temperate forests, *Vegetatio* **33**:153–162.

Grubb, P. J., and Whitmore, T. C., 1966, A comparison of montane and lowland rain forest in Ecuador. II. The climate and its effects on the distribution and physiognomy of the forests, *J. Ecol.* **54**:303–333.

Grubb, P. J., Lloyd, J. R., Pennington, T. D., and Whitmore, T. C., 1963, A comparison of montane and lowland rain forest in Ecuador. I. The forest structure, physiognomy and floristics, *J. Ecol.* **51**:567–601.

Hall, J. B., and Swaine, M. D., 1976, Classification and ecology of closed-canopy forest in Ghana, *J. Ecol.* **64**:913–951.

Hartshorn, G., 1978, Tree falls and tropical forest dynamics, in: *Tropical Trees as Living*

Systems, (P. Tomlinson and M. Zimmerman, eds.), pp. 617–683. Cambridge University Press, London.

Hartshorn, G., 1982, in press, in: *Costa Rican Natural History* (D. Janzen, ed.)

Heinrich, B., and Raven, P. H., 1972, Energetics and pollination ecology, *Science* **176:**597–602.

Heithaus, R., 1974, The role of plant-pollinator interactions in determining community structure, *Ann. Mo. Bot. Gard.* **61:**675–691.

Heithaus, R., Fleming, T. H., and Opler, P. A., 1975, Foraging patterns and resource utilization in seven species of bats in a seasonal tropical forest, *Ecology* **56:**841–854.

Hilty, S. L., 1981, Flowering and fruiting periodicity in a premontane rain forest in Pacific Colombia, *Biotropica* **12:**292–306.

Holdridge, L., 1967, *Life Zone Ecology,* Tropical Science Center, San Jose.

Holdridge, L., Grenke, W. C., Hatheway, W. H., Liang, T., and Tosi, Jr., J. A., 1971, *Forest Environments in Tropical Life Zones, A Pilot Study,* Pergamon Press, Oxford.

Hubbell, S. P., 1979, Tree dispersion, abundance, and diversity in a tropical dry forest, *Science* **203:**1299–1309.

Huston, M., 1979, A general hypothesis of species diversity, *Am. Nat.* **113:**81–101.

Huston, M., 1980, Soil nutrients and tree species richness in Costa Rican forests, *J. Biogeogr.* **7:**147–157.

Janzen, D. H., 1977, Promising directions of study in tropical animal–plant interactions, *Ann. Mo. Bot. Gard.* **64:**706–736.

Janzen, D. H., and Liesner, R., 1980, Annotated checklist of lowland Guanacaste Province, Costa Rica, exclusive of grasses and nonvascular cryptograms, *Brenesia* **18:**15–90.

Johnson, M. P., and Raven, P. H., 1970, Natural regulation of plant species diversity, *Evolutionary Biology,* Volume 4 (T. Dobzhansky, ed.), pp. 127–162, Plenum Press, New York.

Johnson, M. P., and Simberloff, D., 1974, Environmental determinants of island species numbers in the British Isles, *J. Biogeogr.* **1:**149–154.

Jones, E. W., 1955/1956, Ecological studies on the rain forest of southern Nigeria. The plateau forest of the Okomu forest reserve, *J. Ecol.* **43:**564–594; **44:**83–117.

Klinge, H., and Rodrigues, W., 1968, Litter production in an area of Amazonian terra firme forest. Part I. Litter-fall, organic carbon and total nitrogen contents of litter, *Amazoniana* **1:**287–302.

Knight, D. H., 1975, A phytosociological analysis of species-rich tropical forest on Barro Colorado Island, Panama, *Ecol. Monogr.* **45:**259–284.

Lang, G. E., Knight, D. H., and Anderson, D. A., 1971, Sampling the density of tree species with quadrats in a species-rich tropical forest, *Forest Sci.* **17:**395–400.

Loucks, O. L., 1962, Ordinating forest communities by means of environmental scalars and phytosociological indices, *Ecol. Monogr.* **32:**137–166.

Maas, P. J.M., 1971, Floristic observations on forest types in western Surinam I, *Verh. K. Ned. Akad. Wet.* **74:**269–302.

MacArthur, R., 1965, Patterns of species diversity, *Biol. Rev.* **40:**510–533.

MacArthur, R., 1969, Patterns of communities in the tropics, *Biol. J. Linn. Soc.* **1:**19–30.

MacArthur, R., 1972, *Geographical Ecology,* Harper & Row, New York.

MacArthur, R., Wilson, E. O., 1967, *The Theory of Island Biogeography,* Princeton University Press, Princeton, New Jersey.

Madison, M., 1977, Vascular epiphytes: Their systematic occurrence and salient features, *Selbyana* **2:**1–13.

May, R. M., 1975, Patterns of species abundance and diversity, in: *Ecology and Evolution of Communities* (M. L. Cody and J. M. Diamond, eds.), pp. 81–120, Belknap, Cambridge, Massachusetts.

Monasterio, M., 1971, Ecológia de las sabanas de America tropical II. Caracterización ecológica del clima en los Llanos de Calabozo, Venezuela, *Rev. Geogr.* 21:5–38.

Monasterio, M., and Sarmiento, G., 1976, Phenological strategies of plant species in the tropical savanna and the semi-deciduous forest of the venezuelan Llanos, *J. Biogeogr.* 3:325–356.

Monk, C. D., 1967, Tree species diversity in the eastern deciduous forest with particular reference to north central Florida, *Am. Nat.* 101:173–187.

Moral, R., del, 1972, Diversity patterns in forest vegetation of the Wenatchee Mountains, Washington, *Bull. Torrey Bot. Club* 99:57–64.

Myers, C. W., 1969, The ecological geography of cloud forest in Panamá, *Am. Mus. Novit.* 2396:1–52.

Paijmans, K., 1969, An analysis of four tropical rain forest sites in New Guinea, *J. Ecol.* 58:76–101.

Pianka, L., 1966, Latitudinal gradients in species diversity: A review of concepts, *Am. Nat.* 100:33–46.

Pires, J. M., Dobzhansky, Th., and Black, G. A., 1953, An estimate of species of trees in an Amazonian forest community, *Bot. Gaz.* 114:467–477.

Poore, M. E. D., 1963, Problems in the classification of tropical rain forest, *Tropical Geography* 17:12–19.

Prance, G. T., 1968, *Maranthes* (Chrysobalanaceae), a new generic record for America, *Brittonia* 20:203–204.

Prance, G. T., 1975, The history of the INPA capoeira based on ecological studies of Lecythidaceae, *Acta Amazonica* 5:261–263.

Prance, G. T., and Schaller, G. B., 1982, A preliminary study of some vegetation types of the Pantanal, Mato Grosso, Brazil, *Brittonia* 34 (in press).

Prance, G. T., Rodrigues, W. A., and da Silva, Marlene F., 1976, Inventário florestal de um hectare de mata de terra firme km. 30 da Estrada Manaus-Itacoatiara, *Acta Amazonica* 6:9–35.

Preston, F. W., 1948, The commonness, and rarity, of species, *Ecology* 29:254–283.

Preston, F. W., 1962, The canonical distribution of commonness and rarity, *Ecology* 43:185–215, 410–432.

Raven, P. H., 1976, Ethics and attitudes, in: *Conservation of Threatened Plants* (J. Simmons, ed.), pp. 155–179, Plenum, New York.

Regal, P. J., 1977, Ecology and evolution of flowering plant dominance, *Science* 196:622–629.

Richards, R. P., 1952, *The Tropical Rain Forest,* Cambridge.

Richards, R. P., 1963, What the tropics can contribute to ecology, *J. Ecol.* 51:231–241.

Schoener, T. W., 1976, The species–area relation within archipelagos: Models and evidence from island land birds, in: *Proceedings of the 16th International Ornithological Conference* (H. J. Firth and J. H. Calaby, eds.), pp. 629–642, Australian Academy of Science, Canberra, Australia.

Schultz, J. P., 1960, Ecological studies on rain forest in northern Suriname, *Verh. K. Ned. Akad. Wet. Afd. Natuurk.* 53:1–267.

Simberloff, D. S., 1970, Taxonomic diversity of island biotas, *Evolution* 24:23–47.

Simberloff, D. S., 1978, Use of rarefaction and related methods in ecology, in: *Biological Data in Water Pollution Assessment: Quantitative and Statistical Analyses,* (K. Dickson, J. Cairns, Jr., and R. Livingston, eds.), pp. 150–165, American Society of Testing and Materials Special Technical Publication 652.

Simpson, B. B., 1974, Glacial migration of plants: Island biogeographical evidence, *Science* 185:698–700.

Snow, D. W., 1965, A possible selective factor in the evolution of fruiting seasons in tropical forest, *Oikos* 15:274–281.

Sota, E., de la, 1972, Las pteridofitas y el epifitismo en el Departmento del Chocó (Colombia), *An. Soc. Cien. Argent.* **194**:245–278.

Stiles, F. G., 1975, Ecology, flowering phenology and hummingbird pollination of some Costa Rican *Heliconia* species, *Ecology* **56**:285–310.

Stiles, F. G., 1978, Temporal organization of flowering among the hummingbird food plants of a tropical food web, *Biotropica* **10**:194–210.

Tanner, E. V. J., 1977, Four mantane rain forests in Jamaica: A quantitative characterization of the floristics, the soils and the foliar mineral levels, and a discussion of the interrelations, *J. Ecol.* **65**:883–918.

Tramer, E. J., 1974, On latitudinal gradients in avian diversity, *Condor* **76**:123–130.

Troth, R. G., 1979, Vegetational types on a ranch in the central Llanos of Venezuela, in: *Vertebrate Ecology in the Northern Neotropics* (J. Eisenberg, ed.), pp. 17–30, Smithsonian Institution Press.

Tschirley, F. N., Dowler, C. C., and Duke, J. A., 1970, Species diversity in two plant communities of Puerto Rico, in: *A Tropical Rain Forest* (H. T. Odum, ed.), pp. B91–B96, U.S. Atomic Energy Commission, Oak Ridge, Tennessee.

Vega, L., 1968, La estructura de los bosques húmedos tropicales del Carare, Colombia, *Turrialba* **18**:416–436.

Vivanco de la Torre, O., Cárdenas C., M., Tosi, Jr., J. A., and Gortaire, I., G., 1963, Mapa Ecológico de la Costa y Sierra del Ecuador, Min. Fomento, Dir. Gen. de Bosques, Instituto Geografico Militar, Quito.

Webb, L. J., Tracy, J. G., Williams, W. T., and Lance, G. N., 1967, Studies in the numerical analysis of complex rain-forest communities, *J. Ecol.* **55**:171–191, 525–538.

Whitmore, T., 1975, *Tropical Rain Forests of the Far East,* Clarendon Press, Oxford.

Whittaker, R. H., 1956, Vegetation of the Great Smoky Mountains, *Ecol. Monogr.* **26**:1–80.

Whittaker, R. H., 1960, Vegetation of the Siskiyou Mountains, Oregon and California, *Ecol. Monogr.* **30**:279–338.

Whittaker, R. H., 1972, Evolution and measurement of species diversity, *Taxon* **21**:213–251.

Whittaker, R. H., 1977, Evolution of species diversity in land communities, *Evol. Biol.* **10**:1–87.

Whittaker, R. H., and Levin, S. A., 1975, *Niche: Theory and Application,* Dowden, Hutchinson and Ross, Stroudsburg, Pennsylvania.

Whittaker, R. H., and Niering, W. A., 1965, Vegetation of the Santa Catalina Mountains, Arizona (II). A gradient analysis of the south slope. *Ecology* **46**:429–452.

Williams, C. B., 1943, Area and number of species, *Nature* **152**:264–267.

Williams, C. B., 1964, *Patterns in the Balance of Nature,* Academic Press, London.

World Meterological Organization, 1975, *Atlas climatico de América del Sur,* Cartographia, Budapest, Hungary.

Zobel, D. B., McKee, A., Hawk, G. M., and Dyrness, C. T., 1976, Relationships of environment to composition, structure, and diversity of forest communities of the central western Cascades of Oregon, *Ecol. Monogr.* **46**:135–156.

2

Evolution on a Petri Dish

The Evolved β-Galactosidase System as a Model for Studying Acquisitive Evolution in the Laboratory

BARRY G. HALL

Microbiology Section
University of Connecticut
Storrs, Connecticut 06268

INTRODUCTION

Several years ago I began a project whose major goal was to understand the variety of ways by which an organism can evolve new physiologic functions. How does an organism which is already well adapted to its environment evolve a new function to cope with an altered environment? The synthetic theory of evolution does not really address the problem of evolution of novel functions; nor does population genetics deal with the appearance of new functions. Although there is general agreement about how genes, having once acquired a particular function, are successively modified to improve the efficiency of that function, little is known about the detailed process by which functions evolve.

That process may be studied in the laboratory by applying known selective pressures to microorganisms in order to direct the evolution of new functions, and monitoring the changes which occur.

I have, for the past several years, been employing the EBG (evolved β-galactosidase) system of *E. coli* as a model system to study evolution experimentally. The system employs the strategy of removing a single specific function (lactose hydrolysis) via the irreversible loss of genetic information and exerting powerful selective pressure for the reacquisition or evolution of that function. In order to more closely mimic nature, I have relied entirely on spontaneous mutations for evolution in the laboratory.

The major goal can best be approached by posing a series of evolutionary questions that can be answered using genetic and biochemical techniques applied to the model system. Some of the principal questions which have been addressed with the EBG model system are: Where do new functions come from? How many new functions can evolve from a single ancestral function? How many mutational steps are required to evolve a particular new function? How many different genes can give rise to the same new function; i.e., what is the evolutionary potential of an organism with respect to any particular function? What is the effect on an old function when a new function is evolved? How do catalytic mechanisms of enzymes change as they acquire new functions? What regulatory functions change during the evolution of new metabolic functions? Can the evolution of new regulatory functions be "directed" in the same sense that evolution of new enzymatic functions can be directed? When mutations conferring new functions first occur in diploid organisms they must necessarily be in the heterozygous state. How does such heterozygosity affect selection for these new functions?

Many of these questions are not new, and it is not anticipated that my studies will greatly revolutionize evolutionary thought. What is important, however, is that the questions are posed in ways that permit experimental tests of evolutionary hypotheses. The ways in which we think about evolution are, to a large extent, conditioned by certain basic, "obvious" assumptions which we make. It is important, and a major function of my studies, to critically test such basic assumptions whenever possible. An an example, it has been generally assumed that any mutation that would alter an enzyme so profoundly that the enzyme could catalyze a new function would also reduce the effectiveness with which the enzyme catalyzes its normal reaction [for example, see Hinegardner (1977)]. I have recently shown that this assumption is invalid, by demonstrating cases in which the "old" function of an enzyme was unimpaired when the enzyme was evolved to carry out a new function (Hall, 1978a).

One of the difficulties of current evolutionary theory is that it does not make clear, testable, predictions. Instead, evolutionary theory seeks to explain how the biologic world *has* evolved. Evolutionary theory has been generated by considering the diversity of life as it is now and comparing it with life as we believe it was in the past. Taken together, these observations have permitted us to generate a sophisticated, internally consistent set of evolutionary hypotheses. These hypotheses, however, have generally not been tested. One problem is that no hypothesis can be tested by employing the same observations which generated that hypothesis. Another problem is that current evolutionary theory is not usually stated in a form which makes predictions. For instance, the rate of amino acid substitutions into certain proteins has been calculated by comparing amino acid sequences of homologous proteins in a wide variety of species, from primitive to highly evolved (Kimura, 1968; Kimura and Ohta, 1971). However, we cannot *predict* with any accuracy the rate for another protein. Studies have clearly shown that the rate of amino acid replacement varies enormously from one protein to another (Clarke, 1970; King and Jukes, 1969; Dickerson, 1971), and we can speculate on the reasons for this (Lewontin, 1974). We may even note relationships between protein functions and rates of amino acid replacement (Dickerson, 1971), but correlations do not demonstrate causality, and the hypotheses advanced remain only reasonable speculations.

By its very nature, evolution is a branch of biology which deals with enormous time spans. For this reason, evolution is usually considered a nonexperimental science. The utilization of microorganisms to deal with very large populations, over many generations, within a relatively short time span has permitted the study of evolution to move into the laboratory.

Although it is clear that mutations are the raw material of evolution and that selection is a major force in evolution, we know far too little about the detailed relationships between selective pressures and evolution. Evolutionary theory does not tell us what pieces of information we need in order to predict the evolutionary consequences of a particular environmental pressure applied to a particular organism. We have no way to assess the probability of an organism's achieving a particular evolutionary destination. Without an understanding of how organisms evolve new functions, we cannot manipulate the environment to direct evolution in desirable ways, nor can we anticipate (and hope to prevent) undesirable evolutionary consequences of our continual alteration of our environment.

APPROACHES TO MOLECULAR EVOLUTION

There are two distinct approaches to understanding the evolution of genes, i.e., the descriptive and the experimental approaches. The descriptive method involves choosing some character, studying it in a group of organisms in which taxonomic relationships are relatively clear, and attempting to deduce the sequence of evolutionary changes in the character. The experimental approach involves applying known selective pressures to an organism and monitoring the changes that occur in an appropriately chosen character.

The Descriptive Approach

The descriptive method has been by far the most widely employed approach to the study of molecular evolution. It has the distinct advantage that the investigator need not wait for evolutionary events to occur. Indeed, until recently, the miniscule rate at which evolutionarily advantageous mutations occur and are fixed in populations has prohibited alternative approaches.

Most current ideas about how genes evolve are derived from descriptive studies. Most useful, from a molecular biologist's point of view, has been the wealth of information gained by comparing amino acid sequences from functionally related proteins in different species. This method has shown, in several cases, that functionally related proteins are "homologous," i.e., probably derived from a common ancestral gene. For a review of such studies, see Nolan and Margoliash (1968). In the now classical case, the comparison of amino acid sequences of cytochrome C protein from 45 species, including all five eukaryotic kingdoms, has allowed the construction of an extensive phylogenetic tree and the elucidation of the probable sequence of the ancestral cytochrome C (McLaughlin and Dayhoff, 1973).

An important concept which has arisen from such comparative structural studies coupled with sophisticated studies of function is that of strong conservation of amino acid sequences about active sites (Hartley et al., 1965). This suggests that when existing enzymes are modified to acquire new functions, there is strong selection against those modifications that extensively alter the existing active sites. Thus, we may expect to find definite relationships between the functions of existing genes and the ancestral genes from which they arose.

Not only are critically important sequences conserved, but tertiary structure also tends to be preserved. Myoglobin and hemoglobin subunits appear very similar in shape (Perutz, 1964; Watson and Kendrew, 1961). There are clearly strong physical constraints upon sequences that are capable of providing stable tertiary structures. This observation has led to the widely held opinion that new genes generally arise from preexisting functional genes, rather than from random "silent" stretches of DNA.

A number of mechanisms have been proposed for the creation of new genes from old, the most prominent being duplication followed by divergence. Evidence supporting the concept of duplication–divergence has been obtained via the comparison of sequences of peptide fragments of various proteins. Trypsin and chymotrypsin, similar in function, probably arose from a common ancestral gene (Walsh and Neurath, 1964), and their coexistence in single species is evidence supporting gene duplication. The hemoglobins and myoglobins clearly arose by gene duplication (Zuckerkandel and Pauling, 1965; Fitch and Margoliash, 1970), as did the NAD-binding enzymes (Rossmann *et al.*, 1974).

On a finer level, the existence of repeated amino acid sequences within immunoglobins (Hill *et al.*, 1966), bacterial ferridoxins (Tanaka *et al.*, 1966), and human haptoglobin α chains (Smithies *et al.*, 1962) is presumed to be the result of serial duplications.

A dramatic case of duplication–divergence is provided by the genes for egg white lysozyme and bovine α-lactalbumin. The sequences of the two proteins are so similar that they are almost certainly homologous (Brew *et al.*, 1967), yet their enzymatic activities are very different. Neither enzyme acts on the substrate of the other, but both work on β 1–4 linkages, α-lactalbumin in synthesis and lysozyme in degradation. Additionally, the isoelectric points are quite different, α-lactalbumin being pH 10 and lysozyme pH 5. One of the most interesting aspects of the α-lactalbumin–egg white lysozyme story is that the two proteins are not immunologically cross-reactive (Arnon and Maron, 1970). This finding, astonishing in view of the clear sequence homology, was verified by four laboratories. It was later shown (Arnon and Maron, 1971) that although the *native* proteins are immunologically distinct, the strongly denatured (reduced, carboxymethylated) proteins *do* cross-react. Thus, an immunologic homology that was concealed by the native proteins was revealed by denaturation. This is particularly exciting because it suggests the possibility of detecting evolutionary homologies for proteins that cannot be sequenced for technical reasons. A similar result was obtained by Arnheim *et al.* (1971), who showed that human and egg white lysozymes that were completely non-cross-reactive in the native form did cross-react when denatured.

An alternative source of new genes from old is fusion of two or more genes to yield a new single gene. There is, to my knowledge, no direct evidence for this process being responsible for the origin of a known functional gene product. However, studies of tryptophan synthetase in *E. coli* and *Neurospora crassa* strongly suggest that such a fusion may have occurred (Bonner *et al.*, 1965). The tryptophan synthetase of *E. coli* is an aggregate of the *trp A* and *trp B* gene products; the tryptophan synthetase of *N. crassa* is the product of a single gene. Yet biochemical, immunologic, and genetic evidence suggest that the *N. crassa* enzyme has distinct regions with functions remarkably similar to the A and B components of the *E. coli* enzyme. Although the evidence is not conclusive, there is the strong possibility that the single chain of *N. crassa* (and yeast) tryptophan synthetase is the result of fusion of two ancestral genes similar to *E. coli* tryptophan synthetase A and B.

One of the major contributions of descriptive molecular studies has been the discovery of the "evolutionary clock." A large number of studies, employing sequencing and also immunologic techniques, have shown that the rate of amino acid substitutions into proteins is much larger than expected from a Poisson process; thus, proteins evolve in a clocklike manner. These studies show that evolution goes on at approximately a constant rate for proteins of the same functional class, but that proteins of different *functional* classes evolve at different rates. For instance, the immunoglobins evolve several hundredfold faster than do the histones. In an outstanding review of the "evolutionary clock" and rates of protein evolution, Wilson *et al.* (1977) point out that, although the rate of *sequence* evolution is about the same in frogs and mammals, the rate of phenotypic evolution is enormously greater in mammals than in frogs (who have changed only slightly in appearance over the last 90 million years). They argue that most of the phenotypic evolution must therefore arise not from alterations in structural genes, but from changes in regulatory functions. Until recently, Wilson stood virtually alone in stressing the importance of regulatory functions in evolution; yet *every* experimental microbial system studied has demonstrated the requirement for regulatory mutations in the evolution of new functions.

This discussion has necessarily been a brief skimming of the surface of an active and exciting field. There are a number of excellent reviews of this approach to evolution; among them are Bryson and Vogel (1965), Dixon (1966), Mandel (1969), Nolan and Margoliash (1968), Wilson *et al.* (1977), and Goodman (1977).

With perhaps the single exception of the derivation of α-lactalbumin from egg white lysozyme (Brew *et al.*, 1967), descriptive studies fail to provide information on the origin of new functions. Even in this case, the details of the evolution of α-lactalbumin from lysozyme are missing; indeed, the conclusion that lysozyme is the progenitor, rather than the descendant, is based solely on the wide taxonomic distribution of lysozyme and the very narrow distribution of α-lactalbumin. The existence of "gaps" between putative ancestral genes and mature genes leaves the evolutionist in much the same predicament as the anthropologist searching for the "missing link"; he knows it must have existed at some time, but it may well be lost now. It is for the purpose of filling the gap between ancestral gene of unknown function and evolved gene of recognizable function that some evolutionists are turning to experimental systems.

The Experimental Approach

The process by which organisms evolve new functions is often termed "acquisitive evolution." It is studied by applying strong selective pressures to microbial populations to direct the evolution of novel metabolic functions. The literature on directed evolution has been reviewed by Hegeman and Rosenberg (1970), Clarke (1974), Lin *et al.* (1976), Riley and Anilions (1978), and most recently by Clarke (1978).

Mutations that permit microorganisms to metabolize novel substrates have been found to affect both regulatory and structural genes. One of the most common ways of acquiring a new function is by constitutive synthesis of an enzyme that is already able to degrade the novel substrate, but for which the novel substrate is not an inducer. The first *lac* constitutive mutant was isolated in this manner by selecting for colonies that could grow on neolactose, which is not an inducer of the *lac* operon (Lederberg, 1951). In other cases the mutations affect primarily the *activity* of enzymes, as is the case for the amidase system studied by Clarke and co-workers (Brown and Clarke, 1970, 1972; Brown *et al.*, 1970; Betz and Clarke, 1972, 1973).

I shall summarize a few of these studies to show the *variety* of mutations that can lead to new metabolic capabilities.

Selection for mutants of *Klebsiella aerogenes* PRL-R3 that could grow on the unnatural pentitol xylitol led initially to mutants that constitutively synthesized high levels of ribitol dehydrogenase (Mortlock and Wood, 1964a,b), although the properties of the dehydrogenase itself were not different from those of the wild-type enzyme (Mortlock *et al.*, 1965). Lin and co-workers observed the same thing with *K. aerogenes* strain 1033 (Learner *et al.*, 1964), except that in this case, further selection led to a second mutation which altered the ribitol dehydrogenase so that it had a lower K_m for xylitol (Wu *et al.*, 1968; Burleigh *et al.*, 1974). Selection for growth at low xylitol concentrations led to a third mutation giving constitutive synthesis of an arabitol permease that could transport xylitol (Wu *et al.*, 1968). This series of mutations thus involved two regulatory mutations and one structural mutation.

Selection of *K. aerogenes* PRL-R3 mutants that could utilize D-arabinose yielded very similar results: the mutants constitutively synthesized an unaltered *L*-fucose isomerase. Selection for a strain that could grow more rapidly on D-arabinose led to an altered L-fucose isomerase with a lower K_m for D-arabinose (Camyre and Mortlock, 1965; Oliver and Mortlock, 1971a,b). Again, efficient utilization of a novel metabolite was evolved via (1) loss of regulation, followed by (2) altered enzyme activity.

Selection for *E. coli* K12 mutants able to use D-arabinose again yielded regulatory mutations. In this case, however, enzyme synthesis was not constitutive; instead, the regulation was altered so that synthesis of the L-fucose pathway enzymes was induced by D-arabinose (LeBlanc and Mortlock, 1971a,b).

In the preceding cases, constitutive strains that synthesized an enzyme with low activity toward the novel metabolite became better adapted to the new metabolite by a second mutation which altered the activity of the enzyme toward the new metabolite. Several studies have demonstrated a different tactic by which cells can evolve to become better able to use a poor metabolite. In each of these cases the selection scheme employed a chemostat to limit the availability of the new metabolite. Continuous culture of $lacI^-$ (constitutive) mutants of *E. coli* in lactose-limited chemostats resulted in the selection of "hyperconstitutive" strains in which the *lac* operon genes were duplicated (Horiuchi *et al.*, 1962, 1963). When Mortlock and Wood (1971) seeded a xylitol-limited chemostat with a ribitol dehydrogenase (RDH) constitutive strain of *K. aerogenes,* they obtained mutants which synthesized fourfold higher levels of RDH than did the parental strain, but the properties of the RDH were unaltered. They suggested that the RDH gene might have been duplicated during the selection process. More extensive experiments by Hartley's group (Hartley *et al.*, 1972; Rigby *et al.*, 1974) supported this hypothesis. That group found that continued selection for strains that could grow faster and faster on limited amounts of xylitol led to gene duplication, reduplication, etc., until the rapidly growing strains had many copies of the RDH gene. Essentially similar results were obtained later by Mortlock's group (Inderlied and Mortlock, 1977). In both cases, careful studies were unable to detect any difference in the properties of the RDH synthesized by the hyperconstitutive strains. In Hartley's hands, repeated gene duplication led to an evolvant which synthesized more than 20% of its total soluble protein as ribitol dehydrogenase (compared with 1% for a normal constitutive) (Rigby *et al.*, 1974; Hartley, 1974). These observations led Hartley (1974) to conclude that gene duplication is the *only* way in which the efficiency of ribitol dehydrogenase can be improved by spontaneous mutations. This strong conclusion is based upon screening 10^{14} organisms with a system that would detect improved RDH enzyme with an efficiency of 10%, and finding no improved strains in which the properties of RDH were altered. Assuming that single amino acid replacements would occur with a frequency of 10^{-9}, Hartley's conclusion seems justified. He explains the improved RDH observed by Wu *et al.* (1968) as resulting from multiple mutations induced by nitrosoguanidine, and has himself employed nitrosoguanidine to get mutants in which the RDH enzyme has a higher affinity for xylitol (Hartley *et al.*, 1972).

The evolution of L-1,2-propanediol utilization by *E. coli* K12 provides an example of still another tactic for acquisitive evolution. Wild-type strains of *E. coli* K12 are unable to use L-1,2-propanediol as a carbon and energy source. Sridhara *et al.* (1969) described the isolation of a mutant (Strain 3) which could grow on propanediol. The mutant constitutively synthesized an NAD-linked dehydrogenase which oxidized propanediol to lactaldehyde; however, because of its wide substrate specificity, its origin and natural function remained a mystery. Several years later Cocks *et al.* (1974) provided a vital clue by demonstrating that the mutation that permitted propanediol utilization was closely linked to the genes for L-fucose utilization. An intimate relationship between the pathways for propanediol and fucose utilization was further indicated by (1) the observation that Strain 3 (propanediol$^+$) was unable to utilize fucose, and (2) the observation that wild-type strains excrete propanediol during fucose fermentation. It was shown that during anaerobic growth on fucose, synthesis of a lactaldehyde dehydrogenase was induced. The lactaldehyde dehydrogenase was an NAD-dependent enzyme which converted lactaldehyde to L-1,2-propanediol. The lactaldehyde dehydrogenase comigrated on DEAE cellulose with the propanediol dehydrogenase of Strain 3, and it was concluded that they were the same enzyme. The pathway worked out by Cocks *et al.* employs the propanediol (lactaldehyde) dehydrogenase to convert L-1,2-propanediol to lactaldehyde, which is metabolized to pyruvate by two dehydrogenases which function during an aerobic metabolism of fucose. Thus, *E. coli* evolved the ability to use propanediol by constitutively synthesizing an enzyme that is normally produced only during anaerobic growth on fucose. It was not clear whether or not the properties of the propanediol dehydrogenase itself had been altered in the evolutionary process. The number and nature of the mutations involved in the evolution were unclear. The possibility of several mutations was indicated by the observation that Strain 3 had become uninducible for the first three enzymes involved in fucose utilization. Hacking and Lin (1977) have shown that it is possible to obtain propanediol-utilizing mutants that can still grow on fucose, but that these mutants give rise to fucose-negative mutants upon prolonged selection on propanediol. Again, the fucose-negative mutants were uninducible for the first three enzymes in the fucose pathway. They also concluded that several mutations were probably involved in the evolution of propanediol utilization. Recently, Hacking *et al.* (1978) isolated derivatives of Strain 3 that were able to grow faster at low propanediol concentrations. One such mutant, Strain 430, was shown to possess a propanediol transport system that permitted rapid uptake at low propanediol concentrations. Transductional analysis showed that the gene for the new transport system was not linked to the genes for fucose utilization. The origin and natural function of the new transport system remain unknown.

Clarke and her co-workers have studied the evolution of an aliphatic amidase of *Pseudomonas aeruginosa* extensively. Their studies have been reviewed very thoroughly (Clarke, 1974, 1978), and I will address only one aspect of their work as an example of the evolution of a regulatory protein. *Pseudomonas aeruginosa* grows well on acetamide and proprion-amide as carbon and nitrogen sources. The substrates are hydrolyzed by an amidase which is the product of the *amiE* gene, and both are good inducers of amidase synthesis. Wild-type cells are unable to utilize form-amide as a nitrogen source, because it is both a poor substrate of the amidase and a poor inducer of enzyme synthesis. When Brammar *et al.* (1967) isolated mutants that could use formamide as a nitrogen source, two kinds of mutants were obtained: (1) constitutive mutants, which pro-duced amidase at a rate equal to fully induced wild-type cells, and (2) formamide-inducible mutants, which were 10–30 times more inducible by formamide than were the wild-type cells. Genetic analysis showed that both kinds were the result of mutations in the *amiR* gene, a regulatory locus tightly linked (90% cotransduction) with the *amiE* locus. This study is particularly important in that it is the first example of the evolution of a regulatory gene so that it responds to a novel substrate as an inducer.

Several other workers have developed experimental evolution sys-tems which, while not as well developed as the foregoing, deserve men-tion. Schaefler and co-workers have investigated the phospho-β-glucos-idase system in *E. coli* with similar results. Wild-type strains of *E. coli* synthesize P-β-glucosidase A constitutively, but lack P-β-glucosidase B and β-glucoside permease I activities, and are therefore unable to utilize arbutin. A mutant capable of utilizing arbutin was selected and found to synthesize both the permease and P-β-glucosidase B inducibly due to a mutation at the regulatory locus *bglB*. Nitrosoguanidine mutagenesis of another regulatory locus, *bglC,* rendered synthesis of both P-β-glu B and the permease constitutive (Prasad *et al.*, 1973; Prasad and Schaefler, 1974; Schaefler and Maas, 1967). Again, the acquisition of a new catabolic activity involved principally regulatory changes.

Francis and Hansche (1972, 1973) have devised a system for directed evolution of metabolic pathways in yeast. Selection was for improved capability of the enzyme acid phosphatase to hydrolyze β-glycerolphos-phate in a chemostat in a medium buffered to pH 6 (a pH which reduces the efficiency of acid phosphatase by about 70%). Continuous screening of the population for 1000 generations revealed a sequence of "adapta-tion." The first, M_1, was a mutation which effected a 30% increase in the efficiency of orthophosphate metabolism. The second, M_2, effected a change in the pH optimum of acid phosphatase and an increase in its activity. This change was shown to be in the acid phosphatase structural gene. The third, M_3, caused cell clumping—an adaptation to the che-mostat itself and unrelated to phosphate metabolism. This procedure was carried out a second time (Francis and Hansche, 1973) with essentially the same results, except that the mutation in the acid phosphatase structural gene yielded an enzyme with an activity at pH 6 almost equal to the activity of wild type at its pH optimum of 4. In the two experiments, comprising 2000 generations, no events were detected that led to an increased amount of the enzyme. Thus, these experiments in yeast give a result exactly the opposite of results in enterobacteria, i.e., in bacteria, regulatory changes are more common; in yeast, structural changes are more common. Hansche (1975) has shown that yeast *can* exploit the "quantitative change" pathway in order to obtain greater acid phosphatase activity. In one case, there were mutations that increased the amount of acid phosphatase associated with the cell wall, and in another case, there was a duplication of the acid phosphatase gene. This latter case is particularly important, as it provides direct support for the idea that the first step in the acquisition of a new metabolic function may be via gene duplication.

Kemper (1974a,b) has developed a system in *Salmonella typhimurium* which provides an anabolic parallel to my *EBG* studies on evolution. Leucine auxotrophs carrying deletions of the *leuD* gene revert to prototrophy by mutations at the *supQ* locus. Kemper proposes a model in which the *supQ* gene and an additional gene, *newD*, code for two different subunits of an enzyme whose function is yet to be determined. The *newD* gene product is able to replace the absent *leuD* in the complex with the *leuC* gene product, which is the active enzyme. The *newD* product, however, has a greater affinity for *supQ* gene product than for *leuC* gene product; thus, mutations in *supQ* are necessary to make sufficient *newD* product available for active enzyme and prototrophy.

In nature, microorganisms do not grow in pure culture; instead, they exist as members of ecologic communities living in close proximity to one another, competing for resources, and also interacting for their mutual benefit. Slater and his co-workers have modeled this complex situation by studying microbial communities that utilize 2,2'-dichloroproprionic acid (2,2DCPA), which, under the name Dalapon, is used as a herbicide.

Continuous culture (chemostat) procedures were used to isolate communities that could grow on 2,2DCPA (Senior *et al.,* 1976). On each occasion, from several different soil sources, a seven-membered community capable of growth on 2,2DCPA was isolated. The communities consisted of two groups of organisms: three primary utilizers (all of which could grow on 2,2DCPA in pure monoculture) and four secondary utilizers (which were unable to utilize 2,2DCPA in monoculture). The primary utilizers included an unidentified gram-negative bacterium, a *Pseudomonad,* and *Trichoderma viride*; while the secondary utilizers included a budding yeast, a *Flavobacterium* species, an unidentified *Pseudomonad,* and *Pseudomonas putida*. The secondary utilizers are presumed to utilize either excreted metabolites produced by the primary utilizers or to prey directly upon the primary utilizers.

These communities were maintained in continuous culture for up to 3700 generations, and their compositions were remarkable stable, with two exceptions: (1) when the dilution rate was increased above $0.2 \ hr^{-1}$ the yeast was eliminated from the community, and (2) in one community after 2900 hr of growth the secondary utilizer *Pseudomonas putida* acquired the ability to utilize 2,2DCPA, i.e., it became a primary utilizer.

To determine whether *P. putida* had acquired this new metabolic ability by mutation or whether it had acquired it by gene transfer from one of the other primary utilizers, the original *P. putida* strain was studied in pure monoculture.

The original *P. putida* strain, unable to utilize 2,2DCPA, was inoculated into a chemostat supplied with an excess of 2,2DCPA and with sodium proprionate as a growth-limiting carbon source. After 1200 hr a 2,2DCPA-utilizing mutant displaced the original strain, indicating that the new capability arose by mutation.

It is particularly interesting that a new primary utilizer arose in the community despite the presence of three primary utilizers already competing for 2,2DCPA. The mutant *P. putida* did not displace the original strain from the mixed community, but instead coexisted with the unevolved strain.

The unevolved strain of *P. putida* has been designated S3, and the evolved strain PP3. Subsequent studies (Slater *et al.*, 1979) showed that both the unevolved strain S3 and the evolved strain PP3 contained dehalogenase activity, but that the activity was 10–40 times greated in PP3 than in S3. Strain PP3 grew on 2-monochloroproprionate (2MCPA) as well as on 2,2DCPA, while S3 was unable to utilize either substrate. Crude extract of both strains exhibited dehalogenase activity toward mono and dichloroacetic acid; however, those substrates did not serve as growth substrates. Weightman *et al.* (1979) showed that strain PP3 synthesizes two dehalogenases, one with a primary specificity for the dechlorination of proprionic acid, the other with a primary specificity for acetic acid.

This study is particularly important, as it represents the first attempt to apply the approach of experimental evolution to a complex community.

This has been a necessarily and intentionally brief review of the experimental evolution literature. Its purpose was to convey the flavor, rather than the detail, of the experimental approach to evolution. In a particularly clear and well-written article Clarke (1978) has reviewed the literature in depth. Clarke's own elegant work is presented in some detail in that article, and the interested reader will notice many paralells between the evolution of the aliphatic amidase in *Pseudomonas aeruginosa* and the evolution of the EBG β-galactosidase in *Escherichia coli*. A recent review by Mortlock (1981) has stressed the role of regulatory mutations in the acquisition of new metabolic pathways, with particular attention to the evolution of unnatural pentose utilization.

MECHANISMS FOR THE ACQUISITION OF NEW GENETIC MATERIAL

The majority of the studies which I have described demonstrate that cells can evolve new functions by exploiting elements of existing metabolic pathways via structural and regulatory mutations. The extent to which an organism can evolve solely by modification of existing genes must be limited by physical constraints upon the number of functions that can be carried out by a single protein. At some point in its evolutionary career, an organism must acquire additional DNA if it is to continue to evolve. Such acquisition can be divided into two categories: addition of genetic material from sources outside the genome, and addition of DNA by duplication of genetic material from within the genome.

The possible role of extrachromosomal elements in evolution has been reviewed by Reanney (1976). He points out that the available gene pool must be considered to extend across species boundaries, since many conjugative plasmids can transfer genetic information interspecifically. This possibility is supported by findings that determinants for sugar fermentation can be found on conjugative plasmids. Johnson *et al.* (1976) have shown that the ability of *Salmonella tennessee* to ferment lactose and sucrose is determined by the presence of a conjugative plasmid, and that the plasmid can be transferred to a Lac⁻ strain of *E. coli,* thereby rendering that strain both Lac⁺ and Sucrose⁺.

Anderson and Roth (1977) have recently reviewed the literature on tandem duplications in bacteria. They point out that duplications (a) generally cause no loss of gene function, (b) may be of unlimited size, (c) may arise spontaneously at frequencies as high as 10^{-3}. These properties lend themselves to an important role for gene duplication in bacterial evolution. However, duplications are highly unstable, and are subject to both loss and amplification by recombination events. In the one study that I described where evolution of a new function involved gene duplication, Hartley (1974) found that the duplicated copies of the gene for RDH were lost when the selective pressure for xylitol utilization was withdrawn.

Recently, Riley (Zipkas and Riley, 1975; Riley and Anilionis, 1978) has argued strongly for the idea of *genome* duplication in evolution. As early as 1967 Hopwood (1967) pointed out that apparently related genes of *Streptomyces coelicolor* were located directly across from each other on the circular map. Zipkas and Riley (1975) extended this idea to *E. coli,* noting that many related genes were located at 90° or 180° on the genetic map. They suggested that *E. coli* may have undergone two genome doublings in its evolutionary history. More recently (Riley *et al.,* 1978) the map has been reexamined and statistical analyses of positions of gene pairs carried out using several different criteria to define "related gene pairs." This recent analysis showed that all of the genes for enzymes concerned with the central pathways for glucose metabolism fall into four clusters located at 90° and 180° from each other. Other sets of gene pairs considered did not show this highly significant departure from random distribution of position on the map. Based upon an entirely different line of evidence, Sparrow and Naumen (1976) also suggested that genome doubling plays an important role in evolution. They have measured the amount of DNA per haploid genome in a wide variety of species, from viroids to mammals, and found that it does not increase smoothly as one goes up the taxonomic scale, but that it distributes as a series of peaks at doubling intervals. The idea of genome doubling in evolution is by no means established, but it is an exciting idea which may be subject to experimental verification.

THE EBG SYSTEM AS A MODEL FOR ACQUISITIVE EVOLUTION

Most of the experimental systems described above involve selection of mutants that are capable of metabolizing a carbon source that is completely novel to the species. Evolution of the new catabolic function in such cases may involve alteration in regulation, permeation, and degradation of the novel metabolite. The EBG system focuses evolutionary pressure more precisely by directing the evolution of a single, specific new function: the ability to hydrolyze lactose. This is accomplished by beginning with a strain of *E. coli* in which the *lacZ* (β-galactosidase) gene is deleted, but the *lacY* (lactose permease) gene and all of the genes for glucose and galactose catabolism are intact. Such a strain is unable to hydrolyze lactose, but is able to carry out all other functions involved in the lactose degradative pathway. For such a strain, lactose is truly a novel carbon source.

Aside from allowing us to focus our attention upon a single, clearly defined new function, the EBG system has other inherent advantages. The organism itself, *E. coli,* possesses the obvious advantages of being subject to simple genetic manipulation and of permitting use of modern molecular biologic techniques. By selecting for a new lactase function it is possible to take advantage of the vast technology (synthetic substrates, inducers, etc.) developed to deal with studies of the *lac* operon. Finally, the system possesses the advantage of permitting extensive comparisons between the newly evolved lactase function and the older *lacZ* β-galactosidase.

The first EBG mutant was isolated by Campbell *et al.* (1973). They began with *E. coli* strain FO5/RV which carried a deletion of the entire *lac* operon, and an episome on which the *lacY*$^+$ gene was expressed constitutively under the control of an unknown promoter. As it turned out, constitutive synthesis of the lactose permease was a key element in the evolution of lactose utilization by that strain. Strain FO5/RV was streaked onto agar plates containing broth, lactose, and the fermentation indicator triphenyl tetrazolium chloride (Lac-TET plates). The colonies that arose were deep red, indicative of a failure to ferment lactose. After several weeks, white (lactose-fermenting) papillae appeared on the surface of the colonies. Cells from the papillae were streaked onto lactose minimal medium, and the largest colony was designated Ebg-1, ebg standing for *evolved* β-*galactosidase.* Colonies of strain Ebg-1 were incubated until papillae appeared on their surfaces, the papillae were restreaked on lactose minimal medium, and the largest colony was designated Ebg-2. Two further rounds of selection gave strain Ebg-4 which grew poorly on lactose at 42°C. Selection for growth at 42°C yielded strain Ebg-5. When cured of the *lacY*$^+$ episome, the strain was Lac$^-$, due to the absence of a lactose permease. Further selection for lactose utilization yielded an "evolved" permease designated Ebp-1. No further work on the evolved permease has been reported, probably because it was a highly unstable characteristic (Campbell, personal communication). At least two genes in *E. coli* can mutate to transport lactose (Messer, 1974; Hobson, 1978), and one of these may have been the Ebp-1 gene. The final strain carrying the Ebg-5 and Ebp-1 genes was designated LC110, and it was this strain that Campbell characterized.

Strain LC110 synthesized a β-galactosidase enzyme constitutively. The enzyme's activity was assayed on the synthetic substrate *o*-nitrophenyl-β-D-galactoside (ONPG) rather than on lactose, for reasons of convenience. The new β-galactosidase was called ebg enzyme, a designation that remains in use. The ebg enzyme was shown to have kinetic characteristics and specific ionic requirements which distinguished it from the classical *lacZ* β-galactosidase. It was physically distinct, in that ebg enzyme sedimented well ahead of *lacZ* β-galactosidase in a sucrose gradient (Table I). Immunologically, ebg enzyme was completely distinct from *lacZ* enzyme. Antisera raised against *lacZ* enzyme did not react with extracts of strain LC110, nor did antisera against ebg enzyme react with *lacZ* enzyme (Arraj and Campbell, 1975). Additional (unpublished) experiments in my own laboratory have failed to detect any immunologic cross-reactivity between ebg and *lacZ* enzymes.

It was shown that the gene responsible for lactose utilization cotransduced at a low frequency with the *metC* gene, on the opposite side of the chromosome from the *lac* operon. The cotransduction frequency suggested that the gene should be very close to either *serA* or *tolC*. Failure to find cotransduction with *serA* led to the tentative conclusion that the gene was near *tolC*. Subsequent mapping (Hall and Hartl, 1974, 1975) confirmed that conclusion. All of the lactose-utilizing transductants expressed the ebg enzyme constitutively, providing solid genetic evidence that the new enzyme activity was responsible for lactose utilization by strain LC110.

Campbell *et al.* thus provided biochemical, immunologic, and genetic evidence that ebg enzyme was unrelated to the classical *lacZ* enzyme, and thus truely represented evolution of a new function.

The recipient strain that was used in the transductions was *lac* $I^+ Z^- Y^+$. The lactose-utilizing (ebg$^+$) transductants were unable to utilize lactose, despite constitutive synthesis of the ebg$^+$ enzyme, unless isopropyl-thio-β-galactoside (IPTG) was included in the medium to induce synthesis of the lactose permease. Lactose is not an inducer of the *lac* operon (Jobe and Bourgeois, 1972), and it was concluded that ebg enzyme probably could not convert lactose into an inducer of the lac operon.

Campbell felt very strongly that the ebg$^+$ enzyme specified by the *ebg-5* allele present in strain LC110 arose via at least five mutations in

TABLE I. Comparison of *lacZ* and ebg β-Galactosidases[a]

Property	*lacZ* enzyme	ebg enzyme
Doubling time of strains making enzyme, min	52	72
β-galactosidase activity of crude extracts, units/mg	1600	21
pH optimum	5.5–6.5	7.5
Monovalent cation	K^+ or Na^+	K^+
K_m (ONPG), mM	0.35	1.4
K_i (lactose), mM	5.0	20
Estimated molecular weight	5.4×10^5	1.0×10^6
Reaction with anti-*lacZ* serum	Yes	No

[a] Data from Campbell *et al.* (1973).
[b] The molecular weight of ebg enzyme is more reliably estimated as 7.2×10^5 (Hall, 1976a).

the ebg gene. This conclusion was based primarily upon the sequential rounds of selection, each involving picking papillae from the surface of colonies growing on lactose plates. It was assumed that each papilla represented a mutation improving the enzyme for lactose hydrolysis, ebg enzyme. Subsequent studies (Hall, 1977) showed that the ebg enzyme present in strain LC110 probably carried only a single mutation, suggesting that most of the rounds of selection involved advantageous mutations in genes unrelated to lactose utilization. The final round of selection for growth at 42°C probably selected for constitutive synthesis of ebg enzyme, since it was later shown (Hall and Hartl, 1974) that only constitutive strains grow well at 42°C.

Campbell *et al.* thus provided the foundation for this entire series of evolutionary studies by providing genetic, biochemical, and immunologic evidence that lactose utilization could be reevolved in *E. coli* by mutations in a locus completely distinct from the classical *lac* operon.

Campbell *et al.* (1973) had shown that at least one gene, located near *tolC,* could evolve to give a new lactase function. My studies were initiated with several questions in mind: (1) would all lactose-utilizing mutants of a *lacZ* deletion strain arise from mutations in the same gene? (2) Would all the evolved lactose-utilizing strains be alike physiologically, or would a number of different classes of mutants arise? (3) Could strains representing each of the steps in the evolution of the new lactase function be isolated and characterized?

We began (Hall and Hartl, 1974) with an HfrC strain of *E. coli* carrying *lacI*$^+$*Z*$^{deletion\ W4680}$*Y*$^+$, strain DS4680A. Strain DS4680A also carried a mutation making it resistant to the drug spectinomycin. Because the *spcA* mutation is rare (Miller, 1972), it serves as an excellent marker to distinguish descendents of DS4680A from possible Lac$^+$ contaminants. The *lacZ* deletion removes about one-third of the *lacZ* gene, but leaves a fully functional lactose permease (*lacY*) gene under the control of the lac repressor. Strain DS4680A was streaked onto Lac-TET plates identical to those used by Campbell, except that 0.2 mM isopropyl-B-galactoside (IPTG) was included to induce synthesis of the lactose permease, thus assuring the presence of lactose inside the cells. As Campbell had observed, white, lactose-fermenting papillae arose within a few days. Cells from the papillae were streaked onto MacConkey lactose plates also containing 0.2 mM IPTG. MacConkey medium is an extremely sensitive fermentation indicator, and red (lactose-fermenting) colonies were easily detected. Two points concerning this sort of selection should be borne in mind: (1) MacConkey medium is an indicator medium, not a selective medium, and thus the lactose-fermenting colonies were subjected to no *further* selection after being picked from the papillae, and (2) by retaining

Barry G. Hall

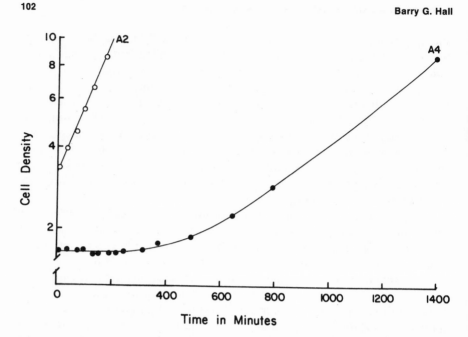

FIG. 1. Growth of constitutive (open circles) and inducible (closed circles) ebg⁺ strains in lactose minimal medium.

only *one* isolate derived from any single colony on the Lac-TET plates we were assured that each isolate was the result of an *independent* spontaneous mutation.

Thirty-four independent lactose-fermenting derivatives of DS4680A were isolated in this way. They fell into two clearly distinguishable classes. One class grew rapidly (0.335 ± 0.039 hr^{-1})* and without delay when shifted from glycerol to lactose miminal medium, while the other class grew more slowly (0.099 ± 0.006 hr^{-1}), and exhibited a several hour delay in growth when shifted from glycerol to lactose minimal medium (Fig. 1).

The three members of the rapidly growing class synthesized a β-galactosidase activity constitutively. The 31 slowly growing strains synthesized β-galactosidase activity *only* when grown in lactose-containing medium. The growth lag exhibited by these strains was thus attributed to an induction lag. Protein synthesis was required for synthesis of β-galactosidase activity in one of the 31 strains, strain A4; thus, the enzyme

---

* Growth rates are given as the first-order rate constant, in reciprocal hours, usually ± the 95% confidence limits. These rates can be converted to the more traditional doubling time in minutes by the equation: doubling time (minutes) = (60 ln 2)/growth rate. The first-order rate constant has the advantage that a larger value indicates faster growth.

was not simply being assembled from preexisting subunits. Induction of the β-galactosidase required lactose. IPTG, a very powerful inducer of the *lac* operon, failed to induce synthesis of the β-galactosidase.

At this point it is important to discuss the use of IPTG in these studies. Unless otherwise indicated, IPTG is *always* included in media containing lactose or other β-galactoside sugars. The sole function of the IPTG is to induce synthesis of the lactose permease, and thus to deliver lactose to the inside of the cell. Neither the constitutive nor the inducible evolved strains grew on lactose in the absence of IPTG. Likewise, induction of β-galactosidase synthesis required *both* lactose and IPTG in the medium. To be sure that induction was not the result of some sort of interaction between lactose and IPTG, we isolated from the inducible strain A4 a mutant that synthesized the *lac* permease constitutively. That mutant synthesized ebg β-galactosidase activity inducibly when grown in the presence of lactose without IPTG. The requirement for IPTG to induce the lac permease has actually been quite a convenience in these studies. Evolved strains can easily be distinguished from wild-type (*lacZ*$^+$) contaminants on the basis of their *lac* phenotypes in the presence and absence of IPTG.

Comparisons of the enzymes from the constitutive and the inducible strains showed that the activities were indistinguishable from each other and, more importantly, indistinguishable from the ebg enzyme described by Campbell *et al.* (1973) as judged by pH optimum, K$^+$ and Mg^{++} requirements, and K_m for ONPG (Table II). Crude extracts of the constitutive strain A2 and of the inducible strain A4 (when grown on lactose) formed precipitin lines with anti-ebg enzyme prepared by J. Campbell. The locus permitting lactose utilization for both strain A2 and A4 was shown to be linked (17–20% recombination) with the *tolC* locus, as pre-

TABLE II. Comparison of Regulated and Constitutive ebg$^+$ Strains^a

Property	Strain A4	Strain A2
Specific activity, glycerol-grown cells	0.16	148
Specific activity, lactose-grown cells	16	127
K_m, (ONPG), no Mg^{++} added, mM	0.88	0.89
K_m, (ONPG), 5 mM Mg^{++}, mM	2.5	2.6
Precipitate with anti-*lacZ* serum	No	No
Precipitate with anti-ebg serum	Yes	Yes
Growth rate in lactose, hr^{-1}	0.11	0.45
pH optimum	7.5	7.5
Monovalent cation	K$^+$	K$^+$

^{a} Data from Hall and Hartl (1974), Hall (1976a) and unpublished experiments.

dicted by Campbell *et al.* It was concluded that lactose utilization had evolved via mutations in the *ebg* gene described by Campbell *et al.* (1973).

There were several differences between the constitutive strains and the inducible strains: (1) constitutive strains grew without a lag upon being shifted to lactose medium, (2) constitutive strains grew much faster than inducible strains, and (3) constitutive strains exhibited much higher levels of ebg β-galactosidase activity than did fully induced inducible strains (Table II). All of the differences could be attributed to the different regulatory states. For most systems examined, it has been observed that constitutive mutants synthesize more enzyme than do fully induced or derepressed regulated strains. Isolation of a constitutive mutant from the lactose-inducible strain A4 provided evidence to support this hypothesis. The constitutive mutant exhibited an enzyme level and growth behavior indistinguishable from those of the constitutive strain A2 (Hall, unpublished observation).

Most surprising, however, was the *homogeneity* of growth rates within each class. For the inducible strains the growth rates were 0.099 ± 0.006 hr^{-1}, a variability no greater than that observed for replicate cultures of the same strain (Hall and Clarke, 1977; Hall, 1978a). This homogeneity led us to conclude that all 34 independent isolates probably synthesized the same evolved enzyme as that studied by Campbell *et al.*

We thus tentatively concluded that in *E. coli* K12 there is only one locus, the *ebg* locus, which can evolve a lactose-hydrolyzing enzyme under our selective conditions. That tentative hypothesis has been borne out over several years. A strain deleted for both the *ebg* gene and the *lacZ* gene has repeatedly been subjected to the standard selective regimin on Lac-TET plates, and has additionally been mutagenized with a variety of chemical mutagens before being subjected to selection. All of those experiments failed to yield any lactose-utilizing mutants (Hall, unpublished experiments). Thus, *E. coli* has a clearly limited evolutionary potential for this particular new function.

While it was clear that we were evolving the same system as that described by Campbell *et al.*, there were some striking differences between their observations and ours. First, our lactose-utilizing strains had been isolated by only a single round of selection, yet the enzyme was indistinguishable from that reported by Campbell *et al.* following five rounds of selection. Our observations seemed inconsistent with their assertation that at least five mutations were required for evolution of the new lactase function. In subsequent experiments, described below, I addressed this question directly.

The second major difference between our results and those of Campbell *et al.* was that the majority of our isolates were regulated, in that they synthesized the ebg β-galactosidase only when grown in the presence of lactose, whereas their strain LC110 synthesized the enzyme constitutively. It seemed unlikely that 31 out of 34 independent isolates had simultaneously evolved a new regulatory system to accompany the newly evolved enzyme; thus, we were forced to conclude that the new β-galactosidase had evolved from a gene that was already subject to induction by lactose.

The Unevolved Enzyme

We had hoped to isolate mutants that represented early steps in the evolution of the new lactase function, but the results described above made that seem unlikely. It still seemed critical, however, to identify the unevolved gene and to isolate and characterize its product. The idea that the expression of the *unevolved* gene might be induced by lactose provided a handle for identification and characterization of the unevolved gene.

We observed that concentrated cell extracts prepared from the *lacZ* deletion strain DS4680A exhibited an extremely low, but detectable, level of β-galactosidase activity as assayed by hydrolysis of ONPG (Hartl and Hall, 1974). When these same cells, which had not been subjected to any selection for lactose utilization, were grown on a non-catabolite repressing carbon source plus lactose and IPTG, the level of β-galactosidase activity was increased over 100-fold (Table III). As was the case for the regulated evolved strains, IPTG alone did not induce synthesis of the enzyme. Two additional "unrelated" $lacZ^\Delta Y^+$ strains exhibited the same behavior: a very low level of β-galactosidase activity when grown in the absence of

TABLE III. Specific Activities of Crude Extracts of *lacZ* Deletion Strains[a]

Strain	Uninduced	Induced
DS4680A	0.28, 0.30[b]	33
W4680	0.11	21
2320M15	0.023	13
1B1	536	—

[a] Data from Hartl and Hall (1974) and Hall and Clark (1977). Specific activities are units/mg of protein assayed on 5 mM ONPG. Uninduced: cells grown in proline or succinate minimal medium. Induced: cells grown in proline or succinate minimal medium containing 0.2 mM IPTG and 0.1% lactose.
[b] Cells grown in succinate minimal medium containing 0.2 mM IPTG.

lactose, and over 100-fold induction when grown (with IPTG) in its presence (Table III).

The fully induced cultures of the unevolved strain DS4680A were unable to utilize lactose as a carbon and energy source, despite having *more* β-galactosidase activity than the fully induced evolved strain A4 (Table II). It seemed likely that the β-galactosidase activity being measured was due to the unevolved ebg enzyme. Taking advantage of an *ebg* deletion mutant isolated from an evolved strain, we mapped the gene for the inducible β-galactosidase activity to the same locus as *ebg,* near *tolC.* *In vitro* assays showed that, although the enzyme was active toward ONPG, it was virtually inactive toward lactose.

It is appropriate at this point to introduce the terminology we use to describe the various states of the ebg enzyme and the *ebg* alleles. The classical nomenclature for bacterial mutants and genes introduced by Demerec (1966) and in general use today is insufficient for dealing with a system where mutants have acquired new activities, rather than having lost normal activities. The structural gene specifying ebg enzyme is designated *ebgA.* In keeping with the terminology developed by Campbell *et al.* (1973), we have reserved the "+" designation for evolved alleles specifying enzyme that allows growth on lactose. As is clear from the preceding section, the unevolved enzyme does not hydrolyze lactose well enough to permit growth. The "unevolved" allele can also be called the "ancestral" or the wild-type allele. We designate the wild type allele $ebgA^o$, the "o" standing for "original." We clearly distinguish the $ebgA^o$ allele from an $ebgA^-$ allele that produces no enzyme or defective enzyme, because the product of the $ebgA^o$ is a β-galactosidase enzyme whose natural function is unknown.

Evolved alleles are designated $ebgA^+$ and differ by one or several mutations from the wild-type allele $ebgA^o$. The various evolved alleles are identified by allele numbers ($ebgA52^+$), and all different alleles are the result of independent spontaneous mutations.

In addition to allele designations, it is often necessary to refer to the enzymes themselves. The wild-type enzyme, the product of the $ebgA^o$ gene, is designated ebg° enzyme, while various evolved enzymes may be designated as ebg^{a+}, ebg^{b+}, etc. We often employ the synonyms "ancestral" or "unevolved" when referring to the wild-type enzyme or allele.

We can now summarize our studies of the unevolved strain DS4680A by saying that synthesis of the unevolved enzyme ebg° is induced by lactose or some lactose derivative. The ebg° enzyme is a β-galactosidase, since it hydrolyzes the synthetic substrate ONPG; however, it is not a lactase, since it does not hydrolyze lactose well enough to support growth.

Regulation of ebg Enzyme Synthesis

The observation that ebg enzyme synthesis is inducible by lactose in both the enevolved strain and in the majority of evolved strains implies the existence of a regulatory gene. It was assumed that the constitutive strains carried a mutation in that regulatory gene, which we designated *ebgR* (Hall and Hartl, 1975). Preliminary experiments in which phage P1 grown on the constitutive *ebgA⁺* strain A2 were used to transduce a *metC tolC* strain showed that the gene order was *ebgA-tolC-metC*. In those experiments all of the *ebgA⁺* transductants were also constitutive, or *ebgR⁻* mutants, suggesting that *ebgR* was tightly linked to *ebgA*.

To test this we selected an *ebgR⁻* mutant of DS4680A that synthesized the unevolved enzyme constitutively. We had observed that phenyl-β-galactoside did not induce ebg enzyme synthesis. We then reasoned that if the unevolved enzyme was active toward ONPG, it might well be active toward phenyl-β-galactoside as well. The *ebgR* mutant was therefore selected by plating DS4680A cells on phenyl-β-galactoside minimal medium. That *ebgR⁻ ebgAº* strain, 1B1, was mated with the F⁻ strain SJ15, which was *argG, tolC, metC,* and carried the inducible evolved gene *ebgA4⁺*. Analysis of the recombinants from that cross yielded the gene order and distances shown in Fig. 2.

To determine how the *ebgR* gene functioned, a series of constitutive *ebgA⁺* alleles were transduced into an F⁻ *argG metC* strain. The level of enzyme synthesis was measured in each of these strains, then F′122, which covers the region from *argG* through *metC,* and therefore carries the wild-type *ebgR* allele, was introduced into each of the strains and the level of enzyme synthesis was again determined. As shown in Table IV, the synthesis of ebg enzyme was strongly repressed in the merodiploid strains. Further analysis showed that about 5% of the cells in the merodiploid populations had lost the wild-type *ebgR* allele and were thus fully constitutive. When corrected for this mutant population, the merodiploid strains were found to be fully repressed by the presence of the wild-type *ebgR* allele on the F′ episome. Thus, by this classical test, the *ebgR⁻* alleles leading to constitutive ebg enzyme synthesis were recessive

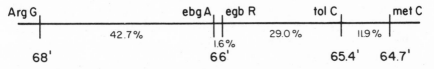

FIG. 2. Map of the *ebg* region of *E. coli.* Numbers below the vertical lines indicate the map positions in minutes. Distances between the loci are in percent recombination.

TABLE IV. Specific Activities of Extracts of Uninduced
Cultures[a]

Strain[b]	F⁻ strain	F 122 merodiploid derivative
SJ5	120	5.7
SJ6	282	7.2
SJ8	193	6.6
SJ9	257	7.2
SJ10	392	15.0
SJ11	444	7.8
SJ12	341	7.4
SJ13	<0.1	—
SJ14	0.1	—

[a] Data from Hall and Hartl (1975).
[b] Strains SJ13 and SJ14 are $ebgR^+$ haploids.

to the wild-type allele. That, together with the ease with which spontaneous constitutive mutants are isolated, led to the conclusion that the $ebgR$ gene specifies a repressor which controls expression of the $ebgA$ gene in a manner directly analogous to the way in which the lac repressor controls expression of the lac operon.

Thus, $ebgR^-$ alleles permit constitutive (unregulated) synthesis of ebg enzyme, while $ebgR^+$ alleles permit ebg enzyme synthesis only in the presence of lactose. The ebg repressor responds to lactose, but *not* to IPTG, as an inducer. It should be noted, however, that when fully induced, the unevolved strain DS4680A synthesizes only 5% as much ebg^o enzyme as does its constitutive derivative strain 1B1 (Table III).

The ebg system thus consists of (at least) two elements: the $ebgA$ gene specifying ebg enzyme, and the $ebgR$ gene specifying ebg repressor. We will turn our attention next to the properties of ebg enzymes, both unevolved and evolved; and then to studies concerning evolution of the regulatory system itself.

Properties of ebg Enzyme

An understanding of the process by which the ancestral protein ebg^o is evolved into an ebg^+ enzyme which hydrolyzes lactose efficiently required biochemical characterization of both the ancestral protein and the evolved proteins (Hall, 1976a).

The ancestral enzyme, ebg^o, was purified from the $ebgR^-$ (constitutive) strain 1B1. Purification was by straightforward classical methods, involving a solubility step (methanol precipitation), a gel filtration step, and, finally, separation on a hydroxyl apatite column. The final product was essentially pure (96% pure). The ebg° enzyme has a subunit molecular weight of 120,000 as judged by SDS-polyacrylamide gel electrophoresis, and a native molecular weight of 720,000 as judged by analysis in the analytical ultracentrifuge. The ebg enzyme thus appears to be a hexameric protein consisting of identical subunits.

Kinetic studies showed that ebg° enzyme had a low affinity for both ONPG and for lactose as judged by apparent K_m values (Table V); however, the maximal velocity V_{max} for ONPG was nearly as high as that of the classical *lacZ* β-galactosidase. Comparisons with *lacZ* β-galactosidase in the same table show that ebg° enzyme must be considered a very inefficient, primitive β-galactosidase.

One of the original goals mentioned was to determine the variety of mutations in the *ebgA* gene that could lead to evolved (ebg$^+$) enzyme. The homogeneity of growth rates in my initial study had suggested that the same mutation had recurred in 34 independent isolates (Hall and Hartl, 1974). A second mutant hunt (Hall *et al.*, 1974), in which 53 mutants were isolated, revealed a new class of *ebgA* $^+$ mutants. While most of the mutants grew exactly like the earlier isolates, three constitutive strains grew at about 25% of the rate of the constitutive strain A2. A fourth strain, 5A1, failed to grow on lactose minimal medium, although it appeared to be weakly lactose positive on MacConkey-lactose indicator plates. When an *ebgR* $^-$ recombinant (strain SJ17) was created from 5A1, it, too, grew at about 25% of the rate of strain A2. It seemed likely that these *ebgA* $^+$ strains represented a new class of mutants. I designated the enzyme from strain A2 as ebg^{+a}, the enzyme from one of the new constitutive mutant strains 5A2 as ebg^{+b}, and the enzyme from the recombinant strain SJ17 as ebg^{+c}.

I had already been surprised by the observation that crude extracts of the lactose-negative *ebgA° ebgR* $^-$ strain 1B1 had about twice as much ebg enzyme activity as the evolved *ebgA* $^+$ *ebgR* $^-$ strain A2 when activity was assayed using ONPG. With *lacZ* β-galactosidase from a large variety of *E. coli* strains it has been observed that activity toward ONPG correlates well with activity toward lactose (B. Levin, personal communication). The new strains 5A2 and SJ17, although they grew only one-fourth as fast, also had about twice as much ebg enzyme activity as strain A2. It seemed likely that for mutant ebg enzymes there was little correlation between ONPG and lactose-hydrolyzing activity. Purification of the ebg^{+a}, ebg^{+b}, and ebg^{+c} enzymes confirmed this idea (Table V). The newly evolved enzymes certainly exhibited drastically altered activities

TABLE V. Kinetic Properties of Purified ebg Enzymes[a]

Enzyme	Pure specific activity	K_m (ONPG)	V_{max} (ONPG)	V_{max}/K_m (ONPG)	K_i (Lactose)	K_m (Lactose)	V_{max} (Lactose)	V_{max}/K_m (Lactose)	Growth rate[b] hr^{-1}
ebg^o	11,000	156	335,000	2,150	1,800	1,200	1,870	1.6	0
ebg^{+a}	5,500	3.3	10,300	3,090	55	63	4,180	67	0.37
ebg^{+b}	9,300	0.60	10,000	16,660	690	860	11,800	13.6	0.13
ebg^{+c}	11,400	0.76	13,300	17,500	380	240	5,130	21	0.14

[a] Data from Hall (1976a). K_m and K_i values are mM substrate or inhibitor. V_{max} values are in nmole substrate hydrolyzed min^{-1} mg^{-1} protein.
[b] Growth rates of strains synthesizing these enzymes constitutively.

toward ONPG, however, these activities correlated not at all with their growth rates on lactose. On the other hand, activities toward lactose correlated well with growth rates.

While all of the evolved enzymes exhibited increased affinity (decreased apparent K_m) for lactose and an increased V_{max}, it was clear, even from this small sample, that evolution did not act primarily on either of the two major kinetic parameters, substrate affinity or turnover number. The ebg^{+b} enzyme showed only about a 30% improvement in affinity for lactose, but the turnover number (proportional to V_{max}) increased over 600%; on the other hand, the ebg^{+c} enzyme increased nearly 500% in affinity for lactose, while improving only a little over 200% in turnover number. These two enzymes, with very different kinetic parameters, permit similar growth rates on lactose (Table V). This is because at substrate concentrations much below the K_m, the velocity of a reaction is proportional to V_{max}/K_m. This means that at physiologic lactose concentrations (about 53 mM lactose) mutations that affect primarily the affinity or primarily the turnover number can result in the same rate of lactose hydrolysis if the ratio V_{max}/K_m is the same. Thus, from an enzymologist's point of view, evolution is not expected to favor either K_m or V_{max} improvement preferentially, but simply to favor mutations that improve the overall rate at physiologic substrate concentrations. This study bears out that expectation. The evolved enzymes permit constitutive strains to grow on lactose at rates that are roughly proportional to the measured V_{max}/K_m, with ebg^{+a} enzyme yielding the fastest growth rate.

The relationships between enzyme efficiency measured *in vitro* and growth rates will be discussed more thoroughly in a subsequent section, but these data make it very clear that there is more than one mutation in the *ebgA* gene that can lead to evolved enzymes capable of supporting growth on lactose.

How Many Mutations Are Required to Evolve ebg$^+$ Enzyme?

The ebg enzymes can be ranked in the ascending order ebg^o, ebg^{+b}, ebg^{+c}, ebg^{+a} with respect to V_{max}/K_m (lactose), an order that exactly corresponds to the order with which constitutive strains synthesizing these enzymes grow on lactose. Do these three enzymes correspond to three successive steps in the evolution of ebg enzyme, or do they simply represent three alternative single point mutations in the gene? The two alternative models are diagrammed in Fig. 3. The two models make very different predictions: Model I predicts that ebg^{+a} mutants could be recovered as single-step mutants only from ebg^{+b} mutants, while model II

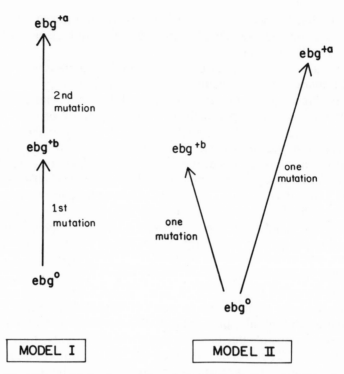

FIG. 3. Alternative models for the evolution of ebg⁺ strains that grow rapidly on lactose.

predicts that ebg^{+a} mutants could be recovered directly by a single-step mutation from the wild-type allele *ebgAᵒ*. Model I is clearly too specific, in that there might be a number of multistep pathways that could lead to enzymes as efficient as ebg^{+a}; model II, however, derives its strength from its specificity. If it can be shown that an enzyme as efficient as ebg^{+a} enzyme can arise via a single mutation in the *ebgA* gene, then model I and all its variants can be rejected as being obligatory for the evolution of ebg enzyme. For the sake of simplicity, it was decided to test the hypothesis that a single mutation in *ebgA* could lead to a strain that grew as rapidly as strain A2, the strain synthesizing ebg^{+a} enzyme constitutively.

To avoid possible regulatory complications, I began with strain 1B1, which synthesizes the unevolved enzyme ebgᵒ constitutively (*ebgAᵒ ebgR⁻*) (Hall, 1977).

Two kinds of experiments were carried out; one involving mutagenesis with ethylmethanesulfonate (EMS), and the other measuring the spontaneous mutation rate. In both cases, the frequency or rate of mutations to *ebgA*⁺ was compared with the frequency or rate of mutations to streptomycin resistence. The comparison was valid because resistance to the high concentrations of streptomycin employed was known to arise via a *single* mutation in the *rpsL* gene, and (like *ebgA*⁺ mutations), the *rpsL* mutation must still permit synthesis of a functional protein.

In four mutagenesis experiments the frequency of lactose-utilizing mutants always exceeded the frequency of streptomycin-resistant mutants, in two of those experiments by a factor of over 100 (Table VI). Several of those mutants were isolated and studied further. All of the mutations studied mapped to the *ebgA* gene as judged by cotransduction with *tolC*. The growth rates of four of those strains were the same as those of strains synthesizing ebg⁺ᵃ enzyme, while a fifth mutant grew at a rate similar to that of strains synthesizing ebg⁺ᵇ enzyme. It seemed very likely that ebg⁺ᵃ enzyme arose via a single mutation in the *ebgA°* gene.

To confirm that hypothesis, I measured the spontaneous mutation rate directly using the Luria and Delbruck (1943) fluctuation test. In two experiments the mutation rate to *ebgA*⁺ was 3.1×10^{-9} and 2.4×10^{-9} per cell division; the rate to streptomycin resistance was 1.1×10^{-10} per cell division. Again, mutations were shown to be in the *ebgA* gene, and isolates were obtained that grew as fast as strain A2.

It was clear that the "best" ebg enzyme, ebg⁺ᵃ, could arise as the consequence of a single point mutation.

At this point, I had obtained answers to several of the questions we posed at the beginning of the study. How many genes can evolve into a lactose-hydrolyzing system? All my data indicate that for *E. coli* K12 the answer is one. How many mutations are required to evolve the new enzyme? Again, the answer is one. How many *different* mutations in the same gene can yield a new lactase enzyme? At least three, leading to enzymes ebg⁺ᵃ, ebg⁺ᵇ, and ebg⁺ᶜ.

These answers were quite satisfying; however, they raised a perplexing problem: why were spontaneous lactose-utilizing mutants never

TABLE VI. Mutation Frequency per Survivor in EMS-Treated
Cultures of Strain 1B1

Expt. no	Lac⁺	Streptomycin resistant	Lac⁺/strep resistant
1	7×10^{-8}	2.4×10^{-8}	2.9
2	140×10^{-8}	1.2×10^{-8}	117
3	250×10^{-8}	2.3×10^{-8}	108
4	7.4×10^{-8}	0.7×10^{-8}	10.6

obtained simply by plating a *lacZ* deletion mutant on lactose minimal medium? Recall that the measurements of the spontaneous mutation rates had been performed with the constitutive strain 1B1, and measured the rate of mutation in the *ebgA* gene itself. Strain DS4680A, which differed from 1B1 only in that it was *ebgR*$^+$, never gave rise to spontaneous lactose-utilizing mutants. Regulated evolved strains, such as A4, grew more slowly than their constitutive counterparts (see p. 104), and I considered the possibility that they might have an insufficient selective advantage when dense cultures were plated onto lactose minimal medium. Reconstruction experiments in which strain A4 was mixed with strain DS4680A at a ratio of 10^{-9} showed that the evolved strain could be recovered, and that it was still regulated after the selection on minimal plates. The paradox, then, was that spontaneous, regulated *ebgA*$^+$ mutants were easily isolated as papillae, but they could not be isolated on minimal medium. Since the ability to select spontaneous *ebgA*$^+$ mutants seemed to depend upon the state of the *ebgR* gene (ebgR$^-$ worked, but *ebgR*$^+$ did not), we decided to investigate the properties of *ebgR* carefully (Hall and Clarke, 1977).

Evolution of the Regulatory Gene *ebgR*

Just as it was important to characterize the unevolved enzyme in order to understand its evolution, it seemed reasonable to begin at the beginning and characterize the unevolved repressor of strain DS4680A.

We began by determining the efficiency with which various β-galactoside compounds induced synthesis of ebg enzyme (Table VII). Of all the compounds tested, only lactose is an effective inducer. A cautionary point should be mentioned: *in vivo* induction experiments such as those described here cannot determine the nature of the actual inducer or effector molecule that interacts with the repressor. For instance, although lactose acts as an *in vivo* inducer of the *lac* operon, *in vitro* studies of effector–repressor interactions have shown that lactose is actually an anti-inducer of the *lac* operon (Barkley *et al.*, 1975), and that it is converted into the true inducer allolactose by the *lacZ* β-galactosidase itself (Jobe and Bourgeois, 1972). Thus, it cannot be concluded that lactose itself interacts with the *ebg* repressor. On the other hand, we *can* conclude that those compounds listed as noninducers or weak inducers do not interact effectively with the *ebg* repressor, because all of those compounds are known to be transported into the cell either by the *lac* permease or other permeases. Those compounds indicated by an asterisk are known to be effective inducers of the *lac* operon.

TABLE VII. Induction of ebg° Enzyme in Strain DS4680A (*ebgA° ebgR*[+])[a]

Additions	Crude extract specific activity, mg[−1] ± 95% confidence interval
Noninducers	
None	0.28 ± 0.058
0.2 mM Isopropyl thiogalactoside*	0.30 ± 0.110
0.1% Melibiose*	0.20 ± 0.045
0.1% Galactose	0.20 ± 0.052
0.1% Lactobionate	0.20 ± 0.077
1.0 mM Glycerol-β-D-galactoside*	0.20 ± 0.059
1.0 mM β-Methyl thiogalactoside*	0.34 ± 0.047
0.1% Phenyl-β-galactoside	0.13 ± 0.080
0.1% Galacturonic acid	0.22 ± 0.060
Weak inducers	
0.1% Methyl-β-galactoside	1.4 ± 0.3
1.0 mM Thiodigalactoside	1.3 ± 0.6
0.1% Lactulose	3 ± 1.9
0.1% Galactosyl-β-1,3-D-arabinose	2.3 ± 0.6
Strong inducers	
0.1% α-Lactose	29 ± 5.4
0.1% β-Lactose	33 ± 3.1
Strain 1B1 (*ebgA° ebgR*[−]) constitutive	
None	536 ± 43

[a] Data from Hall and Clarke (1977). Asterisk indicates a compound known to be an effective inducer of the *lac* operon.

It is worth noting that there seems to be no interaction between the repressors and operators of *ebg* and of *lac*. The *lacI*[+] *ebgR*[−] strains express the *ebgA* gene constitutively; hence the *lac* repressor does not bind the *ebg* operator; and *lacI*[−] *ebgR*[+] strains express the *lac* operon constitutively, indicating that *ebg* repressor does not bind to *lac* operator. Likewise, powerful inducers of the *lac* operon do not induce the *ebgA* gene.

To study regulation in the evolved strains it was necessary to have a valid basis for comparing strains that synthesize different ebg enzymes. The values in Table VII were obtained by measuring the activities of crude extracts toward ONPG. The comparisons were valid because all extracts contained the same enzyme; however, direct comparison of a strain synthesizing ebg° enzyme with a strain that synthesized ebg[+a] enzyme at exactly the same rate would not be valid because ebg[+a] enzyme is only one-half as active toward ONPG as is ebg° enzyme (see Table V). To facilitate comparisons specific activity values were converted into specific synthesis values by dividing the specific activity of the crude extract by the specific activity of the purified enzyme synthesized by that strain. Specific synthesis, then, is a dimensionless number which is the fraction of total soluble protein that is ebg enzyme; i.e., specific synthesis directly reflects the number of molecules of ebg enzyme present in an extract.

Table VIII shows the specific synthesis of ebg enzyme for strain DS4680A and for the regulated evolved strain A4. The two strains have the same uninduced, or basal, level of enzyme synthesis; however, lactose induces the evolved strain A4 four times greater than it induces the unevolved strain DS4680A.

Why is lactose a better inducer in the evolved than in the unevolved strain? Again, two alternative models can be offered to explain the observation.

Model I would suggest that the difference is a property of the *ebgA* allele present. Perhaps the evolved enzyme present in strain A4 converts lactose to an inducer more efficiently than does the unevolved enzyme present in strain DS4680A. According to this model, there is no difference in the *ebgR* alleles present in the two strains.

Model II would suggest that the difference is a property of the *ebgR* allele present. According to this model, strain A4 would carry a mutant *ebgR* allele (in addition to its mutant *ebgA* allele) which specified a repressor that was more sensitive to lactose as an inducer. According to this model, then, strain A4 is actually a double mutant.

To distinguish between the two models, we constructed a recombinant between the two strains that carried the *ebgA*⁺ allele of strain A4 and the *ebgR*⁺ allele of strain DS4680A. This was accomplished by first selecting an *ebgR*⁻ derivative of A4, then crossing it with DS4680A and recovering a regulated lactose-positive recombinant. The recombinant strain SJ18 appeared only weakly lactose-positive on MacConkey-lactose indicator plates, but was easily distinguished from the dark red strain A4

TABLE VIII. Inducers of Strains DS4680A and A4[a]

| | Specific synthesis $\times$ 10^2 $\pm$ 95% C.I. | |
Inducer	DS4680A ($ebgR^+$)	A4 ($ebgR^{+u}$)
None	0.0022 ± 0.0003	0.0030 ± 0.0007
Thiodigalactoside	0.012 ± 0.0054	0.015 ± 0.0036
Methyl-β-galactoside	0.012 ± 0.0025	0.014 ± 0.0013
Lactulose	0.027 ± 0.017	0.056 ± 0.028
Lactose	0.26 ± 0.05	1.23 ± 0.25

[a] Data from Hall and Clarke (1977).

and the dead white strain DS4680A. Strain SJ18 had a specific synthesis similar to that of strain DS4680A, indicating that the level of expression of the *ebgA* gene is determined by the *ebgR* allele present, i.e. supporting model II (Table VIII).

While model II is not the more economical model (it requires that strain A4 evolved via *two* spontaneous mutations, one in *ebgA*, the other in *ebgR*), it does offer a solution to the paradox that began this section. Spontaneous mutants of the A4 type are not detected on lactose minimal medium because, although we can detect mutants present at a frequency of 10^{-9}, we certainly could not expect to detect mutants present at the square of that frequency (10^{-18}).

In strain A4, the repressor evolved along with the enzyme whose synthesis it controls. Why is coevolution of the repressor and the enzyme obligatory? To understand this coevolution we must return briefly to the properties of the evolved enzymes themselves.

The evolved enzymes ebg^{+a}, ebg^{+b}, and ebg^{+c} had been characterized in terms of their kinetics of lactose hydrolysis *in vitro*. Thus, from the Michaelis–Menton equation, the rate of lactose hydrolysis at the physiologic internal lactose concentration could be calculated if that concentration were known. Since lactose is transported by the *lac* permease in these studies, the data of Winkler and Wilson (1966) were used to estimate that in our growing cultures the internal lactose concentration would be 53 mM. This emphasizes the advantage of using the lactose permease for lactose transport, rather than replying upon an uncharacterized "evolved" permease such as Campbell's ebp. Table IX shows the rate at which various pure enzymes would hydrolyze 53 mM lactose (pure enzyme velocity). The specific synthesis of ebg enzyme was measured for each of these strains as described above. We then calculated the *in vivo* velocity as the specific synthesis times the velocity at 53 mM lactose. This *in vivo* lactase activity, then, is the rate at which cells are expected to hydrolyze lactose in terms of nanomoles hydrolyzed per minute per

TABLE IX. Estimated *in Vivo* Lactose Activities[a]

Strain	Enzyme	Velocity at 53 mM lactose, units mg^{-1}	Specific synthesis $\times 10^2$	Estimated velocity *in vivo*, units mg^{-1}	Growth rate, hr^{-1}
A4	ebg^{+a}	1920	1.22	23.5	0.10
SJ12	ebg^{+b}	681	5.1	34.7	0.13
SJ17	ebg^{+c}	916	4.85	42.1	0.15
A2	ebg^{+a}	1920	4.61	84.4	0.37

[a] Data from Hall and Clarke (1977).

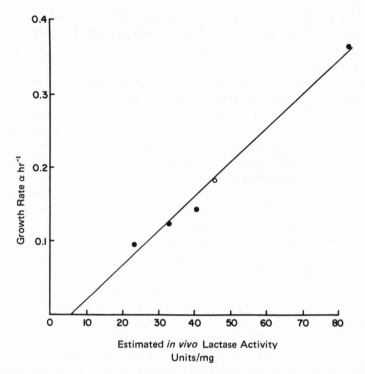

FIG. 4. Growth rate on lactose versus *in vivo* lactase activity for five *ebgA*⁺ strains.

miligram of soluble protein. Figure 4 (solid circles) shows a plot of *in vivo* lactase activity versus observed growth rates. The line shown is the least squares fit to the points, has a corelation coefficient of 0.99, and has the equation: growth rate $\alpha = (0.0046)(in\ vivo\ \text{velocity}) - 0.027$.

The line does not pass through the origin, indicating that there is a minimal lactase activity required for growth, about 5.8 units/mg. Micro-biologists refer to the energy derived from that threshold activity as the "maintenance energy." The equation predicts that strains with less than 5.8 units/mg of *in vivo* lactase activity will not grow on lactose as a sole carbon and energy source. Table X shows several examples bearing out that prediction. Fully induced strain DS4680A has the least activity, and fails to grow. The *ebgA*° constitutive strain fails to grow because, even at its enormous specific synthesis, the enzyme simply has too little lactase activity. Strain 5A1 was mentioned earlier as the original source of the ebg⁺ᶜ enzyme mutation; it failed to grow on lactose until made consti-tutive. Its specific synthesis is characteristic of the unevolved *ebgR* allele, leading to an *in vivo* lactase activity well below the threshold required for growth. Last, the recombinant strain SJ18, carrying the "best" evolved enzyme allele, failed to grow on lactose because it carries the unevolved *ebgR* allele of DS4680A.

The *ebgR* allele present in strain A4 was clearly not wild type; thus, it should not be called *ebgR*$^+$. We decided to designate such alleles *ebgR*$^{+u}$, to indicate a fully functional repressor that is more sensitive to lactose as an inducer than is wild-type repressor. All of the regulated strains isolated in the first two mutant hunts grew at rates very similar to that of strain A4, indicating that they carried *ebgR*$^{+u}$ alleles, and *ebgA*$^+$ alleles specifying ebg enzyme with lactase activity comparable to that of ebg^{+a}. Why were no regulated strains synthesizing enzymes like ebg^{+b} or ebg^{+c} detected in those studies? If they carried *ebgR*$^{+u}$ alleles, their growth rates would be 0.012 and 0.026 (doubling times of 57 and 27 hr), respectively. It does not seem unreasonable to expect that isolation of such slowly growing mutants would be quite rare.

The equation predicts that any regulated evolved strains that grow more rapidly than does strain A4 must have a more efficient enzyme (at 53 mM lactose), or must have a repressor more sensitive to lactose than is *ebgR*$^{+u}$ repressor. In another mutant hunt a series of regulated evolved strains, selected exactly as before, was screened for rapid growers. (That screening was greatly facilitated by the use of XGAL medium, upon which *ebgR*$^-$ colonies are blue.) One isolate, strain PW16-8, clearly not constitutive, had a growth rate of 0.18 hr^{-1}, 80% faster than strain A4. From its growth rate on lactose we predicted that the *in vivo* lactase activity would be 44.7 units/mg. We purified and characterized that enzyme, which we designated ebg^{+d}. The ebg^{+d} enzyme had a specific activity on ONPG of 7150 units/mg, a K_m (lactose) of 49 mM, and a V_{max} (lactose) of 5400 units/mg, yielding an estimate of 2800 units/mg for pure enzyme acting on 53 mM lactose. The specific synthesis of strain PW16-8 was 1.58%, indicating an *ebgR*$^{+u}$ allele. From these biochemical measurements, the *in vivo* lactase activity was estimated as 44.4 units/mg (open

TABLE X. Predicted and Observed Growth Rates[a]

Strain	Enzyme	Velocity at 53 mM lactose, units mg^{-1}	Specific synthesis × 10^2	Estimated velocity *in vivo*, units mg^{-1}	Growth rate, hr^{-1} Predicted	Growth rate, hr^{-1} Observed
DS4680A	ebg^o	82	0.27	0.22	0	0
1B1	ebg^o	82	4.6	3.85	0	0
5A1	ebg^{+c}	916	0.21	1.92	0	0
SJ18	ebg^{+a}	1920	0.31	5.91	0.00036	0

[a] Data from Hall and Clarke (1977).

circle, Fig. 4), in excellent agreement with the predicted value. Enzyme ebg^{+d} thus is the most efficient lactase yet isolated, but even that enzyme would permit a growth rate of only 0.007 hr^{-1} (100 hr doubling time) on lactose if its synthesis were regulated by the wild-type repressor rather than by an evolved ($ebgR^{+u}$) repressor.

We concluded that coevolution of the regulatory and structural genes is an obligatory feature of the ebg system. We must now make a distinction between evolution of an enzyme and evolution of a function. Evolution of the lactose-hydrolyzing enzyme requires only one mutation in $ebgA$, but evolution of the function "lactose utilization" requires two mutations: one in $ebgA$ to improve the efficiency of the enzyme, and the other in $ebgR$ to increase the *amount* of enzyme that is synthesized. The $ebgR$ mutation may be either an $ebgR^{-}$ mutation, which leads to uncontrolled gene expression, or it may be $ebgR^{+u}$, which leads to more effective control of gene expression. (The problem of two simultaneous, spontaneous mutations is discussed later.)

A priori, one would expect that mutations leading to loss of repressor activity ($ebgR^{-}$) would be more common that mutations leading to increased sensitivity to lactose while retaining full repressor function ($ebgR^{+u}$). Since 90% of the evolved strains isolated are regulated ($ebgR^{+u}$), it seems likely that unregulated enzyme synthesis must itself be very disadvantageous.

The *ebgB* Gene

We have recently detected a second gene in the *ebg* operon. As controls for an experiment testing the specificity of antibodies directed against ebg enzyme, we ran crude extracts of the $ebgR^{+}$ strain DS4680A and of the $ebgR^{-}$ strain 1B1 on SDS-polyacrylamide gels (SDS-PAGE). The ebg enzyme band at 120,000 mol. wt. was clearly visible in the strain 1B1 extract; but there was a second band at 79,000 mol. wt. that was much stronger in strain 1B1 than in strain DS4680A. A further series of gels (Fig. 5) showed that both the ebg band and the second band were always present in constitutive strains, but absent in repressed strains. The 79,000 mol. wt. protein was not precipitated by antibodies that precipitated ebg enzyme effectively; thus the 79,000 mol. wt. protein is the product of a second gene under control of the ebg repressor, *ebgB*. Measurements of the peak areas showed that ebg enzyme constitutes about 1.8% of the total protein, while *ebgB* protein constitutes about 1.25%. The figure 1.8% for ebg enzyme is somewhat lower than the earlier estimates based upon enzyme activity; however, those estimates were based

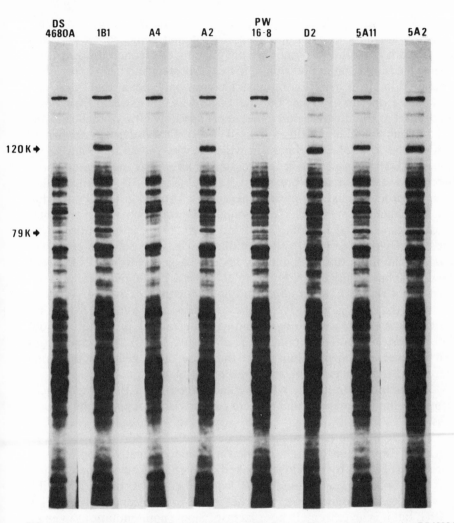

FIG. 5.   Autoradiographs of SDS-polyacrylamide gels. Proteins from uninduced strains DS4680A, A4, and PW 16-8 and from constitutive strains 1B1, A2, 5A11, and 5A2. The 120-kilodalton protein is ebg enzyme; the 79-kilodalton protein is the *ebgB* gene product.

upon soluble protein, not including membrane-bound or nucleic acid-bound proteins. To determine whether or not *ebgB* was in the same operon as *ebgA*, we selected a series of mutants from the constitutive strain A2 in which insertion of a transposable element (Tn5 or Tn9) eliminated expression of the *ebgA* gene. Transposable elements are mobile DNA sequences, several thousand bases long, which include drug resistance genes. When such an element becomes inserted into a gene, it disrupts the gene, not only preventing expression of that gene, but of all other downstream genes in the same operon. When we screened the *ebgA::Tn* strains, we found that expression of the *ebgB* protein had been completely eliminated in all 15 insertion strains. This not only meant that *ebgB* was in the same operon as *ebgA,* but that *ebgB* was downstream from *ebgA* (if *ebgA* were the downstream gene, approximately 60% of the Tn insertion mutants would have still expressed *ebgB*). That order was confirmed by a second kind of experiment in which it was shown that, during ultraviolet (UV) light treatment, expression of *ebgB* was inactivated about three times faster than expression of *ebgA.* (Since RNA transcription does not proceed past the pyrimidine dimers formed during UV irradiation, distal gene expression is inactivated more rapidly than are promoter proximal genes.) Our mapping (Hall and Hartl, 1975), later confirmed by Hartl and Dykhausen (1979), showed that the distance between *ebgA* and *ebgR* was no more than 0.02 min, or about 850 base pairs of DNA. Since about 2000 base pairs are required to code for a 79,000 mol. wt. protein, there is not room between *ebgA* and *ebgR* for the *ebgB* gene. We thus conclude that the gene order in the *ebg* region is *tolC-ebgR-(ebgA$_p$, ebgA$_o$)-ebgA-ebgB,* where *ebgA$_p$* and *ebgA$_o$* mean the promoter and operator sites, respectively (Backman and Low, 1980). We cannot, of course, rule out the possibility that there are other genes in the same operon either between *ebgA* and *ebgB* or distal to *ebgB.*

We can now summarize what we know about the EBG system itself, before beginning to apply that system to study some evolutionary questions.

The *ebg* operon is a second "β-galactosidase" operon present in *E. coli* K12. Expression of the *ebg* operon is under negative control by a repressor specified by the closely linked gene *ebgR.* The operon consists of at least two genes, *ebgA* and *ebgB,* transcribed in that order. The wild-type *ebgA* gene product, ebg° enzyme, is a β-galactosidase, hydrolyzing ONPG, but it is an extremely poor lactase. Single point mutations in *ebgA* can alter ebg enzyme so that its lactase activity is greatly enhanced. Different *ebgA* mutations lead to ebg enzymes with differing lactase activities.

The basal (repressed) level of expression of *ebgA* is very low, 0.002% of the soluble protein, or about five molecules per cell. In a wild-type strain, induction by lactose leads to a 100-fold increase in the expression of *ebgA*; however, thio-β-galactosides and most other compounds tested are unable to act as inducers. Lactose is not a very effective inducer, since it leads to only 5% of the maximal level of expression attained by *ebgR$^-$* strains. The maximal level of expression is high, amounting to about 5% of the soluble protein. Taken together, the *ebgA* and *ebgB* gene products account for about 3% of the *total* protein synthesis in a constitutive strain, indicating that the *ebg* operon has a high-efficiency promotor. Despite high levels of synthesis, the wild-type enzyme (ebg°) hydrolyzes lactose too poorly to permit an *ebgR*⁻ strain to grow on lactose.

Because lactose is not a very effective inducer of the wild-type repressor, mutations in the *ebgR* gene are required if a strain is to grow on lactose. One way to greatly increase the amount of ebg enzyme synthesized is by *ebgR*⁻ mutations, which result in complete loss of repressor function. Such mutations are observed in only about 10% of the evolved strains, suggesting that unregulated expression of the *ebg* operon is disadvantageous under these selective conditions. The remaining 90% of the evolved strains carry *ebgR*⁺ᵘ mutations, which result in a fully functional repressor which exhibits an increased sensitivity to lactose as an inducer.

EVOLUTION OF MULTIPLE NEW FUNCTIONS FOR EBG ENZYME

We had originally hoped to describe a "pathway" for the evolution of lactose utilization by isolating mutants representing steps consisting of increasingly more active ebg enzyme. The realization that a single mutation in *ebgA* was sufficient to convert ebg° enzyme into an efficient lactase was therefore disappointing. As an alternative, I decided to attempt to direct the evolution of ebg enzyme toward effective hydrolysis of additional new substrates. Because most β-galactoside compounds are very ineffective inducers of ebg enzyme synthesis (Hall and Clarke, 1977; see also Table VII), these studies all employed *ebgR*⁻ (constitutive) strains.

I began by directing the evolution of ebg enzyme so that it became an effective β-methylgalactosidase (Hall, 1976b). Strain 1B1 (*ebgA*° *ebgR*⁻) grows very slowly on β-methylgalactoside (MG) (growth rate 0.02 hr⁻¹, or a doubling time of about 31 hr). Strain 1B1 was streaked onto MG-TET plates, and white papillae appeared on the surface of the colonies after 2 weeks. A spontaneous MG-fermenting mutant, strain 2B1, was isolated from one of those papillae. Strain 2B1 had a growth rate of 0.104 hr⁻¹ on MG, fivefold faster than its parental strain. The mutation permitting effective use of MG was mapped to the *ebgA* gene.

When strains synthesizing the enzymes ebg⁺ᵃ, ebg⁺ᵇ, and ebg⁺ᶜ, that had been selected for lactose hydrolysis were examined, they were also found to use MG, with growth rates from 0.077 to 0.091 hr⁻¹. Thus, selection for lactose utilization indirectly selected for improved MG utilization. The converse, however, was not true: strain 2B1, selected for MG utilization, was unable to utilize lactose at all. Thus, the mutation in strain 2B1 was not merely another example of a standard ebg$^+$ mutation, but instead it represented a truly alternative new function, or evolutionary destination, for the *ebgA* gene. This was the first piece of evidence that the evolutionary potential of the *ebgA*° gene was not limited to a single new function.

Because of technical difficulties in assaying methylgalactoside-hydrolyzing activity *in vitro*, I turned to other new substrates for further studies on the evolutionary potential of ebg enzyme.

Four disaccharides were chosen for the study (Hall, 1978a), principally on the basis of cost and commercial availability: lactose (4-O-β-D-galactopyranosyl-D-glucose), lactulose (4-O-β-D-galactopyranosyl-D-fructose), galactosylarabinose (3-O-β-D-galactopyranosyl-D-arabinose), and lactobionic acid (4-O-β-D-galactopyranosyl-D-gluconic acid). The structures of these sugars are shown in Fig. 6.

I began by considering the behavior of strains synthesizing the already studied enzymes ebg^{+a}, ebg^{+b}, ebg^{+c}, and ebg^{+d} toward the various sugars. Table XI shows that the strains synthesizing the enzymes ebg^{+a} and ebg^{+d} (strains A2 and D2) grew rapidly on lactose, extremely slowly on galactosylarabinose, and not at all on lactulose or on lactobionate; while the strains synthesizing enzymes ebg^{+b} and ebg^{+c} (strains 5A2 and SJ17) grew more slowly on lactose, extremely slowly on galactosylarabinose, at a moderate rate on lactulose, and not at all on lactobionate. All four of these strains had been selected for growth on lactose, so that lactulose utilization by strains 5A2 and SJ17 represented indirect selection of a new function, comparable to methylgalactoside utilization exhibited by lactose-utilizing evolvants (Hall, 1976b). *In vitro* assays of ebg enzyme activities confirmed the differences implied by the growth rate measurements: enzymes ebg^{+a} and ebg^{+d} were very active toward lactose, but inactive toward lactulose; enzymes ebg^{+b} and ebg^{+c} were less active toward lactose, and more active toward lactulose than they were toward lactose. On the basis of these differences, the enzymes ebg^{+a} and ebg^{+d} were designated class I, and the enzymes ebg^{+b} and ebg^{+c} were designated class II.

Because lactulose is a poor inducer of ebg enzyme synthesis, classification on the basis of growth rates can only be done in *ebgR*$^-$ strains. Of the 36 independent *ebgR*$^-$ strains selected for growth on lactose, 31 were class I and five were class II strains. This does not imply that class I mutations in *ebgA* arise more frequently. Because class I strains grow more than twice as fast as class II strains on lactose, it is not surprising

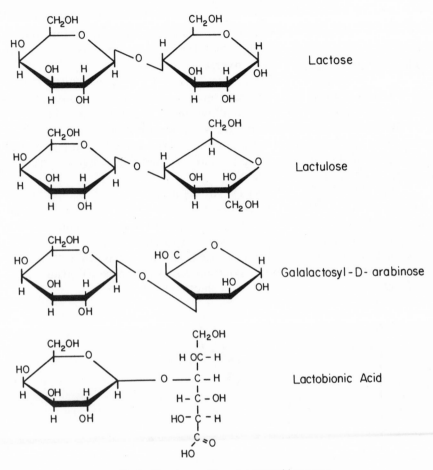

FIG. 6.   Structures of four β-galactoside sugars.

that they should be recovered more often from lactose fermenting papillae. Both class I and class II mutations arise as the consequence of single point mutations in the *ebgA°* gene (Hall, 1977).

The above strains were all selected for growth on lactose. What happens when selection is for lactulose utilization? Thirty-three independent spontaneous lactulose-fermenting mutants were isolated from papillae on the surface of strain 1B1 colonies growing on lactulose-TET plates. Those strains were originally designated class III because they were selected on lactulose; however, their growth rates are indistinguishable from those of class II strains and they are now designated class II, irrespective of the method of selection (Hall and Zuzel, 1980). Two of those strains, C1 and C2, are shown in Table XI. All of the mutants selected for lactulose utilization were class II strains, implying that an active site that will accommodate lactulose will also necessarily accommodate lactose, although the converse is not true.

Up to this point, all of the mutants described have resulted from a single mutation in the *ebgA* gene. I next asked about the consequences of a second mutation which would allow class I strains to utilize lactulose.

The class I strains A2 and D2 were streaked onto lactulose-TET plates, and lactulose-fermenting mutants were isolated from papillae on the surface of the colonies. The lactulose-fermenting mutants were designated class IV strains (Table XI).

Like class II strains, class IV strains utilize both lactose and lactulose. They differ from class II strains, however, in several important ways: (1) class IV strains carry *two* mutations in the *ebgA* gene; (2) class IV strains grow faster on lactose than on lactulose, while class II strains grow faster on lactulose than on lactose (Fig. 7); (3) class IV strains grow well ($\alpha > 0.1$ hr^{-1}) on galactosylarabinose, while class II strains grow extremely slowly, or not at all, on galactosylarabinose.

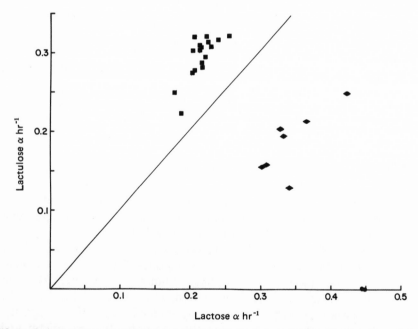

FIG. 7. Growth rates on lactulose versus growth rates on lactose. Circles: class I strains; squares: class II strains; diamonds: class IV strains.

At this juncture, it should be pointed out that all of the mutations have been mapped to the *ebgA* gene.

Galactosylarabinose utilization by class IV strains is another example of indirect selection of a new function, in that selection was for lactulose utilization. More importantly, the class IV mutations considerably increased the range of β-galactoside sugars that could be used. Thus, class IV mutants represent a true second step in an evolutionary pathway leading to ebg enzymes that allow the host organism to exploit a wide range of β-galactosides as carbon and energy sources. The requirement for two mutations in *ebgA* in order to hydrolyze galactosylarabinose seems to be quite strict: all attempts to isolate galactosylarabinose-utilizing mutants directly from the ancestral strain 1B1 have failed.

The existence of galactosylarabinose-hydrolyzing class IV strains encouraged us to use this new function as a selective tool (Hall and Zuzel, 1980). We already knew that class I strains could give rise to the galactosylarabinose-utilizing class IV strains; we therefore deferred applying the selection to class I strains. Four class II strains, two of which had been selected on lactose (5A2 and 5A11, which carries the *ebgA* allele present in SJ17) and two of which had been selected on lactulose (C1 and C2), were subjected to selection for galactosylarabinose utilization. Because the class II strain already grew, albeit extremely slowly, on galactosylarabinose, the usual sort of selection was ineffective. Instead, we employed a serial transfer selection in liquid culture, which resulted in the isolation of galactosylarabinose-utilizing mutants from each of the class II strains. Only one mutant was retained from each culture, and the mutants were designated "GA" for galactosylarabinose use.

The growth rates of the GA strains were strikingly similar to those of class IV strains (Table XI), suggesting the interesting possibility that the *second* mutation in the GA strains might be equivalent to a class I mutation, and implying that the *second* mutation in a class IV strain might be equivalent to a class II mutation.

If the class IV and the GA strains are simply the result of a class I mutation plus a class II mutation within the same gene, then it should be possible to recover both class I and class II recombinants from crosses between strains carrying the wild-type gene *ebgA°* and either class IV or GA strains. When such crosses were performed, indeed, both class I and class II recombinants were recovered. From the class IV strain A23 the recombinants were designated R41 (class I) and R42 (class II); likewise, from strain 5A2GA the recombinants were R61 and R62. The growth rates of the recombinant strains (Table XII) provide strong evidence that both the class IV alleles and the GA alleles are simply the sum of a class I and

TABLE XI. Growth Rates on Four β-Galactoside Sugars $(hr^{-1})^a$

Strain	Class	Lactose	Lactulose	Galactosylarabinose	Lactobionate
1B1	0	0	0	0	0
A2	I	0.446	0	0.029	0
SJ8	I	0.365	0	ND	ND
D2	I	0.451	0	0.032	0
5A2	II	0.188	0.255	0.027	0
5A11	II	0.171	0.233	0	ND
SJ17	II	0.178	0.254	0.017	0
SJ12	II	0.128	0.252	ND	ND
C1	II	0.181	0.235	0	ND
C2	II	0.227	0.312	0.034	0
A23	IV	0.360	0.196	0.139	0
A27	IV	0.424	0.244	0.160	0
D21	IV	0.362	0.225	0.113	0
D22	IV	0.368	0.130	0.137	0
5A2GA	IV	0.355	0.145	0.102	ND
5A11GA	IV	0.352	0.159	0.136	ND
C1GA	IV	0.376	0.181	0.145	ND
C2GA	IV	0.360	0.147	0.102	ND
A231	V	0.328	0.114	0.055	0.149
A232	V	0.459	0.186	0.059	0.144
A233	V	0.435	0.126	0.040	0.146
A234	V	0.393	0.146	0.052	0.136
A271	V	0.358	0.120	0.119	0.101
A272	V	0.183	0.103	0.066	0.199
A273	V	0.237	0.239	0.142	0.094
D211	V	0.382	0.180	0.142	0.084
D212	V	0.384	0.178	0.125	0.091
D213	V	0.376	0.151	0.123	0.073

^{a} Data from Hall (1978a) and Hall and Zuzel (1980).

TABLE XII. Growth Rates of Recombinant Strains $(hr^{-1})^a$

Strain	Class	Lactose	Lactulose	Galactosylarabinose
R41	I	0.377	0	0
R42	II	0.140	0.167	0.031
R61	I	0.389	0	0
R62	II	0.162	0.212	0
RT512	IV	0.201	0.092	0.165
RE522	IV	0.220	0.090	0.112
RT51168	IV	0.230	0.070	0.101
RT52168	IV	0.222	0.069	0.100

^{a} Data from Hall and Zuzel (1980).

a class II mutation. The alleles present in the GA strains are therefore designated class IV also.

We mapped the class I and class II sites within the *ebgA* gene with respect to the tightly linked *ebgR* gene. The order of the sites is shown in Fig. 8. The distance between the class I and class II sites was 1.0–1.5% recombination in several experiments, corresponding to about 650–1000 base pairs in the DNA. The placement of these sites within the *ebgA* gene is somewhat arbitrary, since the class I site is mapped with respect to a mutation an unknown distance into the *ebgR* gene. The exact placement and size of these sites await the results of cloning and sequencing studies that are in progress. It should be pointed out, however, that the distance between the sites corresponds to 25–35% of the length of the *ebgA* gene.

None of the *ebgA* alleles described up to this point permit utilization of lactobionic acid as a carbon and energy source. Attempts to select lactobionate-utilizing mutants on lactobionate-TET plates failed, for reasons that are not yet understood. As it turned out, the failure was fortunate, in that the alternative method chosen allowed quantitation of the frequency of mutations to lactobionate utilization.

I attempted to select lactobionate-utilizing mutants from the unevolved strain 1B1 and from a variety of class I, class II, and class IV strains (Hall, 1978a) simply by plating unmutagenized cultures on lactobionate minimal medium. As shown in Table XI, lactobionate-positive mutants were obtained *only* from class IV strains. If the spontaneous mutation rate is 10^{-9} (Hall, 1977), then the probability of failing to detect lactobionate-utilizing clones for each of the class 0, class I, and class II strains was less than 0.001. It was therefore concluded that lactobionate utilization required *three* mutations in *ebgA*. The lactobionate-utilizing strains were designated class V. Our present understanding of the pathway for the evolution of lactobionate utilization is diagrammed in Fig. 9.

From our understanding of the nature of class IV alleles (Hall and Zuzel, 1980), it is apparent that class V alleles consist of a class I mutation, a class II mutation, and a third x-mutation all in the same gene. The

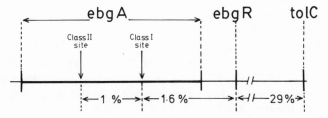

FIG. 8. Map of sites within *ebgA* gene. Distances are in percent recombination.

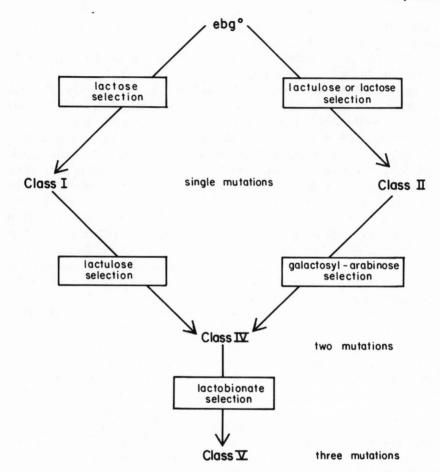

FIG. 9. Diagram of evolutionary pathways for lactobionate utilization via the *ebgA* gene product.

pathway shown simply reflects the results of the selection schemes employed in this laboratory, and does not rule out the possibility of other routes to the same end. Initially (Hall, 1978a), it was believed that the left-hand branch of the pathway was the only route to class IV alleles, and it was only later that the alternate selection was discovered (Hall and Zuzel, 1980). Thus, there may be other selection schemes which would permit isolation of the x-mutations from the wild type, or from class I or class II alleles. The failure to find those selection schemes probably reflects the limitations of this experimenter's imagination more than the limitations of nature.

The foregoing results may leave the reader (and sometimes the experimenter) with the heady feeling that virtually *any* new β-galactosidase function can be selected for ebg enzyme. Sad to say, that is not quite true. Although three of the class IV strains tested yielded class V mutants at frequencies greater than 10^{-9}, strain D22 has never yielded class V mutants. This means that the *ebgA* alleles present in strains D21 and D22 do *not* have the same evolutionary potential. The two alleles carry the same class I mutation (that present in strain D2), but different class II mutations, as reflected in the respective growth rates of D21 and D22 on lactulose (Table XI). To be precise, then, the class II mutation present in strain D22 is not compatible with lactobionate hydrolysis when paired with the strain D2 class I mutation. It may be the case that that particular class II mutation prohibits the evolution of lactobionate utilization, in which case it represents an evolutionary roadblock. On the other hand, it may be the particular combination of class I and class II mutations that prevents the evolution of a class V allele. Experiments to distinguish between these possibilities are in progress. Whatever the reason, the fact remains that despite very similar phenotypes and ancestries, the two alleles do not have the same evolutionary potential.

One of the evolutionary questions we wanted to answer with this system was: How many new functions can be evolved from a single ancestral gene? The answer is at least five: hydrolysis of (1) lactose, (2) lactulose, (3) β-methylgalactoside, (4) galactosylarabinose, and (5) lactobionic acid.

EVOLUTION BY RECOMBINATION WITHIN A GENE

Galactosylarabinose utilization requires two mutations in the *ebgA* gene, which may be selected in either order (Fig. 9). In both cases the evolution is linear, the second mutation being selected in a background containing the first mutation. The alternative scheme would be to have the class I mutation and the class II mutation selected in different (wild-type) individuals in different environments, then brought together by intragenic recombination. The observation that class I and II mutations could be recovered from a class IV allele by recombination within the gene suggested that the second scheme was plausible.

To test the second alternative, strains carrying the class I alleles specifying ebg[+a] or ebg[+d] enzyme were crossed with strains carrying the class II alleles specifying the enzymes ebg[+b] or ebg[+c] and galactosylarabinose-utilizing recombinants were selected (Hall and Zuzel, 1980). From those crosses galactosylarabinose-utilizing recombinants arose at the expected frequency of 1%. Control crosses of class I by class I, or class II by class II alleles yielded no galactosylarabinose-positive recombinants.

These results demonstrated that a new metabolic function, one possessed by neither parent, could arise by intragenic recombination. Class I and class II alleles are alleles of the same gene which have, under different selective pressures, evolved in different directions; i.e., in classical terms, they have diverged. When these divergent populations remixed, rare intragenic recombinants arose that were able to exploit a resource (galactosylarabinose) available to neither parent, nor to the parent's common ancestor. It is not unreasonable to expect that such intragenic recombination may play an important role in evolution in nature, as well as in the laboratory.

Although somewhat outside the scope of this article, and at a somewhat early stage, studies on the catalytic mechanisms of ebg enzyme are in progress in my laboratory and in the laboratory of my collaborator, Dr. M. Sinnott at the University of Bristol, England. The ability to separate or to join class I and class II mutations by recombination will permit us to study in some detail the effects of various mutations singly and together in the same polypeptide chain. Those studies are expected to contribute to our understanding of the ways in which catalytic functions are modified during the evolution of new functions.

EFFECTS ON OLD FUNCTIONS WHEN NEW FUNCTIONS ARE EVOLVED

In a real organism that is well adapted to its environment, a wild-type gene is expected to be the end product of millions of years of evolution, and therefore its product is expected to be nearly maximally efficient at carrying out its particular function. That gene product—enzyme in this case—is the result of selection and chance acting on whatever mutations occurred in its history. If that enzyme catalyzes a vital function under frequently encountered conditions, the gene will neccessarily be under intense selection. If an enzyme is "optimized" for one function, it is generally assumed that any mutation that could generate a new function for the enzyme would, of necessity, impair the original vital function of the enzyme, and thus be selected out of the population (Hindgardner, 1976). The way around this evolutionary roadblock is, of course, gene duplication and subsequent divergence, a well-documented evolutionary mechanism. The *necessity* of gene duplication for evolutionary divergence, however, lies squarely upon the assumption that mutations that confer new functions must impair old functions.

Ideally, that assumption would be examined by considering the effects of various mutations on the natural function of ebg° enzyme. That natural function, however, remains unknown. As a valid alternative, we can compare a series of class V strains with their immediate class IV ancestors, and ask how the new function (growth on lactobionate) has affected the old functions (growth on lactose and on lactulose). Those comparisons are shown in Fig. 10, where open symbols are for class IV strains and closed symbols of the same shape are for their class V descendents. It is immediately apparent that most mutations permitting lactobionate hydrolysis do impair either lactose or lactulose hydrolysis or both; however, this is not always the case. Strain A232 grows 25% *faster* on lactose than its parent, strain A23, and its growth rate on lactulose is unimpaired. While Fig. 10 shows only the effects of the new function on lactose and lactulose hydrolysis, the argument can also be extended to

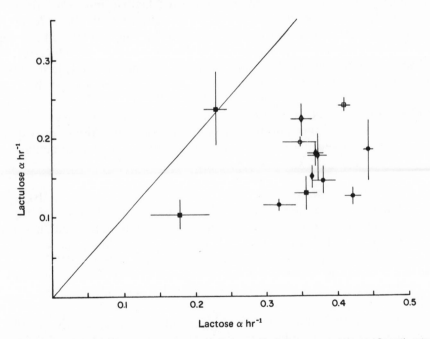

FIG. 10.   Growth rates of class IV and class V strains on lactulose versus lastose. Growth rates are displayed as a cross in which the intersection of the two lines is the mean growth rate on lactose and lactulose, and in which the lengths of the lines indicate the 95% confidence intervals about those means. Symbols at the intersections of lines: (○) A23; (●) class V strains derived from A23; (□) A27; (■) class V strains derived from A27; (◇) strain D2; (◆) class V strains derived from D2.

include galactosylarabinose hydrolysis. Strain D212 is, within the 95% confidence limits, as effective at utilizing lactose, lactulose, and galactosylarabinose as is its parent, strain D21. These results make it clear that evolution of a new function does not always affect the old function(s) adversely. Thus, functional divergence need not always be preceded by gene duplication.

ANOTHER NEW FUNCTION: CONVERSION OF LACTOSE INTO A *LAC* OPERON INDUCER

It was mentioned earlier that lactose is not an inducer of the *lac* operon, and that ebg enzyme does not convert lactose into an inducer, thus necessitating the inclusion of IPTG in media if β-galactoside sugars are to be used by $ebgA^+$ strains. We (Hall and Hartl, 1974) considered the possibility that ebg^{+a} enzyme converted lactose into a *lac* operon inducer, but did so inefficiently. If this were the case, the basal level of the lactose permease might not bring enough lactose into the cell for ebg enzyme to make effective amounts of inducer; however, the induced level of permease might bring in enough lactose to allow ebg enzyme to make enough inducer to maintain induction of the *lac* operon. To test this, we transferred a culture of strain A2 ($ebgA^+$ $ebgR^-$) in which the *lac* operon had been fully induced with IPTG into lactose minimal medium without IPTG. Growth of the culture ceased after four generations, providing evidence that even under the best of conditions ebg enzyme could not convert lactose into effective concentrations of an inducer of the *lac* operon.

The class II strain 5A2 is unable to utilize lactose unless IPTG is included in the medium. After 300 generations of selection for rapid growth on both lactose and lactulose an unstable mutant population arose from strain 5A2. The unstable population segregated clones that were red (lactose-positive) on MacConkey indicator plates without IPTG; while the parental strain 5A2 was white (lactose-negative) on the same plates. One such stable clone was designated strain 4R1 and subjected to further study (Rolseth *et al.*, 1980).

Strain 4R1 exhibited an IPTG-independent lactose-positive phenotype both on indicator plates and in lactose minimal medium. It differed from the parental strain 5A2 in an additional way: it was galactosylarabinose-positive. Direct measurements of *lac* permease activity (Table XIII) showed that the *lac* permease was not expressed constitutively in

TABLE XIII. Induction of *lac* Operon Genes[a]

Strain	Addition[b]	*lac* permease activity[c] (*lacY*)	Transacetylase activity[d] (*lacA*)
5A2	None	<0.3	—
	IPTG	5.1	—
	Lactose	<0.3	—
4R1	None	0.7	<0.02
	IPTG	11.2	3.00
	Lactose	13.6	2.43
SJ60	None	<0.3	<0.02
	IPTG	4.5	3.00
	Lactose	5.9	2.26
SJ61	None	<0.3	<0.02
	IPTG	4.6	2.71
	Lactose	5.6	1.88

[a] Data from Rolseth *et al.* (1980).
[b] Additions to glycerol minimal growth medium were 0.2 mM IPTG, or 0.1% lactose.
[c] *lac* permease activity is μmole of TMG transported $min^{-1} g^{-1}$ dry weight of cells.
[d] Transacetylase activity is nanomole of CoASH produced min^{-1} per 3 $\times 10^8$ cells.

strain 4R1; and simultaneously confirmed that lactose was not an inducer of the *lac* permease in strain 5A2.

Transduction experiments established that the gene responsible for IPTG-independent lactose utilization was in the *ebg* region. Like strain 4R1, the transductants (strains SJ60 and SJ61) expressed the *lac* permease when induced with either lactose or IPTG (Table XIII). Those experiments ruled out the possibilities that 4R1 was really a *lacZ*⁺ contaminant or that the *lac* repressor (*lacI* gene product) in strain 4R1 had become sensitive to lactose as an inducer.

It might still be argued that a gene tightly linked to *ebgA* had evolved into a lactose-TMG-IPTG permease. Two lines of evidence showed that not to be the case: (1) genetic evidence was presented that showed that the *lacY*⁺ gene was necessary for the Lac⁺ phenotype; and (2) it was shown (Table XIII) that synthesis of the thiogalactoside transacetylase (*lacA* gene product) was induced by lactose in strain 4R1 and the transductant strains. This was direct evidence that in strain 4R1, and in strains SJ60 and SJ61, lactose was being converted into an inducer of the *lac* operon.

The best explanation for the observed behavior of strain 4R1 was that ebg enzyme was converting lactose to an inducer of the *lac* operon. Recent experiments have shown that purified ebg enzyme from the transductant strain SJ60 can convert lactose into an inducer of the *lac* operon *in vitro* (Hall, unpublished experiments).

The *lacZ* β-galactosidase converts lactose into the natural inducer allolactose by transgalactosylation (Jobe and Bourgeois, 1972). The identity of the inducer synthesized by the ebg enzyme from strain SJ60 is not yet known; however, studies of this new function for ebg enzyme are continuing.

DIRECTED EVOLUTION OF A REGULATORY PROTEIN

We have been able to direct the evolution of ebg enzyme so that it can carry out a variety of new functions. That directed evolution involved selection of mutations that altered the interactions between ebg enzyme and its substrates. Similarly, evolution of a regulatory protein, a repressor, should involve alterations in the interactions between the repressor and its substrates.

A repressor recognizes two substrates: (1) the operator sequence in the DNA, and (2) small molecule effectors, or inducers. Mutations that alter a repressor so that it no longer recognizes the operator sequence lead to constitutive enzyme synthesis, equivalent to a *loss* of regulatory function rather than evolution of a new function. Mutations that lead to a repressor that binds very tightly to the operator sequence result in a lower level or absence of enzyme synthesis during induction. Again, this is a loss of function, rather than evolution of a new function. On the other hand, just as enzymes can acquire new functions by recognizing new substrates for hydrolysis, repressors should be able to acquire new functions by recognizing and binding new inducers. I decided to attempt to direct the evolution of the *ebg* repressor so that it recognized lactulose as an effective inducer (Hall, 1978b).

Strain 5A1 synthesizes ebg^{+c} enzyme (a class II enzyme) under the control of the wild-type repressor (Hall and Clarke, 1977; Hall, 1978b). Lactulose is not an effective inducer of the wild-type repressor (Table XIV), so strain 5A1 does not grow on lactulose despite the fact that ebg^{+c} enzyme hydrolyzes lactulose effectively.

Strain 5A1 was spread onto lactulose minimal medium to select mutants able to utilize lactulose. Two kinds of spontaneous mutants were expected: (1) $ebgR^-$ mutants, which (like strain 5A11) synthesize ebg^{+c} enzyme constitutively, and (2) mutants that were inducible by lactulose.

TABLE XIV. Induction of ebg Enzyme by Various β-Galactosides[a]

Strain	ebgR allele	None	Lactulose	IPTG	Galactosyl-arabinose	Methyl-galactoside	Lactose
DS4680A	$ebgR^+$	0.28	3	0.30	2.3	1.4	29
5A1	$ebgR^+$	0.23	4.1	0.26	1.2	0.8	23
5A101	$ebgR103^{+L}$	0.24	58	0.47	64	14	75
5A103	$ebgR105^{+L}$	0.27	126	0.48	211	55	230
5A108	$ebgR110^{+L}$	0.30	84	0.45	215	51	199

[a] Data from Hall (1978b). Values shown are units of ebg enzyme activity per mg of protein in crude extracts.

I estimated that the evolved repressors would have to be at least 12 times as sensitive to lactulose as the wild-type repressor in order for growth to occur.

Among over 3000 lactulose-positive colonies screened, only nine mutants were found that were not constitutive. The $ebgR^-$ mutations arose at a frequency of about 4×10^{-8}, while the mutations to lactulose inducibility arose at about 10^{-10}. When the nine nonconstitutive mutants were tested, they all exhibited the normal low basal level in the absence of inducer, and all were induced to 200–500 times the basal level by lactulose.

Three such mutants were subjected to further study. The mutations responsible for lactulose induction were mapped to the $ebgR$ gene, and the new alleles were designated $ebgR^{+L}$, to signify a fully functional repressor that responded to lactulose as an inducer.

Table XIV shows that different $ebgR^{+L}$ alleles have different levels of sensitivity to lactulose as an inducer. Like the wild-type allele $ebgR^+$ and also like the evolved allele $ebgR^{+u}$, $ebgR^{+L}$ repressor is insensitive to IPTG as an inducer. The phenomenon of "indirect selection of a new function," so often observed for ebg enzyme, is also observed for evolution of the repressor. The evolved repressors, selected for increased sensitivity to lactulose, all show greatly increased sensitivity to galactosylarabinose, methyl-β-galactoside, and lactose as inducers. In fact, the inducer specified by the $ebgR105^{+L}$ allele is more sensitive to lactose than any other ebg repressor examined, the level of enzyme synthesis during lactose induction being nearly 50% of the maximal level of expression.

This was the first study deliberately designed to evolve a new specificity for a repressor. The results suggest that evolution of regulatory proteins is not fundamentally different from the evolution of enzymes.

The *ebgA* gene mutations that allowed ebg enzyme to expand its range of substrates to include lactulose, galactosylarabinose, methylgalactoside, and lactobionate all required constitutive enzyme synthesis in order to be useful to the cell. Constitutive synthesis, which leads to about 5% of the soluble protein being ebg enzyme, is very costly to the cell, and would be expected to be disadvantageous in the absence of ebg enzyme substrates as nutrients. With the selection of $ebgR^{+L}$ alleles, however, lactose ceases to be the sole powerful inducer of ebg enzyme synthesis. Strain 5A103 (Table XIV) is able to use both lactose and lactulose without loss of cellular regulation. Further mutations in *ebgA* that would permit hydrolysis of galactosylarabinose or methylgalactoside would be immediately advantageous because the regulatory response to those sugars is already present. Thus, the $ebgR^{+L}$ mutations confer the potential for utilizing a wide variety of β-galactoside sugars without paying the high energetic cost of synthesizing an enzyme when it is not needed.

A MODEL FOR EVOLUTION IN DIPLOID ORGANISMS

We have shown that both regulatory and structural gene mutations are required for the evolution of lactose utilization by the EBG system. While it is clear that haploid *E. coli* cells carrying the required mutations have a strong selective advantage, it is not clear that they would possess a similar advantage were they diploid. Immediately after the required structural and regulatory gene mutations had occurred, the organism would be heterozygous for these evolved alleles. Diploid organisms would remain heterozygous until homozygosis occurred either by sexual matings or by mitotic recombination. Thus, in order for selective pressure to fix the new mutations in the population, the alleles would have to be advantageous in the heterozygous state. This is simply another way of saying that the evolved alleles would have to be dominant. I have examined the dominance of the evolved *ebgA* and *ebgR* alleles separately.

Although alleles specifying functional enzyme are usually dominant over alleles specifying nonfunctional enzyme, it is not obvious that *ebgA*$^+$ would be dominant over *ebgA*^{o}. The *ebgA*^{o} is the wild-type allele, despite the fact that its product is not "functional" with respect to lactose hydrolysis. Intracistronic complementation between mutant and wild-type enzymes can lead to partial restoration of wild-type activity to the mutant enzyme (Hall, 1973). In this case, wild-type activity would mean a loss of lactase activity. The fact that ebg enzyme is a hexamer (Hall, 1976a) means that in an *ebgA*^{o}/*ebgA*$^+$ heterozygote only $\frac{1}{64}$ of the ebg enzyme would contain solely ebg$^+$ subunits. In the absence of direct information about the nature of the active site it is impossible to predict the extent to which the presence of ebg^o subunits would affect the activity of ebg$^+$ subunits.

I constructed a series of F$^-$ *recA* strains that expressed different *ebgA*$^+$ alleles constitutively. Into each of these haploid strains I introduced an F′ episome which expressed *ebgA*^{o} constitutively. Table XV shows the growth rates of these haploid strains and their heterozygous merodiploid derivatives.

It is clear from the fact that the merodiploids all grow on lactose that the *ebgA*$^+$ alleles are dominant. For three of the five strains tested the merodiploids grew faster than the haploid strains, indicating that there was some complementation between the ebg$^+$ and the ebg^o subunits. Note, however, that the interactions between wild-type and evolved subunits never altered the substrate specificities; i.e., class I strains remained lactulose-negative, while class II continued to grow faster on lactose than on lactulose.

It was already known that *ebgR*$^-$ alleles were recessive to *ebgR*$^+$ (Hall and Hartl, 1975). To test the dominance of the functional evolved

TABLE XV. Growth Rates of Constitutive Strains[a]

Strain and ratio[b]	Carbon source		
	Lactose	Lactulose	Glycerol
SJ8R	0.198 ± 0.003	0	0.435 ± 0.017
SJ8R/F′1B1	0.228 ± 0.005	0	0.338 ± 0.005
Ratio	1.48	—	—
SJ12R	0.081 ± 0.010	0.109 ± 0.001	0.367 ± 0.006
SJ12R/F′1B1	0.114 ± 0.006	0.131 ± 0.005	0.321 ± 0.002
Ratio	1.60	1.37	—
SJ24R	0.152 ± 0.003	0.206 ± 0.003	0.458 ± 0.034
SJ24R/F′1B1	0.093 ± 0.003	0.112 ± 0.004	0.339 ± 0.004
Ratio	0.83	0.73	—
SJ470R	0.105 ± 0.010	0.144 ± 0.016	0.317 ± 0.014
SJ470R/F′1B1	0.091 ± 0.002	0.104 ± 0.003	0.287 ± 0.003
Ratio	0.96	0.80	—
SJ480R	0.255 ± 0.003	0	0.403 ± 0.007
SJ480R/F′1B1	0.208 ± 0.003	0	0.295 ± 0.004
Ratio	1.11	—	—

[a] Growth rates are reported as the first-order rate constant in reciprocal hours, ± the 95% confidence limits.
[b] Ratio of haploid to merodiploid growth rates corrected for the difference in growth rate on glycerol.

alleles $ebgR^{+L}$ and $ebgR^{+u}$ (Hall, 1978b; Hall and Clarke, 1977), F^- $recA$ strains were constructed carrying the evolved repressor alleles. These haploid strains were compared with merodiploid derivatives carrying an F′ episome bearing $ebgR^+$ and $ebgA^o$ alleles. Table XVI shows that ebg enzyme synthesis was strongly induced by lactose in the haploid strains, but very poorly induced in the heterozygous merodiploid strains. The evolved repressor alleles were therefore *recessive* to the wild-type $ebgR^+$ allele. As predicted, the $ebgR^+$ $ebgA^o/ebgR^{+uorL}$ $ebgA^+$ double heterozygote merodiploids were lactose-negative.

For all of the bacterial systems examined, several of which were discussed at the beginning of this paper, regulatory mutations were required for evolution of new metabolic functions. These critical regulatory mutations, which involved either constitutive enzyme synthesis or enzyme induction by a novel metabolite, would be expected to be recessive for systems that are under negative control, as was the case for the EBG system. For systems under positive control, on the other hand, it would be expected that both constitutive mutations and mutations that allow the positive effector to respond to a new inducer would be dominant. Farin and Clarke (1978) have provided evidence that the $amiE$ locus in $P.$ $aeruginosa$ is under positive control by the $amiR$ gene product. While $amiR$ mutants that are formamide-inducible have been isolated (Brammar $et\ al.$, 1967), stable diploids which would permit studies of dominance relationships are not available in $P.\ aeruginosa$.

If mutations that allow regulatory proteins to recognize new metabolites as inducers are recessive for negatively controlled systems, then the evolution of new functions for those systems would be very difficult in diploid organisms. If such mutations are dominant for positively controlled systems, then evolution of new functions for positively controlled

TABLE XVI. EBG Enzyme Activities in Whole Cells

| | Units per 10^9 cells[a] | |
Strain	Uninduced[b]	Induced[c]
SJ606R (haploid)	<2	68
SJ606R/F′122 (merodiploid)	<2	2.3
SJ607R (haploid)	<2	20
SJ607R/F′122 (merodiploid)	<2	2.3

[a] Mean of three assays per culture. Values less than two units per 10^9 cells are unreliable.
[b] Uninduced: grown in T-broth + 2×10^{-4} M IPTG.
[c] Induced: grown in T-broth + 2×10^{-4} M IPTG + 0.1% lactose.

systems would be no more difficult in diploid than in haploid organisms. This perhaps partially explains the paucity of examples of simple negative control in the higher eukaryotes, where negatively controlled systems may have been lost as evolutionary dead ends.

CONCLUSIONS AND PROSPECTS

We can draw a few firm conclusions from the studies employing the EBG system for experimental evolution.

First, relying entirely upon spontaneous mutations, coupled with a powerful selective system, we can direct the evolution of new functions for proteins, both enzymatic and regulatory. The utilization of spontaneous mutations makes it seem likely that the kinds of events that we observe during evolution of a new function in the laboratory resemble, at least qualitatively, events in nature.

A variety of new functions can arise from single point mutations within a gene. All of the new functions evolved for ebg enzyme, however, represent variations upon the general function "β-galactosidase," a function already possessed by the ancestral function. We can distinguish between new functions that are of biological significance and new functions that are detectable *in vitro*. The ancestral enzyme ebg° does not hydrolyze lactose, lactulose, galactosylarabinose, or β-methylgalactoside at biologically significant rates; however, hydrolysis of all of those sugars can be detected and measured *in vitro* using purified ebg° enzyme. Thus, in a real sense, the point mutations that generate the evolved enzymes of classes I and II do not result in truly new activities; they simply improve old functions. Hydrolysis of lactobionic acid, however, is not detectable *in vitro* for ebg° enzyme or for enzyme of the evolved classes I and II. Biologically significant lactobionate hydrolysis requires three mutations in the *ebgA* gene; however, *in vitro* lactobionate is hydrolyzed by class IV enzyme, which carries two substitutions (Hall, unpublished results). These results are qualitively similar to those of Clarke and her co-workers (1974, 1978) who found that wild-type amidase of *P. aeruginosa* had no activity toward valeramide, but that a mutant enzyme with enhanced activity toward butyramide exhibited a trace of activity toward valeramide. The butyramide-utilizing mutant did not grow on valeramide, but it was able to give rise to a valeramide-utilizing mutant by a second mutation in the amidase gene. The wild-type gene cannot give rise to valeramide-utilizing mutants directly.

We have shown that there are evolutionary pathways whose individual steps represent single point mutations, each of which confers a selective advantage under some set of conditions. One new function, galactosylarabinose utilization, required a two-step pathway; and there were two alternative routes, because the two required mutations could be selected in either order under different sets of selective pressures.

The above observations led us to experiments that showed that a new function could arise via intragenic recombination. That, and the demonstration that acquisition of a new function did not always impair the old function of an enzyme, probably represent the most important contributions of this project to our understanding of the molecular events in evolution.

This study has fully confirmed the conclusions of other studies (discussed earlier) that regulatory mutations are critical for the evolution of new metabolic functions. *Every single* mutant that has evolved the ability to use a new β-galactoside sugar has required a regulatory gene mutation. The regulatory gene mutations were required because none of the sugars employed for selection were sufficiently powerful inducers of the wild-type repressor to permit enough ebg enzyme synthesis for growth. One kind of regulatory mutation that increased the amount of enzyme synthesised was the $ebgR^-$ mutation, which led to constitutive enzyme synthesis. From the point of view of evolution this was the least interesting kind of regulatory mutation because (1) it led to *loss* of regulatory function, rather than to evolution of new regulatory function, and (2) it led to enzyme synthesis that was wasteful in the absence of the enzyme's substrate. Much more interesting was the second kind of regulatory mutation, the $ebgR^{+u}$ mutation, which led to fully functional repressor with increased sensitivity to lactose as an inducer. Those mutations indeed represented evolution of new regulatory function; however, the $ebgR^{+u}$ repressor showed no increased sensitivity to other β-galactosides as inducers. A second study deliberately directed the evolution of the repressor gene, and resulted in $ebgR^{+L}$ mutations, which increased the sensitivity of the repressor to lactulose, galactosylarabinose, and β-methylgalactoside, as well as to lactose.

In contrast to several other studies in which evolution of a new metabolic function required *only* constitutive enzyme synthesis, evolution of the ability to use β-galactoside sugars required mutations in both the regulatory and structural genes. This requirement for coevolution of the structural and regulatory genes has been one of the most interesting aspects of this study, and it leads to a real paradox. When the *lacZ* deletion strain is streaked onto selective medium, the resulting lactose-utilizing mutants that are isolated some 10 days later all are spontaneous double mutants. When the spontaneous mutation rates are measured independently, they are found to be about 2×10^{-9} for the structural gene $ebgA$ and about 10^{-8} for mutation to $ebgR^-$ (about 10^{-10} for mutation to $ebgR^{+L}$). The double mutants should arise, therefore, at a frequency of about 10^{-18}. Now 10^{18} is about 1 kg of $E. coli$. The colonies from which the spontaneous mutants are isolated consist of about 10^{10} cells. If a mutation to $ebgA^+$ arose when the colony consisted of about 10^9 cells, then when the colony reached its steady state size of about 10^{10} cells there would be about ten $ebgA^+$ cells in the colony. If the required mutation to $ebgR^{+u}$ were independent of the first mutation, it should on average occur when the population of $ebgA^+$ cells had reached about 10^{10} cells, i.e., the entire colony would be $ebgA^+$ $ebgR^+$.

Since $ebgA^+$ $ebgR^+$ cells can not utilize lactose, they can grow no faster than the nonmutant cells in the colony, and their growth must depend upon the turnover (lysis) of other cells in the colony. On the order of 10^9 cell divisions will be required, on average, before a mutation to $ebgR^{+u}$ occurs. If, as we would expect, the mutation to $ebgA^+$ is neutral, then, given a steady state population of 10^{10} cells per colony, including ten $ebgA^+$ cells, some 10^5 yr would be required for 10^9 cell divisions of $ebgA^+$ cells.

If we make the extreme assumptions that $ebgA^+$ cells are somehow protected from turning over and that they reproduce with 100% efficiency, and that under these somewhat anaerobic nutritionally depleted conditions they double in as little as 6 hr, then about 7 days are required for the requisite 10^9 cell divisions. Following the appearance of the first $ebgR^{+u}$ mutation in the $ebgA^+$ population, even if the double mutants grow as fast as they do in highly aerated liquid lactose minimal medium, another 7 days is required for the appearance of a lactose-fermenting papilla consisting of about 10^8 cells.

Thus, the model of two sequential mutations requires *at least* 15 days for papillae to appear; whereas such papillase are usually visible on the majority of colonies within 9 days. On the other hand, 9 days is just about the time that papilla formation is expected to require if the two mutations (to $ebgA^+$ and to $ebgR^{+u}$) occur simultaneously when the colony consists of about 10^9 cells and the double mutant cells grow thereafter at the expense of lactose.

We have shown that the selective medium is not itself mutagenic, and that the strain does not carry mutator mutations, since the colonies exhibit no higher frequency of mutations to drug resistance than expected. We can only conclude that under some conditions spontaneous mutations are not independent events—heresy, I am aware.

Although the major function of the EBG system was to serve as a model for studying evolution, this study and others like it can make contributions to fields other than evolution.

Enzymologists who study enzyme mechanisms utilize substrate analogs to probe the details of catalysis. They often will extend such studies by examining the same enzyme derived from a variety of organisms in order to assess the universality of details of a particular catalytic mechanism. A major limitation of such studies is that the enzymes being compared differ so greatly in primary sequence and structure that it is difficult to relate functional and structural differences. Experimental evolution studies can provide the enzymologist with a wide range of functional mutant enzymes which have different catalytic capabilities, and which differ from each other by single amino acid substitutions. Sequencing of the protein or of the coding DNA for each of the mutants can allow the enzymologist to relate altered catalytic properties to specific alterations in the enzyme. Such studies of ebg enzymes are in progress in collaboration with Dr. M. Sinnott of the University of Bristol. With the aid of synthetic substrates that can also serve as selective agents it may well be possible to direct the evolution of an enzyme to enhance particular aspects of its catalytic mechanism, or to deliberately alter the rate-limiting step in the catalysis of an overall reaction. Experimental evolution systems may therefore serve to test physical-biochemical predictions concerning relationships between protein structure catalytic functions.

The ultimate goal of all experimental evolution studies is to be able to predict, from present knowledge of an organism, the probability that the organism can evolve a particular new function. At present we are very far from that goal. We do not yet even know which properties of an organism are most important for making such predictions. The practical importance of being able to make such predictions is enormous. Novel organic compounds are being generated at increasing rates as byproducts of technological and manufacturing processes. Many of these compounds are highly toxic; nevertheless, they are disposed of in ways that come back to haunt us years later. The recent tragedies that have resulted from residential areas being developed on the sites of old chemical dumps are consequences of the implicit assumption that nature can somehow dispose of all possible compounds harmlessly. That assumption is clearly invalid. With sufficient information, and sufficient understanding of how new functions evolve, we should be able to predict whether a particular compound can be degraded; then we should be able to efficiently direct the evolution of an organism to dispose of the compound under controlled conditions.

REFERENCES

Anderson, P. R., and Roth, J. R., 1977, Tandem genetic duplications in phage and bacteria, *Annu. Rev. Microbiol.* **31**:473–505.

Arnheim, N., Sobel, J., and Canfield, R., 1971, Immunochemical resemblance between human leukemia and hen egg-white lysozyme and their reduced carboxymethyl derivatives, *J. Mol. Biol.* **61**:237–250.

Arnon, R., and Maron, E., 1970, Lack of immunological cross-reaction between bovine α-lactalbumin and hen's egg-white lysozyme, *J. Mol. Biol.* **51**:703–707.

Arnon, R., and Maron, E., 1971, An immunological approach to the structural relationship between hen egg-white lysozyme and bovine α-lactalbumin, *J. Mol. Biol.* **61**:225–235.

Arraj, J.A., and Campbell, J. H., 1975, Isolation and characterization of the newly evolved *ebg* β-galactosidase of *Escherichia coli K12, J. Bacteriol.* **124**:849–856.

Bachmann, B., and Low, K. B., 1980, Linkage map of *Escherichia coli K12,* edition 6, *Microbiol. Rev.* **44**:1–56.

Barkley, M. D., Riggs, A. D., Jobe, A., and Bourgeois, S., 1975, Interaction of effecting ligands with *lac* repressor and repressor–operator complex, *Biochemistry* **14**:1700–1712.

Betz, J. L., and Clarke, P. H., 1972, Selective evolution of phenylacetamide-utilizing strains of *Pseudomonas aeruginosa, J. Gen. Microbiol.* **73**:161–174.

Betz, J. L., and Clarke, P. H., 1973, Growth of *Pseudomonas* species on phenylacetamide, *J. Gen. Microbiol.* **75**:167–177.

Bonner, D. M., DeMoss, J. A., and Mills, S. E., 1965, The evolution of an enzyme, in: *Evolving Genes and Proteins* (V. Bryson and H. J. Vogel, eds.), pp. 305–318, Academic Press, New York.

Brammar, W. J., Clarke, P. H., and Skinner, A. J., 1967, Biochemical and genetic studies with regulator mutants of the *Pseudomonas aeruginosa* 8602 amidase system, *J. Gen. Microbiol.* **47**:87–102.

Brew, K., Vanaman, T. C., and Hill, R. L., 1967, Comparison of the amino acid sequence of bovine α-lactalbumin and hen's egg-white lysozyme, *J. Biol. Chem.* **242**:3747–3749.

Brown, J. E., and Clarke, P. H., 1970, Mutations in a regulator gene allowing *Pseudomonas aeruginosa* 8602 to grow on butyramide, *J. Gen. Microbiol.* **64**:329–342.

Brown, P. R., and Clarke, P. H., 1972, Amino acid substitution in an amidase produced by an acetanilide-utilizing mutant of *Pseudomonas aeruginosa, J. Gen. Microbiol.* **70**:287–298.

Brown, J. E., Brown, P. R., and Clarke, P. H., 1969, Butyramide-utilizing mutants of *Pseudomonas aeruginosa* 8602 which produce an amidase with altered substrate specificity, *J. Gen. Microbiol.* **57**:273–298.

Bryson, V., and Vogel, H. J. (eds.), *Evolving Genes and Proteins,* Academic Press, New York.

Burleigh, B. D., Rigby, P. W. J., and Hartley, B. S., 1974, A comparison of wild-type and mutant ribitol dehydrogenases from *Klebsiella aerogenes, Biochem. J.* **143**:341–352.

Campbell, J. H., Lengyel, J., and Langridge, J., 1973, Evolution of a second gene for β-galactosidase in *Escherichia coli, Proc. Natl. Acad. Sci. USA* **70**:1841–1845.

Camyre, K. P., and Mortlock, R. P., 1965, Growth of *Aerobacter aerogenes* on D-arabinose and L-xylose, *J. Bacteriol.* **90**:1157–1158.

Clarke, B., 1970, Selective constraints on amino-acid substitutions during the evolution of proteins, *Nature* **228**:159–160.

Clarke, P. H., 1974, The evolution of enzymes for the utilisation of novel substrates, in:

Evolution in the Microbial World (M. J. Carlile and J. J. Skehel, eds.), pp. 183–217, Cambridge University Press, London.

Clarke, P. H., 1978, Experiments in microbial evolution, in: *The Bacteria,* Volume VI (L. N. Ornston and J. R. Sokatch, eds.), pp. 137–218, Academic Press, New York.

Cocks, G. T., Aguilar, J., and Lin, E. C. C., 1974, Evolution of L-1,2-propanediol catabolism in *Escherichia coli* by recruitment of enzymes for L-fucose and L-lactate metabolism, *J. Bacteriol.* **118:**83–88.

Demerec, M., Adelberg, E. A., Clark, A. J., and Hartman, P. E., 1966, A proposal for a uniform nomenclature in bacterial genetics, *Genetics* **54:**61–76.

Dickerson, R. E., 1971, The structure of cytochrome C and the rates of molecular evolution, *J. Mol. Evol.* **1:**26–45.

Dixon, G. H., 1966, Mechanisms of protein evolution, in: *Essays in Biochemistry,* Volume 2 (P. N. Campbell and G. D. Greville, eds.), pp. 147–204, Academic Press, New York.

Farin, F., and Clarke, P. H., 1978, Positive regulation of amidase synthesis in *Pseudomonas aeruginosa, J. Bacteriol.* **135:**379–392.

Fitch, W. M., and Margoliash, E., 1970, The usefulness of amino acid and nucleotide sequences in evolutionary studies, *Evol. Biol.* **4:**67–109.

Francis, J. C., and Hansche, P. E., 1972, Directed evolution of metabolic pathways in microbial populations. I. Modification of the acid phosphatase pH optimum in *S. cerevisiae, Genetics* **70:**59–73.

Francis, J. C., and Hansche, P. E., 1973, Directed evolution of metabolic pathways in microbial populations. II. A repeatable adaptation in *Saccharomyces cerevisiae, Genetics* **74:**259–265.

Goodman, M., 1977, Protein sequences in phylogeny, in: *Molecular Evolution* (F. Ayala, ed.), pp. 141–159, Sinauer, Sunderland, Massachusetts.

Hacking, A. J., and Lin, E. C. C., 1977, Regulatory changes in the fucose system associated with the evolution of a catabolic pathway for propanediol in *Escherichia coli, J. Bacteriol.* **130:**832–838.

Hacking, A. J., Aguilar, J., and Lin, E. C. C., 1978, Evolution of propanediol utilization in *Escherichia coli*: Mutant with improved substrate-scavenging power, *J. Bacteriol.* **136:**522–530.

Hall, B. G., 1973, *In vivo* complementation between wild type and mutant β-galactosidase in *Escherichia coli, J. Bacteriol.* **114:**448–450.

Hall, B. G., 1976a, Experimental evolution of a new enzymatic function. Kinetic analysis of the ancestral (ebg^0) and evolved (ebg^+) enzymes, *J. Mol. Biol.* **107:**71–84.

Hall, B. G., 1976b, Methylgalactosidase activity: An alternative evolutionary destination for the $ebgA^0$ gene, *J. Bacteriol.* **126:**536–538.

Hall, B. G., 1977, Number of mutations required to evolve a new lactase function in *Escherichia coli, J. Bacteriol.* **129:**540–543.

Hall, B. G., 1978a, Experimental evolution of a new enzymatic function. II. Evolution of multiple functions for EBG enzyme in *E. coli, Genetics* **89:**453–465.

Hall, B. G., 1978b, Regulation of newly evolved enzymes. IV. Directed evolution of the *ebg* repressor, *Genetics* **90:**673–691.

Hall, B. G., and Clarke, N. D., 1977, Regulation of newly evolved enzymes. III. Evolution of the ebg repressor during selection for enhanced lactase activity, *Genetics* **85:**193–201.

Hall, B. G., and Hartl, D. L., 1974, Regulation of newly evolved enzymes. I. Selection of a novel lactase regulated by lactose in *Escherichia coli, Genetics* **76:**391–400.

Hall, B. G., and Hartl, D. L., 1975, Regulation of newly evolved enzymes. II. The *ebg* repressor, *Genetics* **81:**427–435.

Hall, B. G., and Zuzel, T., 1980, Evolution of a new enzymatic function by recombination within a gene, *Proc. Natl. Acad. Sci. USA* **77**:3529–3533.

Hall, B. G., Hartl, D. L., and Bulbulian, B., 1974, Two pathways for the evolution of a new β-galactosidase (ebg) in *E. coli, Genetics* **77**:s28.

Hansche, P. E., 1975, Gene duplication as a mechanism of genetic adaptation in *Saccharomyces cerevisiae, Genetics* **79**:661–674.

Hartl, D. L., and Dykhausen, D., 1979, Genetic map of *uxaA-ebgA-tolC-metC* region in *E. coli, Genetics* **91**:s44.

Hartl, D. L., and Hall, B. G., 1974, Second naturally occurring β-galactosidase in *E. coli, Nature* **248**:152–153.

Hartley, B. S., Brown, J. R., Kauffman, D. L., and Smillie, L. B., 1965, Evolutionary similarities between pancreatic proteolytic enzymes, *Nature* **207**:1157–1159.

Hartley, B. S., Burleigh, B. D., Midwinter, G. G., Moore, C. H., Morris, H. R., Rigby, P. W. J., Smith, M. J., and Taylor, S. S., 1972, Where do new enzymes come from?, in: *Enzymes: Structure and Function*, 8th FEBS Meeting, Volume 29 (J. Denreth, R. A. Oosterbaan, and C. Veeger, eds.), North-Holland, Amsterdam.

Hartley, B. S., 1974, Enzyme families, in: *Evolution in the Microbial World* (M. J. Carlile, and J. J. Skehel, eds.), pp. 151–182, Cambridge University Press, London.

Hegeman, G. D., and Rosenberg, S. L., 1970, The evolution of bacterial enzyme systems, *Annu. Rev. Microbiol.* **24**:429–462.

Hill, R. L., Delaney, R., Fellows, R. E., and Lebovitz, H. E., 1966, The evolutionary origins of the immunoglobulins, *Proc. Natl. Acad. Sci. USA* **56**:1762–1769.

Hinegardner, R., 1977, Evolution of genome size, in: *Molecular Evolution* (F. J. Ayala, ed.), pp. 179–199, Sinauer, Sunderland, Massachusetts.

Hobson, A. C., 1978, A mutation allowing utilization of lactose by *Escherichia coli lacY* mutants defective in lactose permease, *Mol. Gen. Genet.* **161**:109–110.

Hopwood, D. A., 1967, Genetic analysis and genome structure in *Streptomyces coelicolor, Bacteriol. Rev.* **31**:373–403.

Horiuchi, T., Tomizawa, J., and Novick, A., 1962, Isolation and properties of bacteria capable of high rates of β-galactosidase synthesis, *Biochim. Biophys. Acta* **55**:152.

Horiuchi, T., Horiuchi, S., and Novick, A., 1963, The genetic basis of hyper-synthesis of β-galactosidase, *Genetics* **48**:157–169.

Inderlied, C. B., and Mortlock, R. P., 1977, Growth of *Klebsiella aerogenes* on xylitol: Implications for bacterial enzyme evolution, *J. Mol. Evol.* **9**:181–190.

Jobe, A., and Bourgeois, S., 1972, *Lac* repressor–operator interaction. VI. The natural inducer of the *lac* operon, *J. Mol. Biol.* **69**:397–408.

Johnson, E. M., Wohlhieter, J. A., Placek, B. P., Sleet, R. B., and Baron, L. S., 1976, Plasmid-determined ability of a *Salmonella tennessee* strain to ferment lactose and sucrose, *J. Bacteriol.* **125**:385–386.

Kemper, J., 1974a, Gene order and co-transduction in the *leu-ara-fol-pyrA* region of the *Salmonella typhimurium* linkage map, *J. Bacteriol.* **117**:94–99.

Kemper, J., 1974b, Evolution of a new gene substituting for the *leuD* gene of *Salmonella typhimurium*: Origin and nature of *supQ* and *newD* mutations, *J. Bacteriol.* **120**:1176–1185.

Kimura, M., 1968, Evolutionary rate at the molecular level, *Nature* **217**:624–626.

Kimura, M., and Ohta, T., 1971, On the rate of molecular evolution, *J. Mol. Evol.* **1**:1–17.

King, J. L., and Jukes, T. H., 1969, Non-Darwinian evolution, *Science* **164**:788–798.

Learner, S. A., Wu, T. T., and Lin, E. C. C., 1964, Evolution of a catabolic pathway in bacteria, *Science* **146**:1313–1315.

LeBlanc, D. J., and Mortlock, R. P., 1971a, Metabolism of D-arabinose: Origin of a D-ribulokinase activity in *Escherichia coli, J. Bacteriol.* **106:**82–89.

LeBlanc, D. J., and Mortlock, R. P., 1971b, Metabolism of D-arabinose: A New pathway in *Escherichia coli, J. Bacteriol.* **106:**90–96.

Lederberg, J., 1951, Genetic studies with bacteria, in: *Genetics in the Twentieth Century* (L. C. Dunn, ed.), pp. 263–289, Macmillan, New York.

Lewontin, R. C., 1974, *The Genetic Basis of Evolutionary Change,* Columbia University Press, New York.

Lin, E. C. C., Hacking, A. J., and Aguilar, J., 1976, Experimental models of acquisitive evolution, *BioScience* **26:**548–555.

Luria, S. E., and Delbruck, M., 1943, Mutations of bacteria from virus sensitivity to virus resistance, *Genetics* **28:**491–511.

Mandel, M., 1969, New approaches to bacterial taxonomy: Perspective and prospects, *Annu. Rev. Microbiol.* **23:**239–274.

McLaughlin, P. J., and Dayhoff, M. O., 1973, Eukaryote evolution: A view based on cytochrome C sequence data, *J. Mol. Evol.* **2:**99–116.

Messer, A., 1974, Lactose permeation via the arabinose transport system in *Escherichia coli* K12, *J. Bacteriol.* **120:**266–272.

Miller, J. H., 1972, *Experiments in Molecular Genetics,* p. 228, Cold Spring Harbor Laboratory, New York.

Mortlock, R. P., 1981, Regulatory mutations and the development of new metabolic pathways by bacteria, *Evol. Biol.* **14:**205–267.

Mortlock, R. P., and Wood, W. A., 1964a, Metabolism of pentoses and pentitols by *Aerobacter aerogenes.* I. Demonstration of pentose isomerase, pentulokinase, and pentitol dehydrogenase enzyme families, *J. Bacteriol.* **88:**838–844.

Mortlock, R. P., and Wood, W. A., 1964b, Metabolism of pentoses and pentitols by *Aerobacter aerogenes.* II. Mechanism of acquisition of kinase, isomerase, and dehydrogenase activity, *J. Bacteriol.* **88:**845–849.

Mortlock, R. P., and Wood, W. A., 1971, Genetic and enzymatic mechanisms for the accommodation to novel substrate by *Aerobacter aerogenes,* in: *Biochemical Responses to Environmental Stress* (I. A. Bernstein, ed.), pp. 1–14, Plenum Press, New York.

Mortlock, R. P., Fossitt, D. D., and Wood, W. A., 1965, A basis for utilization of unnatural pentoses and pentitols by *Aerobacter aerogenes, Proc. Natl. Acad. Sci. USA* **54:**572–579.

Nolan, C., and Margoliash, E., 1968, Comparative aspects of primary structures of proteins, *Annu. Rev. Biochem.* **37:**727–790.

Oliver, E. J., and Mortlock, R. P., 1971a, Growth of *Aerobacter aerogenes* on D-arabinose: Origin of the enzyme activities, *J. Bacteriol.* **108:**287–292.

Oliver, E. J., and Mortlock, R. P., 1971b, Metabolism of D-arabinose by *Aerobacter aerogenes*: Purification of the isomerase, *J. Bacteriol.* **108:**293–299.

Perutz, M. F., Bolton, W., Diamond, R., Muirhead, H., Watson, H. C., 1964, Structure of haemoglobin. An X-ray examination of reduced horse haemoglobin, *Nature* **203:**687–690.

Prasad, I., and Schaefler, S., 1974, Regulation of the β-glucoside system in *Escherichia coli* K-12, *J. Bacteriol.* **120:**638–650.

Prasad, I., Young, B., and Schaefler, S., 1973, Genetic determination of the constitutive biosynthesis of phospho-β-glucosidase A in *Escherichia coli* K-12, *J. Bacteriol.* **114:**909–915.

Reanney, D. C., 1976, Extrachromosomal elements as possible agents of adaptation and development, *Bacteriol. Rev.* **40:**552–590.

Rigby, P. W. J., Burleigh, B. D., and Hartley, B. S., 1974, Gene duplication in experimental enzyme evolution, *Nature* **251**:200–204.

Riley, M., and Anilionis, A., 1978, Evolution of the bacterial genome, *Annu. Rev. Microbiol.* **32**:519–560.

Riley, M., Solomon, L., and Zipkas, D., 1978, Relationship between gene function and gene location in *Escherichia coli, J. Mol. Evol.* **11**:47–56.

Rolseth, S. J., Fried, V. A., and Hall, B. G., 1980, A mutant *ebg* enzyme which converts lactose into an inducer of the *lac* operon, *J. Bacteriol.* **142**:1036–1039.

Rossmann, M. G., Moras, D., and Olsen, K. W., 1974, Chemical and biological evolution of a nucleotide-binding protein, *Nature* **250**:194–199.

Schaefler, S., and Maas, W. K., 1967, Inducible system for the utilization of β-glucosides in *Escherichia coli.* II. Description of mutant types and genetic analysis, *J. Bacteriol.* **93**:264–272.

Senior, E., Bull, A. T., and Slater, J. H., 1976, Enzyme evolution in a microbial community growing on the herbicide Dalapon, *Nature* **268**:476–479.

Sinnott, M. L., Withers, S. G., and Viratelle, O. M., 1978, The necessity of magnesium cation for acid assistance of aglycone departure in catalysis by *Escherichia coli* (*lacZ*) β-galactosidase, *Biochem. J.* **175**:539–546.

Slater, J. H., Lovatt, D., Weightman, A. J., Senior, E., and Bull, A. T., 1979, The growth of *Pseudomonas putida* on chlorinated aliphatic acids and its dehalogenase activity, *J. Gen. Microbiol.* **114**:125–136.

Smithies, O., Connell, G. E., and Dixon, G. H., 1962, Chromosomal rearrangements and the evolution of haptoglobin genes, *Nature* **196**:232–236.

Sparrow, A., and Nauman, A., 1976, Evolution of genome size by DNA doublings. Minimum genome size in major taxonomic groups suggests an evolutionary series of DNA doublings, *Science* **192**:524–529.

Sridhara, S., Wu, T. T., Chused, T. M., and Lin, E. C. C., 1969, Ferrous-activated nicotinamide adenine dinucleotide-linked dehydrogenase from a mutant of *Escherichia coli* capable of growth on 1,2-propanediol, *J. Bacteriol.* **98**:87–95.

Tanaka, M., Nakashima, T., Benson, A., Mower, H., and Yasunobu, K. T., 1966, The amino acid sequence of *Clostridium pasteurianum* ferredoxin, *Biochemistry* **5**:1666–1681.

Tenu, J. P., Viratelle, O. M., Garnier, J., and Yon, J., 1971, pH dependence of the activity of β-galactosidase from *Escherichia coli, Eur. J. Biochem.* **20**:363–370.

Walsh, K. A., and Neurath, H., 1964, Trypsinogen and chymotrypsinogen as homologous proteins, *Proc. Natl. Acad. Sci. USA* **52**:884–889.

Warren, R. A. J., 1972, Lactose utilizing mutants of *lac* deletion strains of *Escherichia coli, Can. J. Microbiol.* **18**:1439–1444.

Watson, H. C., and Kendrew, J. C., 1961, Comparison between the amino acid sequences of sperm whate myoglobin and of human haemoglobin, *Nature* **190**:670–672.

Weightman, A. J., Slater, J. H., and Bull, A. T., 1979, The partial purification of two dehalogenases from *Pseudomonas putida* PP3, *FEMS Microbiol. Lett.* **6**:231–234.

Wilson, A. C., Carlson, S. S., and White, T. J., 1977, Biochemical evolution, *Annu. Rev. Biochem.* **46**:573–639.

Winkler, H. H., and Wilson, T. H., 1966, The role of energy coupling in the transport of β-galactosides by *Escherichia coli, J. Biol. Chem.* **241**:2200–2211.

Withers, S. G., Jullien, M., Sinnott, M. L., Viratelle, O. M., and Yon, J. M., 1978, Dependence upon pH of steady-state parameters for the β-galactosidase catalyzed hydrolysis of β-D-galactopyranosyl derivatives of different chemical types, *Eur. J. Biochem.* **87**:249–256.

Wu, T. T., Lin, E. C. C., and Tanaka, S., 1968, Mutants of *Aerobacter aerogenes* capable of utilizing xylitol as a novel carbon source, *J. Bacteriol.* **96**:447–456.

Zipkas, D., and Riley, M., 1975, Proposal concerning mechanism of evolution of the genome of *Escherichia coli, Proc. Natl. Acad. Sci. USA* **72**:1354–1358.

Zuckerkandel, E., and Pauling, L., 1965, Evolutionary divergence and convergence in proteins, in: *Evolving Genes and Proteins* (V. Bryson and H. J. Vogel, eds.), pp. 97–166, Academic Press, New York.

3

A Comparative Summary of Genetic Distances in the Vertebrates

Patterns and Correlations

JOHN C. AVISE

and

CHARLES F. AQUADRO

Department of Molecular and Population Genetics
University of Georgia
Athens, Georgia 30602

INTRODUCTION

For studies in systematic and evolutionary biology, many kinds of molecular data have the unusual distinction of providing "*common yardsticks*" for quantitatively comparing genetic distances in phylogenetically distinct arrays of species. Yet after more than two decades of study, the genetic information residing in DNAs and proteins has only barely been tapped and employed in a comparative framework. Some tantalizing preliminary conclusions have emerged. For example, when different kinds of organisms, such as frogs and mammals, are contrasted, it appears that rates of evolution in structural genes (those that encode proteins) can proceed independently of rates of evolution in organismal morphology and way of life (Cherry *et al.*, 1978; King and Wilson, 1975).

The purpose of this paper is to review specifically the literature of genetic distances between vertebrate species based on conventional electrophoretic analyses of proteins. There are at least two reasons why such a review is appropriate at this time: (1) since the mid-1960s, beginning with the seminal study of Hubby and Throckmorton (1965), multilocus protein electrophoresis has been the most widely employed molecular tool to assay genetic distances between species. Recent reports have summarized levels of within-population variability (heterozygosity) in allozymes (Powell, 1975; Nevo, 1978), but apart from summaries of "genetic differentiation during speciation" (Ayala, 1975; Avise, 1976), no comprehensive reviews of genetic distances between species and genera have appeared since 1975 (Avise, 1974; Nei, 1975); (2) with the advent of new techniques for gene isolation, cloning, and nucleotide sequence determination (Gingeras and Roberts, 1980; Leder *et al.*, 1980; Nathans, 1979), the era in which protein electrophoresis was a "state-of-the-art" survey technique is rapidly coming to an end. Thus now may be a good time to take stock of earlier studies, both for interest in results in their own right, and as a stimulus for identification of problems in need of more refined study. (Nonetheless, it should also be emphasized that because protein electrophoresis is relatively simple and inexpensive, it will continue to be a valuable approach for many projects appropriate to limits of its resolving power.)

METHODS AND MATERIALS—LITERATURE SURVEY AND TAXA STUDIED

The literature review for this report ended in December 1980. We relied heavily upon listings in a bibliography of electrophoretic literature compiled by Smith *et al.* (1982). This bibliography includes publications through 1979, and lists a total of well over 1000 papers dealing with protein electrophoresis and taken from more than 250 journals and books. We also include in this review papers published in 1980, plus a few unpublished data sets of which we happened to be aware. Inevitably, given lags in publication time, important studies may have appeared subsequent to this compilation. Furthermore, because of the highly diverse array of journal outlets for electrophoretic surveys, some pertinent literature may have been overlooked. We apologize for these omissions.

To be included in this review, a study had to satisfy the following arbitrary criteria: (1) calculated genetic distances must be based on information from 14 or more genetic loci; (2) at least three species from a genus must have been examined (or, in comparisons among genera, at least three genera per family). These standards, although fairly lax, elim-

TABLE I. Summary of Protein Electrophoretic Surveys of Genetic Differentiation within Vertebrate Genera

Group	Mean number of individuals per species	Number of loci	Number of species examined	Total number of species in genus[a]	Similarity coefficient[b]	Reference code	Reference
Osteichthyes							
Salmoniformes							
Coregonus	250	27	4[c]	23	C	36	Ferguson *et al.* (1978)
Cypriniformes							
Notropis	4	14	47	101	A	5	Avise (1977)
Campostoma	78	19	3	3	A	26	Buth and Burr (1978)
Hypentelium	43	38	3	3	A	25	Buth (1980)
Thoburnia	21	34	3	3	A	24	Buth (1979)
Atheriniformes							
Cyprinodon	10	31	5	13	C	89	Turner (1974)
Menidia	85	24	5	6	C	51	Johnson (1975)
Perciformes							
Bathygobius	23	26	3	10	A	42	Gorman *et al.* (1976)
Lepomis	110	14	10	11	A	9	Avise and Smith (1977)
Etheostoma	35	23	3	88	A	27	Buth *et al.* (1980)
Amphibia							
Urodela							
Plethodon	5	29	26	~26	A	45	Highton and Larson (1979)
Taricha	121	18	3	3	A	44	Hedgecock and Ayala (1974)
Hydromantes	14	19	5	5	A	92	Wake *et al.* (1978)
Anura							
Hyla	30	19	7	289	A	35	Etges (1979)
Litoria	8	14	16	95	B	34	Dessauer *et al.* (1977)
Rana	40	20	7	248	A	29	Case (1978)
Scaphiopus	64	21	5	7	A	80	Sattler (1980)

(continued)

TABLE I. *(Continued)*

Group	Mean number of individuals per species	Number of loci	Number of species examined	Total number of species in genus[a]	Similarity coefficient[b]	Reference code	Reference
Reptilia							
Squamata							
Anolis	~25	~22	33	251	A	95	Wyles and Gorman (1980)
Bipes	18	22	3	3	A	54	Kim et al. (1976)
Crotaphytus	16	27	4	4	C	64	Montanucci et al. (1975)
Lacerta	159	22	3	53	B	41	Gorman et al. (1975)
Uma	38	22	3	3	A	1	Adest (1977)
Aves							
Anseriformes							
Anser	10	19	3	9	A	72	Patton and Avise (1983)
Aythya	10	19	5	12	A	72	Patton and Avise (1983)
Anas	8	19	10	36	A	72	Patton and Avise (1983)
Passeriformes							
Toxostoma	4	23	3	10	A	15	Avise et al. (1982)
Catharus	9	27	4	10	A	12	Avise et al. (1980a)
Zonotrichia	9	21	3	9	A	13	Avise et al. (1980b)
Ammodramus	8	21	3	10	A	13	Avise et al. (1980b)
Geospiza	34	27	6	6	A	96	Yang and Patton (1981)
Vermivora	10	26	4	11	A	14	Avise et al. (1980c)

					[b]		Reference
Dendroica	9	26	12	27	A	14	Avise et al. (1980c)
Vireo	17	23	5	25	A	15	Avise et al. (1982)
Parus	4	19	3	45	A		Aquadro et al. (in preparation)
Mammalia							
Chiroptera							
Lasiurus	13	20	5[d]	8	A		Patton et al. (in preparation)
Rodentia							
Dipodomys	110	18	11	21	A[e]	53	Johnson and Selander (1971)
Geomys	41	22	5	7	B	74	Penney and Zimmerman (1976)
Neotoma	56	20	3	20	A	99	Zimmerman and Nejtek (1977)
Peromyscus	~40	20	20	59	B	11	Avise et al. (1979)
Spermophilus	88	28	3	30	A	33	Cothran et al. (1977)
Thomomys	55	31	5	9	A	71	Nevo et al. (1974)
Primates							
Cercocebus	—	14	4	4	C	61	Lucotte (1979a)
Macaca	19	21	6	13	A	8	Avise and Duvall (1977)
Papio	—	14	4	4[f]	A	62	Lucotte (1979b)

[a] Total numbers of species in a genus were taken from reference lists compiled by the Gainesville, Florida, Field Station of the National Fish and Wildlife Laboratory, and from Morony et al. (1975).

[b] A, Nei's (1972) *I*; B, Rogers' (1972) *S*; C, other miscellaneous coefficients.

[c] Nine formerly listed taxonomic species reduced to four by Ferguson et al. (1978).

[d] May actually represent about 8–9 biologic species.

[e] Recalculated using allele frequencies given in paper.

[f] Considered geographic races of one polytypic species by some authors.

inated some otherwise appealing candidates for inclusion. For example, Utter *et al.*'s (1973) important study of 23 loci in eight species of Salmonidae was not included because genetic distances between species were calculated for only 8–12 loci. In a few cases, two or more papers treated overlapping sets of species in the same taxonomic genus or family [e.g., Avise *et al.* (1980c) and Barrowclough and Corbin (1978) on parulid warblers]. In such situations, genetic distances across laboratories were invariably in the same range. We have included results only of the study that examined the broadest array of species within the taxonomic grouping.

The great majority of studies utilized Nei's (1972) coefficients to summarize genetic similarities (I) and distances (D) (Table I), where $D = -\ln I$, and I can assume values ranging from zero (indicating no alleles shared by species) to one (species identical in allele frequency at assayed loci). A few studies employed Rogers' (1972) or other coefficients of similarity (S), which can also range from zero to one. Because I and various measures of S are strongly correlated, and exhibit similar absolute values in given species comparisons (Avise, 1974), all genetic distances in this paper were calculated from similarity measures by the equations $D = -\ln I$ or $D = -\ln S$. We believe that variance in D values across studies introduced from the use of different similarity measures is trivial compared to other possible sources of variance, such as particular loci chosen for assay or standards of laboratory practice for distinguishing electromorphs on gels.

Studies on a total of 44 vertebrate genera and 16 families, representing over 3800 pairwise comparisons of species, satisfy the criteria advanced above for inclusion in this summary. Relevant background information on these genera and families is presented in Tables I and II.

GENERAL PATTERNS OF GENETIC DISTANCE AMONG VERTEBRATE TAXA

For each genus, means (and ranges) of D across all pairs of assayed species were calculated. Because we are primarily interested in *relative* levels of genetic differentiation in diverse vertebrate groups, we have plotted these results along a common scale of genetic distance in Fig. 1. There is clearly a rich heterogeneity in mean level of D across genera; nearly the entire spectrum of conceivable distances was observed. For example, all assayed species within several avian genera (e.g., *Anser, Aythya, Catharus, Geospiza, Dendroica*) exhibit very small distances,

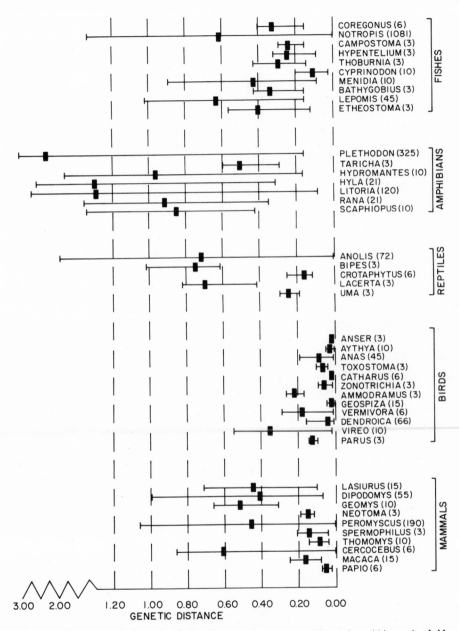

FIG. 1. Means (and ranges) of genetic distance between assayed species within each of 44 vertebrate genera. In parentheses are numbers of pairwise comparisons of species.

TABLE II. Summary of Protein Electrophoretic Surveys of Genetic Differentiation between Vertebrate Genera

Group	Mean number of individuals per species	Number of loci	Number of genera examined	Total number of genera in family[a]	Similarity coefficient[b]	Reference code	Reference
Osteichthyes							
Cypriniformes							
Cyprinidae	4	~15	19	~200	A	5	Avise (1977)
Perciformes							
Cichlidae	25	21	4	83	A	59	Kornfield et al. (1979)
Sciaenidae	6	16	6	46	C	83	Shaw (1970)
Pleuronectiformes							
Pleuronectidae	328	31	3	47	A	93	Ward and Galleguillos (1978)
Aves							
Anseriformes							
Anatidae	8	18	10	43	A	72	Patton and Avise (1983)
Passeriformes							
Mimidae	9	23	3	13	A	15	Avise et al. (1982)
Muscicapidae	8	26	4	~260	A	12	Avise et al. (1980a)

Emberizidae	8	21	7	~130	A	13	Avise et al. (1980b)
Emberizidae	23	27	5	~130	A	96	Yang and Patton (1981)
Parulidae	10	26	12	29	A	14	Avise et al. (1980c)
Icteridae	31	15	6	23	A	85	Smith and Zimmerman (1976)
Troglodytidae	7	19	5	14	A		Aquadro et al. (in preparation)
Hirundinidae	—	19	6	20	A		Zimmerman and Martin (in preparation)
Mammalia							
Chiroptera							
Phyllostomatidae	—	17	6	49	B	19	Baker et al. (1981)
Insectivora							
Soricidae	4	24	3	22	A		Patton et al. (in preparation)
Rodentia							
Cricetidae	3	15	9	113	A	72	Patton et al. (1981)
Primates							
Pongidae	10	22	3	3	A	22	Bruce and Ayala (1978)

[a] See Footnote A to Table I.
[b] See Footnote B to Table I.

attributable to minor allele frequency shifts at a few loci. In contrast, mean distances between species in the amphibian genera *Plethodon, Hyla,* and *Litoria* are huge, due to large allele frequency shifts at the majority of assayed loci.

The range of distances observed within a given genus is also typically very large. (Most studies have focused on exploiting these large ranges within genera to rank-order distances between pairs of species and to estimate phylogenetic relationships within the group.) We have presented ranges of D within a genus, rather than standard errors of D, for two reasons. First, the range is of greater interest in discussions of patterns of taxonomic convention across vertebrate groups (see below). Second, standard errors of D within a genus could be misused in statistical comparisons, because distance values between pairs of species in a matrix are not entirely independent of one another, and are not expected to be normally distributed (Avise, 1978).

Interestingly, some strong trends in magnitudes of genetic differentiation along major taxonomic lines are evident from Fig. 1. Amphibians show by far the greatest genetic distances. Six of the seven assayed amphibian genera exhibit larger mean D values than *any* non-amphibian genus. Even the least differentiated amphibian genus, *Taricha*, shows a larger mean D than do 75% of the non-amphibian genera. Reptiles are represented by few studies, but three of five reptilian genera (*Anolis, Bipes, Crotaphytus*) show larger mean D values than any non-reptilian genera, exclusive of the amphibians. At the other end of the spectrum, birds are highly conservative in level of genetic divergence. To dramatize this point, the average amphibian genus has a mean D about 11 times greater (unweighted values) or 22 times greater (weighted values) than the respective mean D values for avian genera (Table III). In general,

TABLE III. Overall Summary of Mean Genetic Distances in Vertebrate Classes

Comparison	Group	Total number of pairwise species comparisons	Mean D	
			Unweighted	Weighted[a]
Within genera	Osteichthyes	1167	0.36	0.60
	Amphibia	510	1.12	1.75
	Reptilia	87	0.51	0.67
	Aves	173	0.10	0.08
	Mammalia	313	0.30	0.41
Between genera	Osteichthyes	744	1.13	0.76
	Aves	755	0.26	0.24
	Mammalia	114	0.78	1.10

[a] Weighted by number of pairwise species comparisons per genus or per family.

mean genetic distances in fish and mammalian genera fall intermediate to the birds versus the amphibians and some reptiles.

We also calculated means and ranges of D across confamilial vertebrate genera, and results are plotted in Fig. 2. The same general trend is apparent, with avian families exhibiting conservative genetic distances relative to most of the fishes and mammals. We could find no criteria-satisfying estimates of genetic divergence between amphibian or reptilian genera. This is understandable, because for many genera in these classes, essentially the entire observable range of genetic distance had already been exhausted in comparisons among congeners.

Genetic distances in five vertebrate classes are further summarized in Table III. It seems safe to conclude that, for whatever reason, genera and families in the different classes of vertebrates are not equivalent in level of structural gene divergence as currently assayed.

CONSIDERATIONS ON TAXONOMIC CONVENTION

The various patterns discussed above might be attributed to different conventions of taxonomic procedure employed by vertebrate systematists—if, for example, by some standard, avian systematists were "splitters" at the generic level and amphibian systematists were "lumpers." Since to this point our summary has taken all taxonomic assignments at face value, this possibility deserves consideration.

In general, if all else were equal, oversplitting at the species level (i.e., the taxonomic assignment of conspecific populations to different species) would result in a shift in D values to the right in Fig. 1, toward lower distances. Fortunately, at the species level the independent criterion of reproductive isolation in nature can be invoked to ask whether species splitting has been greater in one class of assayed vertebrates than another.

The assayed birds are definitely *not* oversplit at the species level. Virtually all the taxonomic species also represent unambiguous biologic species, retaining separate identities even where sympatric in nature. Evidence substantiating this point has been explicitly presented for the Parulidae (Avise *et al.*, 1980c), and similar evidence could be advanced for the other avian groups as well. However, in some of the non-avian vertebrates, species oversplitting may have contributed to a lowering of D values. For example, among the mammals the four assayed *Papio* taxonomic species are considered geographic races of a single biologic species by some authors (Lucotte, 1979b). Similarly, surveyed allopatric

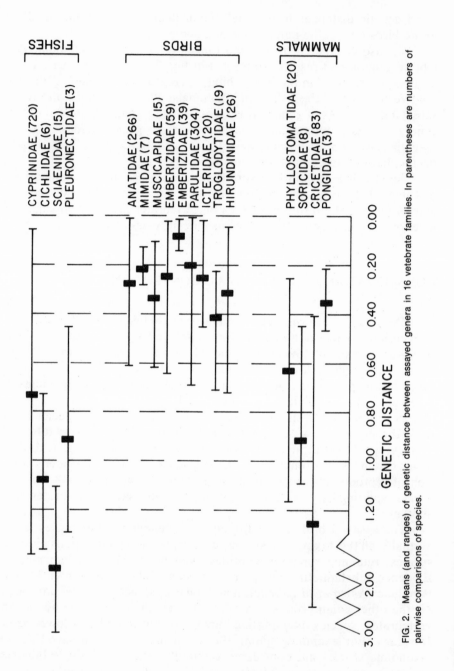

FIG. 2. Means (and ranges) of genetic distance between assayed genera in 16 vetebrate families. In parentheses are numbers of pairwise comparisons of species.

"species" of *Neotoma* and of *Spermophilus* might just as validly be considered subspecies or semispecies (Cothran *et al.*, 1977; Zimmerman and Nejtek, 1977). The five chromosomally defined species of *Thomomys* are very closely related and were originally included within *Thomomys talpoides*. According to Nevo *et al.* (1974), the forms represent "early to late stages" of the speciation process. Among fishes, the *Cyprinodon* species inhabit disjunct desert drainages of California and Nevada, and since they yield fertile F_1 hybrids in the laboratory, their specific status might conceivably be subject to question (they are, nonetheless, quite different morphologically and behaviorally). Similar reservations might be expressed about the validity of the specific status of certain other geographically separate forms, for example of *Thoburnia*, as well as of a few scattered representatives of other assayed fish, amphibian, and reptilian genera. If these other groups have been oversplit to any appreciable extent at the species level, this would serve to accentuate even more the significance of the already apparent conservative pattern of genetic differentiation in birds.

Conversely, it is likely that "good" biologic species in some vertebrate groups have remained unrecognized because of morphologic conservatism. It is also conceivable that such sibling species are more common among amphibians, for example, than among the birds, where optical cues and associated morphologic changes often play an important role in reproductive isolation. If closely related, unrecognized sibling species are especially abundant among certain non-avian vertebrates, the apparent genetic distances in these groups would be artificially inflated for this reason.

At the generic level, if all else were equal, taxonomic oversplitting (the taxonomic assignment of very closely related species to different genera) would be reflected in a truncation of observed genetic distances at a fairly low level within a genus. Since avian genera (and to some extent fish and mammalian genera) do exhibit this truncated pattern relative to most amphibians and some reptiles, is this evidence of generic oversplitting? Conversely, excessive lumping (for example, the artificial placement of cladistically unrelated species into a genus) would lead to unrealistic inflation of genetic distances for a genus. Since amphibian (and to some extent reptilian) genera do exhibit large distances relative to birds, is this evidence that these genera represent to some extent artificial amalgamations of species? These questions are difficult to answer, because, as noted in the introduction, in the past there have been no uniform criteria for recognition of taxonomic categories which could be applied evenly in organisms as different as amphibians and birds. With respect to the standard of structural gene divergence as reflected in proteins, the avian genera assayed *are* oversplit relative to amphibians (or, equivalently, amphibians are "undersplit" relative to birds).

We have employed a taxonomic list, "Vertebrates of the World" (1978, unpublished; compiled by the National Fish and Wildlife Laboratory, Gainesville, Florida), to count numbers of extant species in various vertebrate taxa. Results are summarized in Table IV. Genera and families of amphibians and reptiles contain, on the average, roughly 2–5 times as many recognized species as do genera and families, respectively, of fishes, birds, and mammals. This is true for the particular taxa for which genetic distance estimates are available, as well as for the vertebrate classes as a whole (Table IV). Thus, strictly in terms of census species numbers, fishes, birds, and mammals are somewhat oversplit relative to the reptiles and amphibians, a result which roughly parallels the conclusions reached from genetic distance estimates. However, the degree of "census" oversplitting may not be entirely sufficient to account for the huge disparity in genetic distances between, for example, birds and amphibians. Suppose, for sake of argument, that an assayed family of birds were to be reduced to a "genus." That "genus" would contain, on the average, more than three times as many species as an average assayed amphibian genus (Table IV), yet genetic distances in that "genus" would remain several-fold lower than distances within typical amphibian genera (Table III; also compare Figs. 1 and 2).

These conclusions implicitly assume that the estimates of genetic distance in genera of the vertebrate classes have been based on adequate representations of species within those genera. This assumption is justified in Table IV (right-hand side), where the mean proportions of extant species assayed per genus are fairly large and uniform (range 0.40–0.64) across classes. Nonetheless, some of the scatter in Figs. 1 and 2 is certainly attributable to differential representation of assayed species within genera or families. For example, *Neotoma* is represented by only three of the most closely related of 20 species in the genus, while *Geomys* is represented by five of the seven total species within that genus.

Even if it should prove true that practices of taxonomic procedure could account for the observed patterns of *D* across vertebrate classes, this result might nonetheless be of considerable biologic significance if it could also be shown that the taxonomic practices reflect general biologic characteristics of the groups, such as different rates of morphologic divergence. For example, it appears that amphibians such as frogs have evolved very slowly in morphology relative to mammals (Wilson *et al.*, 1977), and this conservative morphologic differentiation is probably reflected, in part, in the larger average size of amphibian genera. Cherry *et al.* (1978) have recently made valiant preliminary attempts to stand-

TABLE IV. Taxonomic Characteristics of the Vertebrate Classes

Group	Entire class worldwide			Particular taxa in this report (Tables I and II)			
	Total no. species	Mean no. species per genus	Mean no. species per family	Total no. species	Mean no. species per genus	Mean no. species per family	Mean proportion species per genus actually assayed
Fishes	13,732	4.1	32.8	261	26.1	431.8	0.61
Amphibians	3,373	8.7	108.8	671	95.8	—	0.56
Reptiles	6,508	7.1	135.6	314	62.8	—	0.64
Birds	9,031	4.4	56.4	210	17.5	315.1	0.40
Mammals	3,993	4.0	32.4	182	18.2	258.2	0.55

ardize and quantify levels of morphologic differentiation across vertebrate classes, and these approaches introduce a potentially fruitful field for future comparative studies.

PROTEINS ASSAYED

Fortunately for purposes of this comparative review, multilocus electrophoretic techniques have become fairly standardized across laboratories. The detailed electrophoretic procedures most often followed are those of Shaw and Prasad (1970), Selander *et al.* (1971), and Ayala *et al.* (1972). Table V lists the proteins most commonly assayed, and Table VI shows the monitored distributions of loci encoding these and other proteins in the 44 vertebrate genera assayed. Several proteins, such as GOT, IDH, LDH, MDH, PGM, 6-PGD, EST's, and PT's, were included in almost every study.

Gillespie and Kojima (1968) and Kojima *et al.* (1970) distinguish two general classes of proteins, glucose-metabolizing enzymes (group I) and nonspecific enzymes (group II), on the basis of mean level of polymorphism within species. A higher level of heterozygosity in group II enzymes was attributed to the greater diversity of environmental substrates for reactions catalyzed by these enzymes. Several other authors have made similar attempts to classify proteins by function, and have noted some general trends in heterozygosity levels [Gillespie and Langley (1974), Johnson (1973, 1974); but see Selander (1976)]. Since correlations across loci between heterozygosity levels and level of genetic divergence between species have been reported (Johnson and Mickevich, 1977; Koehn and Eanes, 1978), an examination of the loci used for D estimates is warranted.

Our primary concern here is not whether different arrays of proteins have evolved at different rates in the vertebrates, but rather whether a sampling bias in loci assayed could potentially account for the huge differences in mean D values across vertebrate classes. Relevant results are summarized in Table VII. When the proteins are arranged by metabolic function according to Gillespie and Kojima's (1968) [and Selander's (1976)] classification, it is apparent that very similar sets of loci have been assayed in all the major vertebrate groups. For example, surveys of the Amphibia and Aves included an average of 16% versus 13%, respectively, of group II enzymes (and 8% versus 4%, respectively, of esterases). The other vertebrates exhibit comparable figures.

Sarich (1977) has argued that "a rather striking bimodality describes

TABLE V. Names and Abbreviations of Proteins Most Commonly Included in Electrophoretic Surveys of Genetic Distance[a]

Name	Abbreviation	Enzyme commission (E.C.) number
Group I enzymes		
Adenylate kinase	ADKIN	2.7.4.3
Creatine kinase	CK	2.7.3.2
Glucose-6-phosphate dehydrogenase	G-6-PDH	1.1.1.49
Glutamate-oxaloacetate aminotransferase (Aspartate aminotransferase)	GOT	2.6.1.1
α-Glycerophosphate dehydrogenase	α-GPDH	1.1.1.8
Isocitrate dehydrogenase	IDH	1.1.1.42
Lactate dehydrogenase	LDH	1.1.1.28
Malate dehydrogenase	MDH	1.1.1.37
Malic enzyme	ME	1.1.1.40
Mannose phosphate isomerase	MPI	5.3.1.7
Phosphoglucomutase	PGM	2.7.5.1.
6-Phosphogluconate dehydrogenase	6-PGDH	1.1.1.43
Phosphoglucoisomerase	PGI	5.3.1.9
Xanthine dehydrogenase	XDH	1.2.3.2
Group II enzymes		
Acid phosphatase	ACPH	3.1.3.2
Alcohol dehydrogenase	ADH	1.1.1.1
Esterases	EST	3.1.1.2
Peptidases	PEP	3.4.1.1
Sorbitol dehydrogenase	SDH	1.1.1.14
Group III proteins		
Nonenzymatic proteins	PT	—
Unknown function		
Tetrazolium oxidase[b]	TO	—

[a] Scored in six or more of the 44 vertebrate genera of Fig. 1.
[b] Probably represents superoxide dismutase (E.C. 1.15.1.1) in at least some studies.

the distribution of electrophoretically measured rates of evolution at the various protein loci. In other words, there are two sets of proteins, and one of them is accumulating electrophoretically detectable substitutions at a rate tenfold greater than the other." The rapidly evolving set supposedly includes the nonspecific esterases and the various nonenzymatic blood proteins (group III proteins of Table V). There is now abundant evidence that different proteins evolve on the average at different rates, but Sarich's specific conclusions are less than convincing for at least two reasons: (1) they ignore the possibility of variance in rates among proteins within each of the "bimodal" classes, as well as among different lineages, and (2) empirical data do not always support the assignment of proteins

TABLE VI. Numbers of Protein-Coding Loci from Which Genetic Distances Were Estimated in Various Vertebrate Genera

Protein	Reference[a]																					
	36	5	26	25	24	89	51	42	9	27	45	44	92	35	34	29	80	95[b]	54	64	41	1
ADKIN	3				1		2				1			1								
CK			1	3	3		2			1				1								2
G-6-PDH						2	1				1											1
GOT		1	2	3	3	2	2	2	2	2	2		2	2	1	1	2	2	2	1	2	2
α-GPDH						1		1		1	1	1	1	1		1	2	1	1	2	1	1
IDH	1	1				1	2		1		2	1	2	1			3		1	1	2	2
LDH	4	2	3	3	3	3	3	2	2	2	2	2	2	2	2	2	1	2	2	2	2	2
MDH	1	1	3	5	5	2	2	3	1	3	2	1	1	2	1	2	2	2	1	2	1	2
ME	1										1	1	1	1		2				1		
MPI							1			1	1		1	1								
PGM	2	1	1	2	2		1	1	1	1	1		1	2		2	1	2	1	1	2	2
6-PGDH		1	2	2				2	1		1		1	1		1	1	1		1	1	
PGI	3	2	2	3	3	1	2	2	2	2	1		1	1				1	1		1	1
XDH	1		1	1	1			1		1								1				1
ACPH			1	1						1	2				1					1		
ADH	2		1	1	1	1		2		1	1			1	1			1				
EST	3	2	1	2	2	10	1	3	2		1	4			1	3		2	3	2	4	3
PEP	2		1	2	2	1	2	2	1		2		2				1	2	3	2		2
SDH				1		1	1		1	1	1		1	1			1		1			
PT		3		5	5	2		3		2	4	5			7	5	4	5	4	7	4	3
TO	2		1	2	2	1	1	1		1	1	1	2	1	1	1	1		2		1	2
All others	2		1	3	3	3	1	1		3	3	2	2		1		2			4	1	2

Protein	72	72	72	15	12	13	13	96	14	14	14	16	c	d	53	74	99	11	33	71	61	8	62
ADKIN	2	2	2	2	2	1	1		1	1	1									2		1	1
CK						2	2		2	2	2	2	2	3									
G-6-PDH																		1				1	1
GOT	2	2	2	2	2	2	2	2	2	2	2	2	2	2	2	2	2	2	2	2			
α-GPDH	1	1	1	2	2	2	2	2	1	1	1	1		1	1	1	1	1	1	1			
IDH	2	2	2	2	2	2	2	2	2	2	2	2	1	1	2	2	2	2	2	2		1	
LDH	2	2	2	2	2	2	2	2	2	2	2	2	2	2	2	3	3	2	3	2	2	2	2
MDH	2	2	2	2	2	2	2	2	2	2	2	2	2	2	2	2	2	2	2	2		1	1
ME				1	2			1				1	1	1									
MPI												1											
PGM	1	1	1	1	1	1	1	2	3	3	3	1	1	1	1		1	1	2	1		2	2
6-PGDH	1	1	1	1	1	1	1	1	1	1	1	1	1	1	1	1		1	1	1	1	1	1
PGI	1	1	1	1	1	1	1	1	1	1	1	1	1	1			1	1	1	1	1	1	1
XDH																							3
ACPH																				1		1	
ADH								1	1	1	1									1			
EST	2		2	2	3	2		3	3	1	1	1	1			6	4		6	4	4		
PEP		2	2	1	3	1		3	3	3	3	1	1	3						1			
SDH								1										1					
PT	2	2	2	3	3	3	3	2	3	3	3	3	2		6	4	3	5	6	7	7	4	
TO	1	1	1	1	1	1	1	2	1	1	1	1	1	2	1	1	1	1	1	1			
All others					1				1	1		1	1							1		5	2

_a_ Numbers refer to reference code numbers in Tables I and II.
_b_ From Gorman and Kim (1976).
_c_ From Aquadro et al. (in preparation).
_d_ From Patton et al. (in preparation).

TABLE VII. Mean (and Range) Proportion of Loci of Various Groups Assayed per Study in the Vertebrates[a]

	Group I	Metabolic function[b]		Unknown function	Rapidly evolving[c]	Esterases
		Group II	Group III			
Osteichthyes	0.65 (0.43–0.78)	0.22 (0.13–0.46)	0.08 (0.00–0.21)	0.05 (0.00–0.08)	0.19 (0.04–0.43)	0.11 (0.00–0.35)
Amphibia	0.60 (0.38–0.89)	0.16 (0.06–0.25)	0.22 (0.00–0.54)	0.02 (0.00–0.06)	0.29 (0.00–0.62)	0.08 (0.00–0.25)
Reptilia	0.51 (0.41–0.57)	0.24 (0.19–0.32)	0.20 (0.13–0.30)	0.05 (0.00–0.09)	0.34 (0.27–0.39)	0.13 (0.09–0.19)
Aves	0.71 (0.55–0.76)	0.13 (0.05–0.30)	0.12 (0.08–0.14)	0.04 (0.00–0.07)	0.16 (0.10–0.23)	0.04 (0.00–0.11)
Mammalia	0.57 (0.22–0.75)	0.18 (0.00–0.28)	0.21 (0.00–0.50)	0.04 (0.00–0.10)	0.32 (0.00–0.78)	0.11 (0.00–0.28)

[a] Excludes from consideration "all other" proteins of Table VI.
[b] As classified by Gillespie and Kojima (1968) and Selander (1976); see Table V.
[c] As argued by Sarich (1977).

to the two classes described by Sarich. For example, three of the general proteins assayed in 20 species of *Peromyscus* are *less* divergent electrophoretically than are any of the other 17 loci examined. Nonetheless, for sake of argument we can accept Sarich's conclusions and examine the proportion of "rapidly evolving" loci assayed in the vertebrates (Table VII). These range from 16% (Aves) to 34% (Reptilia). If we standardize estimates of genetic distance in Amphibia and Aves to values expected had identical proportions of "fast-evolving" loci been assayed, we are still left with no less than a sixfold difference in level of mean D between these groups which would remain unexplained.

The conservative pattern of protein differentiation in Aves has been reported independently by at least five electrophoretic research groups, and the nonconservative pattern in Amphibia has been reported by at least seven laboratories. All of these studies have employed roughly similar sets of loci. Thus, in summary, Selander's (1976) conclusion that "the major determinant of the span of variation in estimates of polymorphism is the laboratory in which the survey was conducted" is probably not a valid explanation to account for the widely different levels of genetic distance reviewed in this paper.

EVIDENCE FOR A UNIFORM PROTEIN CLOCK

In a general review of biochemical differentiation, Wilson *et al.* (1977) argue that "the discovery of the evolutionary clock stands out as the most significant result of research in molecular evolution." The concept that genetic divergence between reproductively isolated populations proceeds at stochastically constant rates follows directly from the neutral mutation theory (Nei, 1975), but in fact when long spans of time are involved and averages are made across many proteins, a correlation between genetic distance and elapsed time since speciation is expected under many selectionist scenarios as well (Avise and Ayala, 1975).

Under neutral mutation pressure, low and intermediate values of Nei's D statistic are linearly related to time of divergence of two populations; hence calibrations of the electrophoretic clock against absolute time, and comparisons of these calibrations in different organisms, become possible provided independent information about times of speciation is available. In our opinion, *the major obstacle to critical tests of the electrophoretic protein clock is the almost total lack of reliable independent information about times of speciation.* For most of the 44 genera and 16 families reviewed in the paper, we are unaware of any critical and detailed nonmolecular information about mean times of species splitting. As Wilson *et al.* (1977) admit for electrophoretic data, "Although there is indirect evidence that *m* (genetic distance) is correlated with time elapsed since divergence, direct evidence is fragmentary."

Nonetheless, the literature conveys a very strong impression that genetic distances (electrophoretically assayed) can be reliably used to date speciations. Many papers conclude with statements such as the following: "The predicted divergence time (from electrophoretic data) is about 2.5 million years, which is in accord with geological evidence" (Gorman *et al.*, 1976) and "fossil data from pocket gophers of the genera *Thomomys* . . . and *Geomys*. . . and deer mice, genus *Peromyscus* . . . have been shown to correspond well with estimates of divergence times from biochemical evidence" (Zimmerman and Nejtek, 1977). If such conclusions are to be convincing, one would expect them to be based on at least roughly similar electrophoretic clocks. But in fact the two studies quoted above employed two distinct clocks, calibrated more than 22-fold different from one another!

Such flexibility in clock calibration is not atypical. Table VIII provides examples of some of the various electrophoretic clock calibrations to be found in the literature. These range from a high rate of $D = 1$ accumulating in only 0.8 million years, to a low rate of $D = 1$ accumulating in 18 million years. Many intermediate rates can be found as well, yet without exception all of the references in Table VIII that addressed the issue concluded that molecular-based estimates of absolute divergence time were compatible with fossil or geologic evidence. Figure 3 plots two

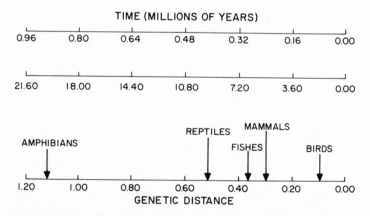

FIG. 3.　Two time scales commonly employed in the literature, plotted against genetic distance (Nei's *D*). Mean distances between congeneric species of the vertebrate classes (taken from Table III) are also plotted for reference.

TABLE VIII. Examples of Various Electrophoretic Protein-Clock Calibrations Found in the Literature

Group	Nei's D	Time, millions of years	Reference
Fishes			
Hesperoleucus	1	1.0	Avise *et al.* (1975)
Astyanax	1	5.0	Chakraborty and Nei (1974)
Bathygobius	1	18.0	Gorman *et al.* (1976)
Amphibians			
Plethodon	1	14.0	Highton and Larson (1979)
Reptiles			
Bipes	1	5.0–18.0	Kim *et al.* (1976)
Anolis	1	18.0	Yang *et al.* (1974)
Uma	1	18.0	Adest (1977)
Birds			
Geospiza, etc.	1	5.0	Yang and Patton (1981)
Mammals			
Peromyscus	1	0.7	Zimmerman *et al.* (1975)
Thomomys	1	1.5	Nevo *et al.* (1974)
Homo	1	3.8	Nei and Roychoudhury (1982)

distinct but commonly employed protein clocks against genetic distance. Given the huge range of time estimates available from different clocks that might be chosen, it is hard to imagine a genetic distance estimate that would not be "compatible" with almost any fossil or geologic data. For example, from Fig. 3 an average pair of reptile species could be estimated to have shared a common ancestor anywhere from 400,000 to 9 million years ago.

A linear relationship between D and time is predicted by the "infinite-allele" model (Fig. 3), but when backward and parallel mutations are taken into consideration, estimated divergence times do not increase linearly when D's are large (Nei, 1981). However, the direction of the relationship is such that, if these stepwise mutation models actually hold, amphibians and reptiles would be older than pictured in Fig. 3, and hence the discrepancy with birds would become even greater.

Even the large range of possible time estimates shown in Fig. 3 may not be wide enough to accommodate all of the observed range in D. For example, species within genera of Parulidae exhibit a mean genetic distance of $D = 0.06$ (Avise *et al.*, 1980c). From Fig. 3 we can estimate that an average pair of parulid species diverged anywhere from about 5000 to about one million years ago. Yet, Mengel (1964) has presented a detailed, convincing, and widely accepted zoogeographic reconstruction for the parulids which would date the mean time of splitting of the assayed species at well more than two million years ago. If Mengel's scenario is correct, the proteins of these birds would appear to be evolving even more slowly than the slowest previously calibrated clock. Additional evidence for a possible slowdown in protein evolution in birds is advanced elsewhere (Patton and Avise, 1983). Interestingly, the calibrations exemplified in Table VIII are generally opposite *even in direction* to this suggestion— thus Yang and Patton (1981) employed a severalfold more rapid calibration for protein divergence in birds than did Yang *et al.* (1974) for *Anolis* lizards.

If, on the other hand, a single protein clock does exist for all vertebrates, the data in Fig. 1 and Table III argue that an average pair of assayed amphibian congeners last shared a common ancestor somewhere between 11 and 22 times as long ago as did an average assayed pair of avian congeners, and that mean speciation times for most fishes, reptiles, and mammals are generally intermediate to these. It is a measure of our ignorance about absolute times of species divergence that we cannot critically eliminate these as viable possibilities.

In summary, the available evidence for (or against) a single, multi-locus protein-electrophoretic clock in the vertebrates is far less compelling than a casual reading of the literature would suggest. It is entirely possible that electrophoretic clocks will have to be calibrated separately for different classes or even lower taxa of vertebrates. Of course as more calibrations are defined, the meaningfulness of any clock concept rapidly diminishes. These comments are not meant to be a general indictment of evidence for molecular clocks—we have dealt solely with electrophoretic data. But for these studies, the protein-clock theory has clearly and unduly molded the interpretation of data, rather than vice versa.

OTHER FACTORS INFLUENCING LEVELS OF GENETIC DIVERGENCE

A simultaneous fascination yet frustration of evolutionary biology is the fact that theory to account for empirical observations may often be advanced from different levels of biologic organization, or from seemingly widely disparate fields of study. This is certainly the case in attempts to explain composite empirical attributes such as mean heterozygosity H or genetic distance. For example, in a recent book summarizing protein variation (Ayala, 1976), potential explanations for varying levels of H were sought from considerations ranging from regulatory control of flux through metabolic pathways (Johnson, 1976) and enzyme function (Selander, 1976), to population size and time since last bottleneck (Soulé, 1976), to global patterns of trophic stability and resource utilization (Valentine, 1976). Such explanations are not necessarily mutually exclusive, but this does not imply that their possible connections are always clear.

A comparable ensemble of considerations, ranging from physiologic and genetic to ecologic and historical factors, might be addressed with respect to the issue of level of genetic divergence as well. An abbreviated list of some such considerations is presented in Table IX. It is well beyond the scope of this report to examine all of these possibilities in detail. Some, such as mutation rate, are simply unknown. Others, such as fecundity, might show general trends across the vertebrate classes roughly correlated with mean levels of D (Hamrick *et al.*, 1979), but even if so, possible causal links, if any, between the two are unclear.

We will briefly discuss only two characteristics of vertebrates roughly correlated with D. These characteristics warrant consideration because plausible (though certainly unproven) causal connections might associate them with D, through the following hypotheses.

1. *Heterozygosity and genetic divergence.* Hypothesis: Mean level of heterozygosity within species is functionally related to level of diver-

TABLE IX. Examples of Biologic Characteristics Which Might Conceivably Influence Levels of Genetic Divergence between Species[a]

I.	Genetic characteristics
	a. Heterozygosity
	b. Mutation rate
	c. Total amount of DNA
	d. Chromosome number
II.	Physiologic characteristics
	a. Body temperature (stability and absolute level)
	b. Blood pH
	c. Parent–offspring immunology
III.	Ecologic characteristics (demographic and reproductive)
	a. Population size (including fluctuations in size)
	b. Generation length
	c. Fecundity
	d. Care of young
	e. Habitat type
IV.	Evolutionary characteristics
	a. Mode of fertilization
	b. Mating system
	c. Form of reproductive isolation
	d. Mode of speciation
	e. Time since speciation

[a] Many characteristics are arbitrary in placement by category.

gence between closely related species, all else being equal. Under this hypothesis, genera exhibiting larger mean D's do so because of increased levels of within-species genetic variability, which is then available for conversion to species differences during speciation. A corollary is that level of genetic divergence between closely related species would be limited in phylads whose species exhibit low genetic variation.

Table X summarizes the most comprehensive available estimates of H per species in five vertebrate classes. Two of the estimates come from reviews by Powell (1975) and Nevo (1978) and are based on totals of 71 and 134 species, respectively; the third estimate is for 284 of the assayed species and genera included in this review (Table I). In each case, amphibians exhibit the highest levels of H, and in two of the three summaries, birds show the lowest H. [For serious reservations concerning tests of statistical significance of these differences, see Powell (1975).] Estimates of H for fishes, reptiles, and mammals are remarkably consistent in all summaries, and generally fall intermediate to mean values for amphibians and birds. Nonetheless, the difference in mean H between birds and amphibians is only about $1\frac{1}{2}$- to 3-fold in the various summaries, and there is tremendous overlap in H's for particular species in the two groups.

Even if it does prove true that the level of heterozygosity within species is correlated with and influences mean D between closely related species, reasons for the difference in H across taxa would remain conjectural. Furthermore, since H represents only the within-population component of within-species genetic variation, the between-population component might be more relevant in influencing the magnitude of genetic divergence during speciation. If this is true, the hypothesis advanced above could be shifted slightly to propose that the magnitude of between-population divergence is related to within-population variation. Finally, it is conceivable that H and D are correlated, but not causally or biologically linked, because they are both determined by a third underlying factor, such as taxon-specific abilities of particular electrophoretic conditions (i.e., buffer concentration or pH) to distinguish different allelic products.

2. *Body temperature and genetic divergence.* Hypothesis: The temperature regimes in which proteins function and evolve causally influence magnitudes of protein variation and divergence. Such a general hypothesis could have several aspects. Perhaps extremes of environmental temperature, particularly high temperatures, might limit the number of amino acid substitutions that would yield a new protein with proper biochemical function at those elevated temperatures. Another possibility is that temporal *stability* in temperature regimes might select for reduced genetic

variation in loci encoding proteins (or, conversely, variable temperatures might select for increased protein diversity).

This hypothesis is motivated by the important effect that temperature is known to have on biochemical processes of cells (Bennett and Ruben, 1979; Hochachka and Somero, 1973; Somero, 1978), in particular in influencing rates of chemical reactions, and in influencing higher orders of protein structure through effects on weak chemical bonds. Thus, for example, Hochachka and Somero (1973) suggest that "in birds and mammals, with body temperatures near 40°C, the increased thermal energy present in the system, relative to ectotherms, may render the ordering or structuring of the enzyme–substrate system more difficult," and "mammalian and avian enzymes . . . may have more rigid higher order structures than ectothermic enzymes."

Among the vertebrates, birds exhibit the highest and most uniform body temperatures; these range from about 38 to 43°C across species, and remain temporally stable within most species (King and Farner, 1961). Associated with the high body temperatures are a host of related physiologic conditions [e.g., metabolic rate, values of blood P_{CO_2} and blood pH (Calder and Schmidt-Nielsen, 1968; Somero, 1978)]. Several of these characteristics might conspire to influence protein evolution, conceivably inhibiting rates of amino acid substitution. If, as we have suggested elsewhere, protein evolution is indeed decelerated in birds (Avise *et al.*, 1980c; Patton and Avise, 1983), part of the explanation may lie in the nature of the selection regime imposed by the internal physiologic environment.

Mammals are also endothermic, with mean body temperatures ranging from about 31 to 40°C across species (Altman and Dittmer, 1974). If the considerations discussed above are valid, we might also expect mammals to be rather conservative in level of protein variation and divergence. There are some suggestions that this is indeed the case (Table X and Fig. 1), at least relative to most amphibians and reptiles. However, many mammals (and a few birds) also undergo various forms of hibernation, ranging from lethargy to deep torpor, with associated, variable lowerings of body temperature (Altman and Dittmer, 1974). Thus, some mammals may experience fluctuating intracellular temperature regimes during their lifetimes. Among the nonendothermic vertebrates, body temperatures are also likely to fluctuate widely both within and among species since the temperatures depend to a large extent on the external environment. It is not at all likely that amphibians as a group inhabit lower or more variable temperature habitats than do fishes. Thus, even if temperature does significantly influence level of protein variation and divergence, it cannot

TABLE X. Estimates of Mean Heterozygosity per Species in the Vertebrate Classes[a]

	Mean D per genus (present study)	Powell (1975)		Nevo (1978)		Present study	
		Number of species	H	Number of species	H	Number of species	H
Fishes	0.36	31	0.058	51	0.051	77	0.054
Amphibians	1.12	3	0.105	13	0.078	69	0.058
Reptiles	0.51	9	0.043	17	0.047	27	0.041
Birds	0.10	3	0.031	7	0.047	61	0.037
Mammals	0.30	25	0.039	46	0.036	50	0.039

[a] Product-moment correlations between mean H and mean D across the vertebrates are $r = 0.94$ ($p < 0.05$), $r = 0.88$ ($p < 0.05$), and $r = 0.76$ (NS) in the summaries of Powell, Nevo, and present study, respectively.

by itself explain all of the observed heterogeneity in D across the vertebrate classes.

CONCLUSIONS AND SUMMARY

Considering the multivarious factors which might be expected to influence mean genetic distances in different genera, it would be surprising if any consistent trends in magnitude of D per genus were apparent among the vertebrates. Nonetheless, our summary of the multilocus protein-electrophoretic literature, including over 3800 pairwise comparisons of species, has disclosed some strong and consistent patterns of genetic differentiation. In particular, avian congeners are extremely conservative in magnitude of protein divergence relative to all amphibian congeners assayed to date. Few reptiles have been assayed, but three of five genera exhibit a nonconservative pattern of divergence approaching that of the amphibians. Fish and mammalian genera are highly variable in D, but generally fall intermediate in level of divergence to the birds and amphibians.

Several possible explanations for these trends are considered:

1. Taxonomic convention: The conservative pattern of protein differentiation in birds cannot be attributed to "oversplitting" at the species level. All assayed avian taxa are "good" biologic species. Oversplitting at the species level in some of the other vertebrate groups does undoubtedly contribute to a lowering of D within various genera, but this consideration serves to emphasize even more the already apparent conser-

vative pattern of genetic divergence in birds. At the generic level, "lumping" (as gauged by census counts of numbers of species per genus) has been about twofold greater in amphibians and reptiles than in other vertebrates, and this convention is a contributor to the explanation of the trend in D's across the vertebrates.

2. Proteins assayed: A consideration of the particular proteins examined in various studies leads to the conclusion that the trends in D across the vertebrates are not attributable to laboratory- or study-dependent sets of proteins assayed. This is true no matter whether the proteins are classified by function (i.e., glucose versus non-glucose metabolizing), or by presumed rate of evolution.

3. Protein clock: Notwithstanding many comments to the contrary in the literature, there is little solid evidence for (or against) a common multilocus electrophoretic clock for the vertebrates. In various studies, clock calibrations have varied more than 20-fold even though common sets of proteins were assayed. The existing 20-fold range of clock calibrations still may not be sufficiently wide to account for all of the observed D's. In particular, it is possible that avian proteins are evolving, on the average, even more slowly than predicted by the slowest previously calibrated clock. On the other hand, given our general ignorance about absolute times of species separation, we cannot yet, by hard criteria, eliminate the possibility of a single electrophoretic clock.

4. Other explanations: A variety of factors, alone or in combination, and ranging from genetic to physiologic to ecologic and evolutionary characteristics, might be responsible for the observed trends in D. We focus on two characteristics which are roughly correlated with D across vertebrates, and for which plausible causal connections with D can be hypothesized: heterozygosity within species, and temperature regime. Heterozygosity could directly influence D because between-species differences are built upon within-species variation, or it could be that D and H are correlated because both are independently influenced by other underlying factors. Given the known significance of temperature to protein structure and function, it seems likely that temperature regimes could play an important role in protein evolution. In particular, the unusually high and stable body temperatures of most birds might account in part for their conservative pattern of protein differentiation.

Whatever the explanations for the heterogeneity in D's among the vertebrates, the results are of significance to interpretations of taxonomic assignment. If the protein-electrophoretic results are corroborated by studies of DNA, it will be abundantly clear that an average genus of amphibians, for example, is by no means equivalent to an average genus of birds from a genetic point of view. In fact, particular amphibian genera probably encompass a genetic diversity characteristic of entire arrays of avian families.

We can conclude by providing an example of how these and related considerations importantly influence views of comparative organismal relationship. King and Wilson (1975) demonstrate that man and chimp are very similar genetically, with an average human polypeptide more than 99% identical to its chimpanzee counterpart. This conclusion was based upon a wealth of molecular data, including inferences based on a protein-electrophoretic distance of $D = 0.62$ [a value of $D = 0.39$ was reported by Bruce and Ayala (1978, 1979)]. From a "frog's point of view," this is indeed a very small genetic distance (Cherry *et al.*, 1978; King and Wilson, 1975), and it has raised the question of whether man and chimp warrant their current placement in different taxonomic families (Sarich and Cronin, 1977). (It has also led some authors to be rather forceful about concluding that the human–chimp split occurred only about five million years ago [e.g., Zihlman and Lowenstein (1979).] However, from a bird's perspective, the human–chimp D of 0.62 is quite large—larger than the distance between any pair of avian congeners assayed to date, or between most confamilial avian genera (Figs. 1 and 2). Systematic decisions should of course be based upon many kinds of information, including morphology, physiology, and so forth. But because molecular data often provide quantitative, common scales for comparing disparate organisms, they should play an important role in our perceptions of relationships of organisms summarized in existing taxonomies.

ACKNOWLEDGMENTS

Our work is supported by NSF grants DEB-7814195 and DEB-8022135. CFA was supported by an NIH predoctoral training grant.

REFERENCES

Adest, G. A., 1977, Genetic relationships in the genus *Uma* (Iguanidae), *Copeia* **1977**:47–52.
Altman, P. L., and Dittmer, D. S. (eds.), 1974, *Biology Data Book*, Federation of American Societies for Experimental Biology, Bethesda, Maryland.
Avise, J. C., 1974, Systematic value of electrophoretic data, *Syst. Zool.* **23**:465–481.
Avise, J. C., 1976, Genetic differentiation during speciation, in: *Molecular Evolution* (F. J. Ayala, ed.), pp. 106–122, Sinauer, Sunderland, Massachusetts.

Avise, J. C., 1977, Is evolution gradual or rectangular? Evidence from living fishes, *Proc. Natl. Acad. Sci. USA* **74**:5083–5087.

Avise, J. C., 1978, Variances and frequency distributions of genetic distance in evolutionary phylads, *Heredity* **40**:225–237.

Avise, J. C., and Ayala, F. J., 1975, Genetic change and rates of cladogenesis, *Genetics* **81**:757–773.

Avise, J. C., and Duvall, S. W., 1977, Allelic expression and genetic distance in hybrid macaque monkeys, *J. Hered.* **68**:23–30.

Avise, J. C., and Smith, M. H., 1977, Gene frequency comparisons between sunfish (Centrarchidae) populations at various stages of evolutionary divergence, *Syst. Zool.* **26**:319–334.

Avise, J. C., Smith, J. J., and Ayala, F. J., 1975, Adaptive differentiation with little genic change between two native California minnows, *Evolution* **29**:411–426.

Avise, J. C., Smith, M. H., and Selander, R. K., 1979, Biochemical polymorphism and systematics in the genus *Peromyscus* VII. Geographic differentiation in members of the *truei* and *maniculatus* species groups, *J. Mammal.* **60**:177–192.

Avise, J. C., Patton, J. C., and Aquadro, C. F., 1980a, Evolutionary genetics of birds I. Relationships among North American thrushes and allies, *Auk* **97**:135–147.

Avise, J. C., Patton, J. C., and Aquadro, C. F., 1980b, Evolutionary genetics of birds II. Conservative protein evolution in North American sparrows and relatives, *Syst. Zool.* **29**:323–334.

Avise, J. C., Patton, J. C., and Aquadro, C. F., 1980c, Evolutionary genetics of birds III. Comparative molecular evolution in New World warblers and rodents, *J. Hered.* **71**:303–310.

Avise, J. C., Aquadro, C. F., and Patton, J. C., 1982, Evolutionary genetics of birds V. Genetic distances within Mimidae (mimic thrushes) and Vireonidae (vireos), *Biochem. Genet.* **20**:95–104.

Ayala, F. J., 1975, Genetic differentiation during the speciation process, *Evol. Biol.* **8**:1–78.

Ayala, F. J. (ed.), 1976, *Molecular Evolution*, Sinauer, Sunderland, Massachusetts.

Ayala, F. J., Powell, J. R., Tracey, M. L., Mourão, C. A., and Pérez-Salas, S., 1972, Enzyme variability in the *Drosophila willistoni* group. IV. Genic variation in natural populations of *Drosophila willistoni*, *Genetics* **70**:113–139.

Baker, R. J., Honeycutt, R. L., Arnold, M. L., Sarich, V. M., and Genoways, H. H., 1981, Electrophoretic and immunological studies on the relationship of the Brachyphyllinae and the Glossophaginae, *J. Mamm.* **62**:665–672.

Barrowclough, G. F., and Corbin, K. W., 1978, Genetic variation and differentiation in the Parulidae, *Auk* **95**:691–702.

Bennett, A. F., and Ruben, J. A., 1979, Endothermy and activity in vertebrates, *Science* **206**:649–654.

Bruce, E. J., and Ayala, F. J., 1978, Humans and apes are genetically very similar, *Nature* **276**:264–265.

Bruce, E. J., and Ayala, F. J., 1979, Phylogenetic relationships between man and the apes: Electrophoretic evidence, *Evolution* **33**:1040–1056.

Buth, D. G., 1979, Genetic relationships among the torrent suckers, genus *Thoburnia*, *Biochem. Syst. Ecol.* **7**:311–316.

Buth, D. G., 1980, Evolutionary genetics and systematic relationships in the catostomid genus *Hypentelium*, *Copeia* **1980**:280–290.

Buth, D. G., and Burr, B. M., 1978, Isozyme variability in the cyprinid genus *Campostoma*, *Copeia* **1978**:298–311.

Buth, D. G., Burr, B. M., and Schenck, J. R., 1980, Electrophoretic evidence for relationships and differentiation among members of the percid subgenus *Microperca, Biochem. Syst. Ecol.* **8:**297–304.

Calder, W. A., and Schmidt-Nielsen, K., 1968, Panting and blood carbon dioxide in birds, *Am. J. Physiol.* **215:**477–482.

Case, S. M., 1978, Biochemical systematics of members of the genus *Rana* native to western North America, *Syst. Zool.* **27:**299–311.

Chakraborty, R., and Nei, M., 1974, Dynamics of gene differentiation between incompletely isolated populations of unequal sizes, *Theor. Popul. Biol.* **5:**460–469.

Cherry, L. M., Case, S. M., and Wilson, A. C., 1978, Frog perspective on the morphological difference between humans and chimpanzees, *Science* **200:**209–211.

Cothran, E. G., Zimmerman, E. G., and Nadler, C. F., 1977, Genic differentiation and evolution in the ground squirrel subgenus *Ictidomys* (genus *Spermophilus*), *J. Mammal.* **58:**610–622.

Dessauer, H. C., Gartside, D. F., and Zweifel, R. G., 1977, Protein electrophoresis and the systematics of some New Guinea hylid frogs (genus *Litoria*), *Syst. Zool.* **26:**426–436.

Etges, W. J., 1979, Ecological genetic relationships in selected anurans of the southeastern United States, M. S. Thesis, University of Georgia.

Ferguson, A., Himberg, K. J. M., and Svardson, G., 1978, Systematics of the Irish pollan (*Coregonus pollan* Thompson): An electrophoretic comparison with other Holarctic Coregonidae, *J. Fish Biol.* **12:**221–233.

Gillespie, J. H., and Kojima, K., 1968, The degree of polymorphism in enzymes involved in energy production compared to that in nonspecific enzymes in two *Drosophila ananassae* populations, *Proc. Natl. Acad. Sci. USA* **61:**582–585.

Gillespie, J. H., and Langley, C. H., 1974, A general model to account for enzyme variation in natural populations, *Genetics* **76:**837–884.

Gingeras, T. R., and Roberts, R. J., 1980, Steps toward computer analysis of nucleotide sequences, *Science* **209:**1322–1328.

Gorman, G. C., and Kim, Y. J., 1976, *Anolis* lizards of the eastern Caribbean: a case study in evolution II. Genetic relationships and genetic variation of the *Bimaculatus* group, *Syst. Zool.* **25:**62–77.

Gorman, G. C., Soule, M., Yang, S. Y., and Nevo, E., 1975, Evolutionary genetics of insular adriatic lizards, *Evolution* **29:**52–71.

Gorman, G. C., Kim, Y. J., and Rubinoff, R., 1976, Genetic relationships of three species of *Bathygobius* from the Atlantic and Pacific sides of Panama, *Copeia* **1976:**361–364.

Hamrick, J. L., Linhart, Y. B., and Mitton, J. B., 1979, Relationships between life history characteristics and electrophoretically detectable genetic variation in plants, *Annu. Rev. Ecol. Syst.* **10:**173–200.

Hedgecock, D., and Ayala, F. J., 1974, Evolutionary divergence in the genus *Taricha* (Salamandridae), *Copeia* **1974:**738–747.

Highton, R., and Larson, A., 1979, the genetic relationships of the salamanders of the genus *Plethodon, Syst. Zool.* **28:**579–599.

Hochachka, P. W., and Somero, G. N., 1973, *Strategies of Biochemical Adaptation*, Saunders, Philadelphia, Pennsylvania.

Hubby, J. L., and Throckmorton, L. H., 1965, Protein differences in *Drosophila* II. Comparative species genetics and evolutionary problems, *Genetics* **52:**203–215.

Johnson, G. B., 1973, Importance of substrate variability to enzyme polymorphisms, *Nature New Biol.* **243:**151–153.

Johnson, G. B., 1974, Enzyme polymorphism and metabolism, *Science* **184:**28–37.

Johnson, G. B., 1976, Genetic polymorphism and enzyme function, in: *Molecular Evolution* (F. J. Ayala, ed.), Sinauer, Sunderland, Massachusetts.

Johnson, M. S., 1975, Biochemical systematics of the atherinid genus *Menidia*, *Copeia* **1975**:662–691.

Johnson, M. S., and Mickevich, M. F., 1977, Variability and evolutionary rates of characters, *Evolution* **31**:642–648.

Johnson, W. E., and Selander, R. K., 1971, Protein variation and systematics in kangaroo rats (genus *Dipodomys*), *Syst. Zool.* **20**:377–405.

Kim, Y. J., Gorman, G. C., Papenfuss, T., and Roychoudhury, A. K., 1976, Genetic relationships and genetic variation in the Amphisbaenian genus *Bipes, Copeia* **1976**:120–124.

King, J. R., and Farner, D. S., 1961, Energy metabolism, thermoregulation, and body temperature, in: *Biology and Physiology of Birds*, Volume II (A. J. Marshall, ed.), pp. 215–288, Academic Press, New York.

King, M. C., and Wilson, A. C., 1975, Evolution at two levels in humans and chimpanzees, *Science* **188**:107–116.

Koehn, R. K., and Eanes, W. F., 1978, Molecular structure and protein variation within and among populations, *Evol. Biol.* **11**:39–100.

Kojima, K., Gillespie, J., and Tobari, Y. N., 1970, A profile of *Drosophila* species' enzymes assayed by electrophoresis I. Number of alleles, heterozygosities, and linkage disequilibrium in glucose-metabolizing systems and some other enzymes, *Biochem. Genet.* **4**:627–637.

Kornfield, I. L., Ritte, U., Richler, C., and Wahrman, J., 1979, Biochemical and cytological differentiation among cichlid fishes of the Sea of Galilee, *Evolution* **33**:1–14.

Leder, P., Hansen, J. N., Konkel, D., Leder, A., Nishioka, Y., and Talkington, C., 1980, Mouse globin system: A functional and evolutionary analysis, *Science* **209**:1336–1342.

Lucotte, G., 1979a, Génétique des populations, spéciation et taxonomie chez les babouins: II. Similitudes génétique comparées entre différentes espéces: *Papio papio, P. anubis, P. cynocephalus*, et *P. hamadryas* basées sur les données du polymorphisme des enzymes erythrocytaires, *Biochem. Syst. Ecol.* **7**:245–251.

Lucotte, G., 1979b, Distances électrophorétiques entre les différentes espéces de singes du groupe des mangabeys, *Ann. Genet.* **22**:85–87.

Mengel, R. N., 1964, The probably history of species formation in some northern wood warblers (Parulidae), *Living Bird* **3**:9–43.

Montanucci, R. R., Axtell, R. A., and Dessauer, H. C., 1975, Evolutionary divergence among collared lizards (*Crotaphytus*), with comments on the status of *Gambelia, Herpetologica* **31**:336–347.

Morony, J. J., Jr., Bock, W. J. and Farrand, J., Jr., 1975, *Reference List of the Birds of the World*, Special Publication of the American Museum of Natatural History, New York.

Nathans, D., 1979, Restriction endonucleases, simian virus 40, and the new genetics, *Science* **206**:903–909.

Nei, M., 1972, Genetic distance between populations, *Am. Nat.* **106**:283–292.

Nei, M., 1975, *Molecular Population Genetics and Evolution*, North-Holland Amsterdam.

Nei, M., 1981, Genetic distance and molecular taxonomy, in *Proc. XIV Int. Congr. Genet.*, Problems in General Genetics (Y. P. Altukhov, ed.) Volume 2, pp. 7–22, MIA Publishers, Moscow.

Nei, M., and Roychoudhury, A. K., 1981, Genetic relationship and evolution of human races, *Evol. Biol.* **14**:1–59.

Nevo, E., 1978, Genetic variation in natural populations: patterns and theory, *Theor. Popul. Biol.* **13**:121–177.

Nevo, E., Kim, Y. J., Shaw, C. R., and Thaeler, C. S., Jr., 1974, Genetic variation, selection and speciation in *Thomomys talpoides* pocket gophers, *Evolution* **28**:1–23.

Patton, J. C., and Avise, J. C., 1983, Evolutionary genetics of birds IV. Rates of protein divergence in waterfowl (Anatidae), Submitted.

Patton, J. C., Baker, R. J., and Avise, J. C., 1981, Phenetic and cladistic analyses of biochemical evolution in peromyscine rodents, in: *Mammalian Population Genetics* (M. H. Smith and J. Joule, eds.), pp. 288–308, University of Georgia Press, Athens, Georgia.

Penney, D. F., and Zimmerman, E. G., 1976, Genic divergence and local population differentiation by random drift in the pocket gopher genus *Geomys*, *Evolution* **30**:473–484.

Powell, J. R., 1975, Protein variation in natural populations of animals, *Evol. Biol.* **3**:79–119.

Prager, E. M., and Wilson, A. C., 1975, Slow evolutionary loss of the potential for interspecific hybridization in birds: A manifestation of slow regulatory evolution, *Proc. Natl. Acad. Sci. USA* **72**:200–204.

Rogers, J. S., 1972, Measures of genetic similarity and genetic distance, in: *Univ. Texas Publ.*, 7213, pp. 145–153.

Sarich, V. M., 1977, Rates, sample sizes, and the neutrality hypothesis for electrophoresis in evolutionary studies, *Nature* **26**:24–28.

Sarich, V. M., and Cronin, J. E., 1977, Molecular systematics of the primates, in: *Molecular Anthropology* (M. H. Goodman, and R. E., Tashian, eds.), pp. 144–170, Plenum, New York.

Sattler, P., 1980, Genetic relationships among selected species of North American *Scaphiopus*, *Copeia* **1980**:605–610.

Selander, R. K., 1976, Genic variation in natural populations, in: *Molecular Evolution* (F. J. Ayala, ed.), pp. 21–45, Sinauer, Sunderland, Massachusetts.

Selander, R. K., Smith, M. H., Yang, S. Y., Johnson, W. E., and Gentry, J. B., 1971, Biochemical polymorphism and systematics in the genus *Peromyscus*, I. Variation in the old-field mouse (*Peromyscus polionotus*), in: *Studies in Genetics VI* (Univ. Texas Publ. 7103), pp. 49–90.

Shaw, C. R., 1970, How many genes evolve? *Biochem. Genet.* **4**:275–283.

Shaw, C. R., and Prasad, R., 1970, Starch gel electrophoresis of enzymes—A compilation of recipes, *Biochem. Genet.* **4**:297–320.

Smith, J. K., and Zimmerman, E. G., 1976, Biochemical genetics and evolution of North American blackbirds, family Icteridae, *Comp. Biochem. Physiol.* **53B**:319–324.

Smith, M. W., Aquadro, C. F., Smith, M. H., Chesser, R. K., and Etges, W. J., 1982, A bibliography for electrophoretic studies of biochemical variation in natural populations of vertebrates. Texas Tech Press, in press.

Somero, G. N., 1978, Temperature adaptation of enzymes, *Annu. Rev. Ecol. Syst.* **9**:1–29.

Soulé, M., 1976, Allozyme variation: Its determinants in space and time, in: *Molecular Evolution*, (F. J. Ayala, ed.), pp. 60–77, Sinauer, Sunderland, Massachusetts.

Turner, B. J., 1974, Genetic divergence of Death Valley pupfish species: Biochemical versus morphological evidence, *Evolution* **28**:281–295.

Utter, F. M., Allendorf, F. W., and Hodgins, H. O., 1973, Genetic variability and relationships in pacific salmon and related trout based on protein variations, *Syst. Zool.* **22**:257–270.

Valentine, J. W., 1976, Genetic strategies of adaptation, in: *Molecular Evolution* (F. J. Ayala, ed.), pp. 78–94, Sinauer, Sunderland, Massachusetts.

Wake, D. B., Maxson, L. R., and Wurst, G. Z., 1978, Genetic differentiation, albumin evolution, and their biogeographic implications in plethodontid salamanders of California and southern Europe, *Evolution* **32**:529–539.

Ward, R. D., and Galleguillos, R. A., 1978, Protein variation in the plaice, dab, and flounder and their genetic relationships, in: *Marine Organisms* (B. Battaglia and J. Beardmore, eds.), pp. 71–93, Plenum Press, New York.

Wilson, A. C., Carlson, S. S., and White, T. J., 1977, Biochemical evolution, *Annu. Rev. Biochem.* **46:**573–639.

Wyles, J. S., and Gorman, G. C., 1980, The albumin immunological and Nei electrophoretic distance correlation: A calibration for the Saurian genus *Anolis* (Iguanidae), *Copeia* **1980:**66–71.

Yang, S. Y., and Patton, J. L., 1981, Genic variability and differentiation in the Galapagos finches, *Auk* **98:**230–242.

Yang, S. Y., Soule, M., and Gorman, G. C., 1974, *Anolis* lizards of the eastern Caribbean: A case study in evolution I. Gentic relationships, phylogeny, and colonization sequence of the *roquet* group, *Syst. Zool.* **23:**387–399.

Zihlman, A. L., and Lowenstein, J. M., 1979, False start of the human parade, *Nat. Hist.* **88:**86–91.

Zimmerman, E. G., and Nejtek, M. E., 1977, Genetics and speciation of three semispecies of *Neotoma*, *J. Mammal.* **58:**391–402.

Zimmerman, E. G., Hart, B. J., and Kilpatrick, C. W., 1975, Biochemical genetics of the *truei* and *boylei* groups of the genus *Peromyscus* (Rodentia), *Comp. Biochem. Physiol.* **52B:**541–545.

4

The Alcohol Dehydrogenase Polymorphism in *Drosophila melanogaster*

Selection at an Enzyme Locus

W. VAN DELDEN

Department of Genetics
University of Groningen
9751 NN Haren, The Netherlands

INTRODUCTION

In the late sixties the neo-Darwinian theory of evolution, hitherto generally accepted by biologists, was confronted with a new, revolutionary view: the theory of neutral or non-Darwinian evolution. In the neutralist view amino acid and nucleotide changes in the course of evolution are mainly due to random fixation of selectively neutral mutants (Kimura, 1968; King and Jukes, 1969). The approximately constant rate of evolution in terms of amino acid substitutions per site per year for various lineages, as claimed by neutralists, forms one of the main arguments for the neutralist theory. Enzyme polymorphisms in present-day populations are considered as a phase in molecular evolution (Kimura and Ohta, 1971; Kimura, 1977). The first estimates of the extent of enzyme polymorphisms in populations of *Drosophila pseudoobscura* (Lewontin and Hubby, 1966) and humans (Harris, 1966) based on electrophoresis of proteins were soon followed by many others, covering a wide range of animal and plant species. These surveys showed that most species are highly polymorphic [see reviews by Powell (1975), Nevo (1978), A. H. D. Brown (1979), and Hamrick *et al.* (1979)]. In the neutralist versus selectionist controversy the nature of these allozyme polymorphisms is disputed. The neutralist hypothesis states that the observed variation is mainly a product of mutation and drift of selectively neutral genes. The selectionist hypothesis claims that some form of balancing selection is responsible for the maintenance of allozyme polymorphisms.

Many attempts have been made to determine the nature, either neutral or selective, of the allozyme polymorphisms in nature. In many cases these efforts involved rather indirect methods. In the statistical tests devised for this purpose observations were compared with theoretical expectations based on the neutrality hypothesis. Such tests involved expectations as to the effective number of alleles, means, variances, and distributions of heterozygosities, and geographic allele frequency distributions. These attempts, however, have not been very successful and give ambiguous results, partly because of theoretical objections to the tests applied, partly because of the problems in estimating the parameters involved (Ewens and Feldman, 1976).

Apparently there is a need for more direct methods, preferably including estimates of fitness components. In this respect allele frequencies for particular loci have been tracked in polymorphic laboratory populations of sufficiently high numbers to exclude drift. Frequency changes in these experiments have been interpreted as the result of selection. Some of these experiments have shown no change in allele frequencies over time, such as, e.g., in the carefully performed experiments of Yamazaki (1971). MacIntyre (1972) concluded from his data on allele frequency changes that selection coefficients were extremely small. These results, however, need not be decisive, because selective differences between allozyme variants may occur only under particular environmental conditions, not met in the experiments. On the other hand, when allele frequency changes occur they are not necessarily due to selection at the allozyme locus under study, because of the hitch-hiking effect (Kojima and Schaffer, 1967; Maynard Smith and Haigh, 1974). Whether allele frequency changes can be attributed to selection depends on particular conditions, such as, e.g., repeatability in different independent genetic backgrounds.

It then appears that the unambiguous demonstration of selection creates considerable methodologic difficulties. In the case of allozyme variation, however, a very straightforward approach is available (Lewontin, 1974; Clarke, 1975). When the (external) substrate of an allozyme can be identified and the allozyme variants differ in biochemical properties such as activity, stability, or pH optimum, predictions can be made as to the selective differences among the variants under particular conditions. Such quantitative differences among allozyme variants have been reported for a great number of enzyme systems. In humans, e.g., 60%

of the enzymes in a sample of 30 showed biochemical differences of the sort described above (Harris, 1976). Environmental stresses relevant to the enzyme under study can thus provide strong evidence for selection. Still, care is needed because biochemical differences observed *in vitro* may not be characteristic for the situation *in vivo*. Examples demonstrating the value of this method are the amylase polymorphism (De Jong and Scharloo, 1976; Hoorn and Scharloo, 1979) and the alcohol dehydrogenase polymorphism (Clarke, 1975; David, 1977b; Van Delden *et al.*, 1978) in *Drosophila melanogaster*.

It is the purpose of this article to review the present state of knowledge of the alcohol dehydrogenase polymorphism in *D. melanogaster*, with emphasis on its biologic significance. When no species indication is given, *D. melanogaster* is concerned. The nomenclature of Johnson and Denniston (1964) for the description of the alleles is followed.

GENERAL ASPECTS OF THE ALCOHOL DEHYDROGENASE POLYMORPHISM

Genetic, Developmental, and Biochemical Aspects

The alcohol dehydrogenase (*Adh*) polymorphism in *Drosophila melanogaster* was first described by Johnson and Denniston (1964).The *Adh* locus is located on the second chromosome at map position 50.1 (Grell *et al.*, 1965) and is most probably situated within polytene chromosome band 35B2 (O'Donnell *et al.*, 1977; Woodruff and Ashburner, 1979a,b). Natural populations generally contain two common electrophoretic alleles: the *Adh*-Fast (F) and the *Adh*-Slow (S) alleles; only rarely are exceptional alleles with different electrophoretic mobilities found at very low frequency (see Chambers, 1981). The ADH (E.C. 1.1.1.1) of *D. melanogaster* contains two identical subunits, each 254 residues in length and of molecular weight 27,400. The complete amino acid sequences of F and S allozymes are known and differ by a single amino acid substitution: S enzyme differs from F by a threonine-to-lysine substitution at position 192 (Fletcher *et al.*, 1978; Retzios and Thatcher, 1979; Thatcher, 1980). Amino acid substitutions in other ADH variants, including null mutants, are also determined and generally involve single amino acid substitutions. For more detailed reports on purification and amino acid sequencing of *Drosophila* ADH see Schwartz and Jörnvall (1976), Thatcher (1977), Thatcher and Camfield (1977), Thatcher and Retzios (1980), Juan and Gonzales-Duarte (1980), and Chambers *et al.* (1981).

ADH catalyzes the oxidation of alcohols to aldehydes or ketones and concurrently reduces NAD$^+$ to NADH. A variety of both primary and secondary alcohols serve as substrates. *Adh*-null alleles, lacking ADH activity, can be obtained by mutagenic agents and subsequent screening by unsaturated alcohols (e.g., 1-pentene-3-ol) (Grell *et al.*, 1968; Sofer and Hatkoff, 1972; Vigue and Sofer, 1974, 1976; O'Donell *et al.*, 1975; Schwartz and Sofer, 1976a). These unsaturated alcohols are converted to highly toxic ketones in ADH-positive individuals which will die, while ADH-negative flies survive. For at least one ADH-negative mutant it is found that the defect is in the structural gene coding for ADH (Reddy *et al.*, 1980). The frequency of *Adh*-null alleles in natural populations is extremely low: Voelker *et al.* (1980) found a frequency of 0.001. Apparently ADH is not essential for *Drosophila*, because *Adh*-negative mutants are viable and performing well in pure cultures in the laboratory.

The biochemistry and developmental biology of *Drosophila* ADH are reviewed in Dickinson and Sullivan (1975) (see also O'Brien and MacIntyre, 1978) and will be summarized here briefly. The function in insect metabolism, except for the detoxification of environmental alcohols, is unclear. ADH activity in *Drosophila* can be found in many body parts, but it is primarily localized in fat body, intestine, and Malpighian tubules, while in male adults high specific activities are found in parts of the male genital apparatus (Ursprung *et al.*, 1970; Korotchkin *et al.*, 1972). The developmental profile of ADH shows an increase in specific activity in eggs just before hatching. This increase continues during the larval stages and reaches a maximum in late third instar. In pupae activity decreases, but it increases again in adults to reach a plateau (at 25°C) at about 6 days after emergence, while in older flies it declines again (Dunn *et al.*, 1969; Ursprung *et al.*, 1970; McDonald and Avise, 1976). Eggs show only maternal electrophoretic patterns throughout most of the embryonic development; at hatching the pattern belonging to the genotype of the individual appears (Wright and Shaw, 1970; Leibenguth *et al.*, 1979).

The alcohol dehydrogenases of both SS and FF genotypes generally occur in several forms with different electrophoretic mobilities (Johnson and Denniston, 1964; Ursprung and Leone, 1965; Grell *et al.*, 1965). The three most common isozymes are ADH-1, ADH-3, and ADH-5, of which ADH-5 is the most cathodally migrating band and ADH-1 the most anodally migrating band, while ADH-3 is intermediate. The isozymes show *in vitro* differences in a number of characters, such as heat stability and kinetic parameters (Jacobson, 1968; Day and Needham, 1974). The isozymes can be interconverted *in vitro*, e.g., by NAD$^+$ or acetone (Ursprung and Carlin, 1968; Sofer and Ursprung, 1968; Jacobson *et al.*, 1970, 1972; Jacobson and Pfuderer, 1972; Knopp and Jacobson, 1972) or *in vivo* by additions to the food, such as acetone and isopropanol (Schwartz and

Sofer, 1976b; Papel *et al.*, 1979; Fontdevila *et al.*, 1980). Experiments concerning the nature of these interconversions have shown that the ADH enzyme has a noncovalent binding of two (ADH-1), one (ADH-3), or no (ADH-5) molecules of an NAD-carbonyl compound (Schwartz *et al.*, 1975; Schwartz and Sofer, 1976b; Schwartz *et al.*, 1979).

As the isozymes differ considerably in activity and stability (ADH-5 is the most active but the least stable form), the interconversions may be of adaptive importance (Papel *et al.*, 1979; Anderson and McDonald, 1981a,b).

Generally, larvae and adults from homozygous FF strains possess greater *in vitro* ADH activity than those from homozygous SS strains, the heterozygotes (FS) showing intermediate activities (Rasmuson *et al.*, 1966; Gibson, 1970; Vigue and Johnson, 1973; Day *et al.*, 1974a). Considerable variation in ADH activity, however, can be found among homozygous strains of identical electrophoretic mobility (Gibson and Miklovich, 1971; Ward and Hebert, 1972; Birley and Barnes, 1973, 1975; Ward, 1974, 1975; Barnes and Birley, 1975). Many publications deal with *in vitro* biochemical properties of ADH of the three genotypes. Table I summarizes the results obtained by different authors. The survey shows conformity for relative ADH activities: FF exceeds SS, with FS being intermediate. Absolute ADH activities and relative proportions may vary considerably, however, depending on origin of populations, experimental procedures, and substrates. Except for one case (Oakeshott, 1976b), further agreement is found for temperature stability, expressed as the fraction of the enzyme activity remaining after heat treatment, where extracts of SS have the highest stability. It is possible that thermal stability reflects a more general stability of the molecule, such as, e.g., to protease degradation. Anderson and McDonald (1981a) have measured the relative *in vivo* stability of ADH from an FF strain and calculated a half-lifetime value $T_{1/2}$ of about 55 hr. Interestingly, this value was increased to about 290 hr in isopropanol-treated flies (Anderson and McDonald, 1981b).

Both differences in relative catalytic efficiency and differences in the amount of ADH molecules may be involved in the genotypic differences in ADH activity. Most authors who have determined the amount of ADH claim that FF exceeds SS in the number of molecules. Day *et al.* (1974a) found that FF and SS from the same polymorphic population show no difference in the number of ADH molecules, though it was found that in general FF strains have higher amounts of ADH than do SS strains. They believe that differences in activity for the greater part are due to differences in catalytic efficiency (Clarke, 1975). Such differences are also inferred, in addition to differences in ADH amount, by other authors. Clarke *et al.* (1979) have shown that environmental factors may influence

TABLE I. Differences in *in Vitro* Biochemical Properties of ADH from Different Alcohol Dehydrogenase Genotypes

Life stage	ADH activity	K_m	V_{max}	ADH amount	pH optimum	Temperature optimum	Temperature stability	Reference
Larval	FF > FS > SS						SS > FS > FF	Gibson (1970)
Larval	FF > FS > SS						SS > FS > FF	Gibson and Miklovich (1971)
Adult	FF > FS > SS			FF > FS > SS				Gibson (1972)
Adult	FF > FS > SS	FF = FS = SS			FF = FS = SS	SS > FS = FF	SS > FS > FF	Vigue and Johnson (1973)
Adult	FF > FS > SS	FF = SS	FF > SS	FF = SS	FF = SS	FF = FS = SS	SS > FS = FF	Day et al. (1974a,b)
Larval/adult	FF > FS > SS				FF = FS = SS		FF = FS > SS	Oakeshott (1976b)
Larval	FF > SS		FF > SS	FF > SS				Lewis and Gibson (1978)
Adult	FF > SS	FF = SS					SS > FF	Maroni (1978)
Adult	FF > FS > SS						SS > FS > FF	Van Delden and Kamping (1980)
Adult	FF > SS	FF > SS		FF > SS				McDonald et al. (1980)

considerably the amount of ADH: by increasing the amount of yeast in the food of the larvae, the ADH amount in adults could be more than doubled. This observation points to the necessity for carefully controlled culture conditions when determinations of absolute and relative ADH amounts are required [see also Birley and Marson (1981)]. At present the data suggest that catalytic efficiency and probably the amount of ADH of FF is greater than that of SS.

Activities measured *in vitro* differ considerably, depending on the alcohol used as substrate: in general, activities found for short-chain alcohols are higher than for longer chain alcohols (>4 C atoms). Secondary alcohols are more quickly converted than are their primary counterparts (Vigue and Johnson, 1973; Day *et al.*, 1974a; Morgan, 1975; Oakeshott, 1976b; Maroni, 1978; Chambers *et al.*, 1978; Grossman, 1980). Further, FF and SS show considerable differences in substrate specificities when activity ratios (activity with a particular alcohol relative to a standard alcohol, e.g., ethanol) are determined.

Thus *Adh* genotypes show several differences in *in vitro* enzyme properties, which may form the starting point for natural selection under appropriate conditions, provided that these differences are also expressed *in vivo*.

Interstrain Differences in ADH Activity

In addition to the great differences in ADH activity generally found between FF and SS individuals from the same polymorphic population, considerable variation in activity is found within each electrophoretic class when different strains or inbred lines are compared (Gibson and Miklovich, 1971; Ward and Hebert, 1972; Birley and Barnes, 1973, 1975; Hewitt *et al.*, 1974; Ward, 1974, 1975; Barnes and Birley, 1975, 1978; Thompson and Kaiser, 1977; Kamping and Van Delden, 1978; Birley *et al.*, 1980; Wilson and McDonald, 1981). In fact, some high-activity SS strains have activities equal to or even higher than low-activity FF strains (Ward, 1974; Gibson *et al.*, 1980). Though one should proceed with caution in interpreting these differences, because environmental factors, such as yeast amount, may influence the amount of ADH (Clarke *et al.*, 1979), it can safely be assumed that a considerable part of this intragenotype variation is heritable. ADH activity may be considered as a quantitative character in this respect (Ward, 1975). Birley and Barnes (1973) estimated the narrow heritability at 21% for their material. From an analysis of crosses between SS inbred lines differing in ADH activity, it was concluded that the variation in activity is controlled by additive gene action;

no evidence of directional dominance or of nonallelic interactions was found (Barnes and Birley, 1975). Crosses between SS and FF inbred lines showed dominance for low-ADH activity (Birley and Barnes, 1975). Further analysis of ADH activity at 25 and 35°C by a half-diallel mating design showed considerable additive genetic variation, while dominance deviations were ambidirectional, dependent on specific parents. At 35°C most of the genetic variation was additive and mainly due to modifier loci. There was strong genotype–environment interaction for heat stability. In a thorough analysis Birley and Marson (1981) found genotype–environment interaction for ADH activity among a set of highly inbred lines. This interaction was not wholly associated with genotype–environment interaction for body weight or total protein level (see also Clarke *et al.*, 1979).

Not surprisingly, then, selection for increased and decreased ADH activity has been successful (Ward and Hebert, 1972; Ward, 1975). It has been shown that modifiers of ADH activity are located on the X, second, and third chromosomes (Pipkin and Hewitt, 1972a; Hewitt *et al.*, 1974; Ward, 1975; Barnes and Birley, 1975, 1978). Assays of ADH activity in strains homozygous for various combinations of second and third chromosomes derived from natural populations (McDonald and Ayala, 1978; Ayala and McDonald, 1980) showed considerable variation for modifiers on the second and third chromosomes. The effect of modifying genes at the third chromosome varies as they interact with different second chromosomes. Third chromosomes with high-activity modifiers are partially or completely dominant over chromosomes with low-activity genes. It was found that a low-activity strain has fewer ADH molecules than does a high-activity strain. Laurie-Ahlberg *et al.* (1980) also showed the existence of unlinked ADH-activity modifiers.

Thompson *et al.* (1977) have found evidence that in a low-activity SS strain a control mutation is located proximal and very close to the *Adh* locus (see also Hisey *et al.*, 1979). In this case the low-activity strain had only half as many ADH molecules as did regular SS strains (Thompson and Kaiser, 1977). The work of Grossmann (1980) also suggests the occurrence of a modifier of ADH activity very close to the *Adh* locus. Pipkin and Hewitt (1972a,b) concluded from their work on interspecies hybrids and triploids that the X chromosome has a modifying effect on ADH activity and found a gene dosage effect.

The general idea emerging from these kinds of experiments is that the presence of modifiers in all major chromosomes has unambiguously been proven, though allelic variation at the *Adh* locus itself responsible for activity differences cannot fully be excluded.

Heterogeneity within Electrophoretic Classes

Electrophoretic variants differ from each other by at least one amino acid substitution causing a charge difference of the protein. Substitutions lacking such an effect, e.g., when neutral amino acids are involved, are not detected by electrophoresis. Each electrophoretic class may thus be heterogeneous and genetic variation may be underestimated. In fact considerable hidden variation has been revealed for several enzyme loci in a number of *Drosophila* species by use of such methods as heat denaturation techniques and gel sieving analysis (Bernstein *et al.*, 1973; Singh *et al.*, 1976; G. B. Johnson, 1977).

Heat denaturation studies have revealed additional variation at the *Adh* locus also: Thörig *et al.* (1975) found a variant (Adh ^{71k}) in a laboratory Notch strain with electrophoretic mobility identical to the F variant, but heat-stable at 40°C. In natural populations from the U.S., Sampsell and Milkman discovered three additional thermostability variants, either more or less temperature sensitive than regular alleles. These variants were found in the F as well as in the S mobility classes (Milkman, 1976; Sampsell, 1977; Sampsell and Milkman, 1978). Gibson *et al.* (1980, 1981) discovered an electrophoretic F variant in Australian populations, which had an ADH activity and a secondary/primary alcohol activity ratio characteristic of S variants. This variant was distinguished from both S and F variants by much higher thermostability. There is good evidence that all these variants are allelic to the *Adh* locus. It was found that the Australian thermostable F variant differs from the common F by at least a single amino acid substitution (Chambers *et al.*, 1981). Sampsell (1977) surveyed populations throughout the U.S. and found that the thermostability variants had a collective frequency ranging from 0 to 7%. Frequencies of the heat-stable variant in Australian populations ranged from 0 to 5% (Wilks *et al.*, 1980).

Kreitman (1980) performed an extensive electrophoretic screening of isochromosomal lines derived from a natural population for unusual mobility variants, but was unable to uncover such variants. Still, his method was sensitive enough to distinguish two of Sampsell's (1977) thermostability variants, previously thought to be indistinguishable in electrophoretic mobility from regular F and S alleles. It was concluded that the *Adh* locus in *D. melanogaster*, unlike highly variable loci, such as the *Xdh* locus and *Est* loci in *D. pseudoobscura*, has no substantial hidden variability. A considerable number of alleles at an *Adh* locus was, however, found by Coyne and Felton (1977) in *D. pseudoobscura* and *D. persimilis* and by G. B. Johnson (1978) in *D. mojavensis*. In the latter case it was assumed that considerable variation at the *Adh* locus was disguised by cofactor and subunit binding. Kreitman's (1980) study, however, stresses the importance of genetic analysis for the unambiguous demonstration of the genetic nature of mobility variation. Methodologic artefacts, such as caused by differences in enzyme concentration among strains, may wrongly suggest the existence of genetic variation.

RELATION WITH ALCOHOLS

Drosophila Species and Ethanol Tolerance

Alcohols, the substrates for ADH, form a toxic component in the environment of *Drosophila* (David *et al.*, 1976). *Drosophila* species differ considerably in their tolerance to alcohols, of which ethanol is of special ecologic importance. Though various authors have used different methods for determining tolerance to ethanol, the general conclusion is that *D. melanogaster* belongs to the group of species with high tolerance compared to most other species tested. Other tolerant species are *D. lebanonensis*, a species abundant in wine cellars, which has an even higher tolerance to ethanol than does *D. melanogaster* (David *et al.*, 1979), and *D. rubrostriata* (M. J. Brown, 1934). David and Bocquet (1976) determined the toxicity of several alcohols for the sibling species *D. melanogaster* and *D. simulans*. The toxicity of primary alcohols was found to increase with the number of C atoms from ethanol on. Secondary alcohols are more toxic than primary alcohols with the same number of C atoms. In both species ethanol is the best tolerated alcohol, but the tolerance of *D. melanogaster* is generally greater than that of *D. simulans*. David *et al.* (1974) exposed flies belonging to six different sibling species of the *melanogaster* subgroup to ethanol. The species were classified in the following order of decreasing tolerance: *melanogaster* > *simulans* > *erecta* > *yakuba* > *mauritiana* > *teissieri* (see also Parsons, 1980c). In David's study *D. melanogaster* had an LD_{50} of about 10%, while the LD_{50} values of the other species never exceeded 5%. Some *Drosophila* species are extremely sensitive to ethanol, such as *D. bromeliae*, with an LD_{50} of about 1.5% (David, 1973).

Species differences in tolerance to ethanol are reflected in their distribution. Studies on the microdistribution of *D. melanogaster* and *D. simulans* show that the former, much more tolerant, species is found nearly exclusively within maturation cellars of vineyards, while both species occur outside the cellars (McKenzie and Parsons, 1972, 1974). Laboratory experiments showed that in choice situations larvae of *D. melanogaster* migrate to agar containing ethanol instead of pure agar, while *D. simulans* larvae do not show a preference (Parsons and King, 1977). Compared to other alcohols, the ethanol content will generally be high, though in some cases other alcohols will be present in plant tissues at high concentration, such as isopropanol in necrotic tissue of certain cacti (Heed, 1978).

Apparently *Drosophila* species can cope with alcohols in their environment by means of ADH. Chambers *et al.* (1978) used electrophoresis to investigate the substrate specificities of alcohol-oxidizing enzymes in 13 *Drosophila* species, belonging to four different systematic groups. The species all possessed the same basic complement of enzymes able to oxidize a variety of alcohols and had similar specificities, though they showed substantial differences in activities. Daggard (1981) studied survival in the presence of ethanol and activities of ADH and ALDOX for four *Drosophila* species and found them to be related. McDonald and Avise (1976) have compared the ADH and α-glycerophosphate (α-GPDH) activity levels of nine *Drosophila* species belonging to the *melanogaster*, *obscura*, and *willistoni* groups. For any given life stage interspecific variability was much greater for ADH than for α-GPDH. Also, the ontogenetic profiles of ADH were found to be more variable among species. *D. melanogaster* possessed throughout the highest ADH activity. An important finding from this study is the positive correlation between ADH activity and adult survival on isopropanol for the nine species. The occurrence of *Drosophila* species in particular habitats may thus be associated with ADH activity. Species very sensitive to ethanol, such as *D. bromeliae* (David, 1973), may have niches where no fermenting microorganisms occur. Some *Drosophila* species apparently have no ADH activity at all (Courtright *et al.*, 1966).

Various alcohols, both primary and secondary, occur in fruits (Hulme, 1970, 1971), though in general ethanol will quantitatively dominate, especially in fermenting fruits. This certainly holds for wine cellars and grape musts, where some *Drosophila* species, and notably *D. melanogaster*, occur in great numbers and sometimes lay eggs. The ethanol tolerance of yeast strains involved in wine fermentation can be very high: volume percentages up to 19% are observed (Castelli, 1954). In general, ethanol concentrations in residues of grapes or fermenting liquor will be lower. McKenzie and McKechnie (1979), for example, measured ethanol concentrations from 0.3 to 12.6% in different zones of piles of grape residues. Briscoe *et al.* (1975) reported large populations of *D. melanogaster* which bred in the fermenting liquor (12–15% ethanol) in wine jars. Ainsley and Kitto (1975) found ethanol concentrations between 2 and 4%

in fermenting bananas. In cactus rot ethanol concentrations of about 0.4% were found, and other alcohols, such as isopropanol, may exceed 0.2% by volume (Heed, 1978).

Allele Frequency Changes

In view of the detoxifying properties of ADH in the presence of alcohols, the *Adh* polymorphism forms an appropriate example to study the action of selection, if any, on an allozyme polymorphism as noted in the introduction. Because the three common genotypes differ in *in vitro* ADH activity (FF > FS > SS), it may be assumed that differential detoxifying ability, positively correlated with ADH activity, will result in differences in fitness among genotypes. David *et al.* (1976) have shown in this respect that ADH-negative flies show much higher mortality than do ADH-positive flies on exposure to ethanol.

A common approach to explore fitness differences is to track allele frequencies in experimental polymorphic populations in the presence of ethanol. Table II summarizes the outcome of experiments from various authors. In all cases ethanol was a supplement to regular food medium and all life stages were continuously exposed to ethanol. Listed are deviations of F frequency on ethanol medium from F frequency on control medium over an initial period of 10–20 generations exposure. With one exception it was found that the frequency of the F allele increases significantly on ethanol compared with controls. It should be kept in mind that the figures present net increase: control populations may also have experienced changes in frequencies. The Australian populations (Gibson *et al.*, 1979; see also Oakeshott, 1979) are exceptional both in methods used and outcome. Here populations were selected by exposure to increasing ethanol concentrations in the course of generations. There was no unambiguous rise in F frequency and much variation among subpopulations of the same population. These deviating results may, however, be caused by genetic drift, because only 20 pairs of parents were used each generation. The majority of the results obtained by different authors with populations from quite different origin are thus in full agreement with the expectation that FF, the high-activity genotype, will survive better on ethanol, consequently leading to higher F frequencies. A more detailed inspection of the data shows very rapid changes in F frequency during the first five generations; Bijlsma-Meeles and Van Delden (1974), for example, report a mean Δq of 0.05 per generation. In later generations the rise in F frequency slows down, partly because of the low frequency of SS homozygotes, but perhaps also because of adaptation to ethanol

TABLE II. Changes in *Adh*F Frequencies in Populations Kept on Ethanol-Supplemented Food

Origin of population	Number of subpopulations	Initial *Adh*F frequency	Percent ethanol	Number of generations	Δq^F	Δq^F per generation	Reference
Laboratory strain	2	0.50	6	18	0.275	0.015	Gibson (1970)
Groningen (The Netherlands)	1	0.50	15	19	0.230	0.012	Van Delden *et al.* (1975)
Groningen (The Netherlands)	1	0.50	15	20	0.320	0.016	Van Delden *et al.* (1978)
Haren (The Netherlands)	1	0.50	15	20	0.220	0.011	Van Delden *et al.* (1978)
Curaçao (Dutch Antilles)	1	0.50	15	20	0.250	0.013	Van Delden *et al.* (1978)
Craigmoor (Australia)	4	0.50	6→15	10	−0.160	−0.016	Gibson *et al.* (1979)
Chateau Douglas (Australia)	4	0.50	6→15	10	0.140	0.014	Gibson *et al.* (1979)
Weymouth (U.S.)	2	0.33	10	18	0.358	0.020	Cavener and Clegg (1978, 1981)

not dependent on the *Adh* locus. Final fixation of the F allele will occur eventually (Van Delden *et al.*, 1978; Cavener and Clegg, 1978, 1981; Gibson *et al.*, 1979). In the populations studied by Van Delden *et al.* (1978) the increase in F frequency was positively correlated with the *in vitro* ADH activity ratio of FF to SS. The increase in F frequency on ethanol can be reduced by high humidity, which under regular food conditions tends to decrease the frequency of the F allele. Frequency changes are not restricted to ethanol, but also occur on other alcohols (Van Delden *et al.*, 1975, 1978). In populations polymorphic for an *Adh*-null allele (initial frequency 0.5) and either the F or the S allele, complete loss of the null allele was observed on ethanol medium within eight generations (Van Delden and Kamping, 1979a, and in manuscript). On food containing pentene-3-ol (see section on detection of null mutants), however, a rise of the frequency of the Adh-negative alleles was observed.

Fitness Estimates

The cause of the allele frequency changes on ethanol food have been analyzed by the estimation of the fitnesses of the *Adh* genotypes in separate tests. Morgan (1975) determined the survival of newly emerged larvae to adults on food supplied with various alcohols. FF homozygotes survive better than do SS homozygotes. Survival of SS homozygotes relative to FF on the alcohols tested is negatively correlated with the ratio of *in vitro* FF/SS activities with the corresponding alcohols. The expectation that survival will depend on the relative detoxifying properties of ADH is further confirmed by the experiments of Kamping and Van Delden (1978). They exposed flies of eight different strains, electrophoretically identical (FF homozygotes), but varying in ADH activity, to ethanol. Survival was found to be positively correlated with ADH activity. An identical result was obtained by Thompson and Kaiser (1977), who compared two SS strains differing in ADH activity for egg-to-adult survival on food with various alcohols. Ainsley and Kitto (1975) showed that larvae of different *Adh* genotypes differ in their *in vivo* rates of alcohol oxidation when put on food containing different alcohols. At the low ethanol concentrations used, however, it was found that the SS larvae oxided ethanol better than FF larvae.

The relation between survival and ADH activity only holds when the detoxification products are relatively nontoxic. This is illustrated by the reverse in mortality between FF and SS when flies are exposed to the vapour of 1-pentene-3-ol. In that case FF has the highest mortality because a highly toxic ketone (Sofer and Hatkof, 1972) is produced by ADH action (Morgan, 1975; Oakeshott, 1977). Acetaldehyde is the direct product of ADH action on ethanol, and it is plausible to assume that this toxic product is immediately converted into a nontoxic acetate by aldehyde oxidase (AO). David *et al.* (1978), however, found that mutant strains lacking regular AO activity have about the same tolerance to ethanol as do AO-positive strains. They suggested that another locus is also able to produce an aldehyde oxidase, which is responsible for the conversion of acetaldehyde.

Van Delden *et al.* (1978) made a detailed study of both egg-to-adult and adult survival on ethanol-supplemented food. The results unambiguously showed a considerably higher mortality of SS compared to FF and FS in the juvenile and adult life stages. The differences between FF and FS in egg-to-adult survival are comparatively small and generally not significantly different; in some populations FS tends to be superior, in others FF does. Adult survival is in general higher for FF than for FS, but the degree of dominance varies: in some populations FS is intermediate in survival between FF and SS, while in other populations FS equals FF. The degree of dominance also appears to be dependent on concentration. Briscoe *et al.* (1975) found effective dominance of F over S for adult survival also, while SS has the lowest number of survivors. Oakeshott *et al.* (1980) also found much higher mortality of SS compared to FS and FF when ethanol was added to the food. Short exposure to ethanol vapor, however, gave reversed results and the mortalities were in the order FF > FS > SS (Oakeshott and Gibson, 1981). This phenomenon may be connected with the finding that ethanol vapor can be taken up by the respiration system and may be subsequently used as food (Van Herrewege and David, 1978). Oakeshott (1976a) found that FS adults even survived more often than FF on ethanol food, while larval development time differed in the following way: FF < FS < SS.

Kilias *et al.* (1979) have suggested that ethanol may also influence recombination frequency and that the *Adh* genotypes may react differently to alcohols in this respect.

Selection for Alcohol Tolerance

Substantial fitness differences occur when the *Adh* genotypes are exposed to alcohols, leading to frequency changes in polymorphic populations. It may be asked whether long-term exposure of monomorphic populations to alcohols is also accompanied by an increase in tolerance. Such a phenomenon can be envisaged when selection favors the increase in frequency of high-activity alleles in a heterogeneous group of alleles of the same electrophoretic phenotype or when selection increases the frequency of alleles at modifier loci with a positive effect on ADH activity (Ward and Hebert, 1972; Ward, 1975). In addition, there is also the possibility that increased tolerance to alcohols depends on other loci than the *Adh* locus.

Selection for increased tolerance to ethanol has been successful and has led to increases in LD_{50} which were substantially higher than 100% (David and Boquet, 1977; David *et al.*, 1977). A series of biochemical tests was applied to one particular selected strain. No evidence was found for the existence of structual differences in ADH between selected and control strains (McDonald *et al.*, 1977; Ayala and McDonald, 1980). Selected strains, however, contained higher ADH activity and higher amounts of ADH, probably effected by changes in modifier genes. As ADH amount and activity are allometrically related to body weight (Clarke *et al.*, 1979; Van Dijk, 1981; McKay, 1981), an increase in the quantity of ADH may be brought about by an increase in body weight. Selection for increased tolerance to hexanol appeared to be highly successful in both FF and SS lines (Van Delden and Kamping, 1982), while the selected lines also proved to be considerably more resistant to other alcohols, such as ethanol and propanol. ADH activity in these lines was increased, which may be the cause of the increase in tolerance. The rise in ADH activity was, however, accompanied by a rise in body weight. This raises the question of whether selection has acted on body weight, with increased ADH activity as a consequence, or directly on ADH activity. This study makes clear that ADH activity *per se* is not the only factor involved in alcohol tolerance, as SS strains selected for increased tolerance still possessed considerably lower ADH activity than FF controls, which had a much lower tolerance. The finding of McKenzie and Parsons (1974), who measured an increase in tolerance to ethanol in winery populations compared to nonwinery populations but no increase in F frequency, may also point to an additional mechanism for alcohol adaptation, independent of the *Adh* locus. A similar conclusion was reached by Gibson *et al.* (1979), who successfully selected for increased tolerance in polymorphic laboratory populations, but found no consistent increase in F frequency nor in ADH activity. Van Herrewege and David (1980) found improved ethanol utilization, measured by increased life span at low concentrations, in some strains selected for increased ethanol tolerance, but not in all.

A peculiar effect was found by Bijlsma-Meeles (1979), who discovered that exposure of eggs to ethanol induced an increase in ethanol tolerance in larvae. This apparently phenotypic effect was accompanied by an increase in ADH activity in the eggs. Such an increase in activity after exposure to ethanol was also found for larvae (Gibson, 1970) and for embryonic cells (Horikawa *et al.*, 1967). This may be a case of substrate-induced enzyme synthesis (Chakrabartty, 1978; Anderson and McDonald, 1981b).

Alcohols as Food Components

Instead of being toxic compounds, primary alcohols can function as food components. It has been shown (Van Herrewege and David, 1974; Deltombe-Lietaert *et al.*, 1979; Van Herrewege *et al.*, 1980; Anderson *et al.*, 1981) that the life span of flies deprived of other food sources is extended when primary alcohols are provided at low concentrations. Arqués and Duarte (1980) found an increase of larval viability at low ethanol concentrations, but a decrease when isopropanol was added instead. Positive effects on life span can also be obtained by the exposure to ethanol vapor, suggesting that feeding through the respiratory system is possible (Van Herrewege and David, 1978; Starmer *et al.*, 1977). The survival time on ethanol, in the absence of other food components, appears to be positively correlated with ADH activity (Libion-Mannaert *et al.*, 1976; Daly and Clarke, 1981). Life duration first increases with concentration, but then decreases at higher concentrations when the toxic effects become dominant. There are considerable threshold differences among *Drosophila* species when tested for increase in longevity at various ethanol concentrations (Parsons, 1980c; Parsons and Spence, 1981). Primary alcohols are metabolized in the order: butanol > ethanol > propanol. Secondary alcohols do not extend longevity, although ADH activity *in vitro* with secondary alcohols is higher than on the corresponding primary alcohols. It was concluded (Van Herrewege *et al.*, 1980) that no direct relationship exists between the nutritive value of the various alcohols and their toxicity: highly toxic alcohols, such as butanol, can serve better as food when supplied at low concentrations than ethanol. It was further found (Van Herrewege and David, 1980) that increased tolerance to ethanol obtained by directional selection is not in all cases linked to improved utilization of ethanol as food. Alcohol tolerance and utilization, both related to ADH activity, may thus in part be controled by different genetic mechanisms.

Natural Populations

In view of the laboratory experiments, showing consistent fitness differences among the *Adh* genotypes in the presence of alcohols, it is plausible that ethanol, present in the natural environment of *Drosophila*, influences the allele frequencies in natural populations also. Experiments in this context have concentrated on *Drosophila* populations in wineries, where ethanol is probably present at higher concentrations than in any other part of the species habitat. The effects of selection will probably be maximized in this environment. It is to be expected that winery populations will have higher F frequencies than surrounding populations not exposed to ethanol. The work of Briscoe *et al.* (1975) and Hickey and McLean (1980) confirm this expectation for populations from Spanish and Canadian winery populations, respectively. Such a relation, however, was not found by other authors (McKenzie and Parsons, 1974; McKenzie and McKechnie, 1978; Marks *et al.*, 1980) for Australian and Californian populations. The explanation for this discrepancy may be found in differences in the distance observed between winery and nonwinery populations among these studies (Hickey and McLean, 1980). Gene flow (McKenzie, 1975) may blur allele frequency differences over shorter distances. On the other hand, it was found (McKenzie and Parsons, 1974; McKenzie and McKechnie, 1978) that tolerance to ethanol was higher in cellar populations compared to noncellar populations less than 100 m removed from each other, while no differentiation in allele frequency was observed. Increased tolerance to ethanol in these cases may depend on other loci than the *Adh* locus.

On a larger geographic scale it was found that ethanol tolerance in populations of *D. melanogaster* from Europe, America, and Africa increases with latitude (David and Bocquet, 1974, 1975). This cline is paralleled by changes in quantitative characters such as weight and ovariole number. An increase in ethanol tolerance with latitude was also found in Australia (Parsons, 1980a,b; Parsons and Stanley, 1981).

Behavioral Responses to Ethanol

Because of the important role of fermenting fruits in the ecology of *D. melanogaster*, several investigators have studied the behavioral responses of both larvae and adults to ethanol. Flies are attracted by ethanol and the attractability increases to maximum responses at a concentration between 8 and 16%; higher concentrations lead abruptly to repulsion (Fuyama, 1976). A considerable amount of genetic variation for olfactory response seems to exist in natural populations. It is further worth mentioning that females of *Cothonaspis sp.* (*Hymenoptera, Cynipidae*), a parasite of *D. melanogaster* larvae, are increasingly attracted by ethanol concentrations of up to about 8%, after which the reaction rapidly changes to one of repulsion (Carton, 1976). Apparently ethanol serves as an olfactory stimulus for these parasites to find suitable places for oviposition in *Drosophila* larvae. *Drosophila* species differ in their preference for ethanol: larvae of *D. melanogaster* have a preference for agar supplemented with ethanol in moderate concentrations compared to pure agar, while larvae of *D. simulans* initially show no preference (Parsons, 1977; Parsons and King, 1977). Oviposition choice experiments showed a rejection of ethanol-supplemented food by *D. simulans* females, while *D. melanogaster* females showed a preference for ethanol food (McKenzie and Parsons, 1972). As *D. melanogaster* has a much higher tolerance to ethanol than does *D. simulans*, probably based on a higher ADH level, these data suggest that larval discrimination and adult oviposition site preference may be correlated with the presence of ethanol in the natural environment of the species and on their ADH level. A study of oviposition site preferences in several *Drosophila* species (Richmond and Gerking, 1979) provides further evidence for such a correlation.

As far as intraspecific variation in ethanol preference is concerned, it was found by Cavener (1979) that no differences exist between FF and SS females for ethanol oviposition sites in choice experiments; FF larvae, on the other hand, prefer ethanol-containing food, while SS larvae have no preference. Gelfant and McDonald (1980), however, found avoidance behavior toward higher ethanol concentrations, where the strength of the avoidance response was in the following order: FF > SS > null mutant. This discrepancy could be due to differences in the effective ethanol concentration used. When adults were given a choice between ethanol or water it was found that heterozygotes preferred ethanol more than did either of the homozygotes (Soliman and Knight, 1981. In conclusion it seems possible that the *Adh* genotype is of importance for the degree of preference or avoidance toward ethanol.

OTHER FACTORS INFLUENCING THE ADH POLYMORPHISM

Fitness Differences in the Absence of Alcohols

The *Adh* locus is highly polymorphic: in surveys based on a great number of populations from different geographic origins only very rarely is a monomorphic population observed (Vigue and Johnson, 1973; Johnson and Schaffer, 1973; Johnson and Burrows, 1976; Sampsell, 1977; Wilks *et al.*, 1980). As exposure to ethanol eventually would lead to fixation of the F allele, the persistence of the *Adh* polymorphism suggests the occurrence of some form of balancing selection.

Laboratory experiments in which large populations were started with different initial frequencies showed a convergence of frequencies on regular food leading to values equal to those in the base population (Bijlsma-Meeles and Van Delden, 1974; Van Delden *et al.*, 1978; Bijlsma-Meeles, 1982. However, it can be imagined that selection at other loci in linkage disequilibrium with the *Adh* locus instead of selection at the *Adh* locus has been responsible for such a reaction (Leibenguth and Steinmetz, 1976; Leibenguth, 1977). Hitch-hiking of the *Adh* locus with other loci is unlikely, however, as a repetition of the experiment with reisolated lines from the experimental populations gave identical results (Van Delden *et al.*, 1978). Viability tests in this case suggested overdominance for egg-to-adult survival as a possible mechanism for the maintenance of the polymorphism. This is in agreement with the findings of Franklin (1981) for a natural population where heterozygote advantage at the *Adh* locus was observed. Indirect evidence for the occurrence of balancing selection in natural populations is provided by McKenzie (1980), who found similar, stable allele frequencies in different populations in spite of ample opportunities for allele divergence due to drift.

Kojima and Tobari (1969) claimed that egg-to-adult survival of *Adh* genotypes depends upon their genotypic frequencies, leading to frequency-dependent selection and the establishment of stable equilibria. Morgan (1976) also found frequency-dependent effects, though FF had more survivors than SS at all frequencies and no equilibrium could be obtained. The conditioning of the medium by the larvae has been suggested as a mechanism for frequency-dependent selection at the esterase-6 locus (Huang *et al.*, 1971). Such an effect was not observed for the *Adh* locus (Dolan and Robertson, 1975; Van Delden, unpublished). Yoshimaru and Mukai (1979) found no evidence for frequency-dependent effects for the *Adh* locus. Clarke and Allendorf (1979) have proposed a theoretical model for frequency-dependent selection due to kinetic differences between allozymes with external substrates like ADH.

Van Delden and Kamping (1979b) found that *Adh* genotypes differ in developmental times. It was concluded that FF larvae, and to a lesser extent FS larvae, either reach their critical weights for pupation earlier than SS larvae or possess a lower critical weight. This difference was found to be reinforced by increasing levels of crowding. It was concluded that at very high densities FF survives more often than FS, but that at intermediate densities FS has a higher survival than FF. The SS larvae survive less often than the other genotypes under all conditions. Conditions for frequency-dependent equilibria are not fulfilled in such a situation (De Jong, 1976). Evidence for faster development of FF also comes from the experiments of Oakeshott (1976a,b; 1977). Different rates of development for *Adh* genotypes were also found in the seaweed fly, *Coelopa frigida* (Day et al., 1980; Day and Buckley, 1980).

Pot et al. (1980) investigated whether genotypic differences in mating success could be held responsible for the maintenance of the *Adh* polymorphism. No evidence for overdominance or rare genotype mating advantage was found, though there are genotypic differences in mating success in the order FF > FS > SS. The advantage of FF over SS is dependent on environmental conditions, such as temperature and age of females (Knoppien et al., 1980). It has further been reported (McKenzie and Fegent, 1980) that in female choice situations, most matings are performed by FS males. Bijlsma-Meeles (1982) has attempted to estimate various components of fitness under regular laboratory conditions, using the approach devised by Prout (1971a,b). No significant effects for sexual, fecundity, and zygotic selection components could be shown, mainly due to the large interstrain differences. In another approach (Bijlsma-Meeles and Van Delden, 1974) monomorphic and polymorphic populations were exposed to a variety of extreme environments (high concentrations of ethanol supplement in the food, high temperature, low temperature, high humidity, etc.). The number of populations lost in each environment after eight generations was recorded, and thus a measure of "population fitness" was obtained. The extinction percentage varied considerably among the three types of population. Not surprisingly, the monomorphic SS populations showed the highest extinction in the ethanol environment. But in other environments quite differential extinction also occurred; polymorphic populations, e.g., exhibited the lowest extinction rate at high temperature. This experiment suggests that selection acts on the *Adh* polymorphism under several environmental conditions. A serious problem, however, is that, except for alcohols, no functional relationship can be indicated between environmental variables and the *Adh* locus, leaving the possibility of effects of linked loci or inversions (Alahiotis, 1976; Alahiotis and Pelecanos, 1978; Pieragostini et al., 1979).

Differences in Temperature Resistance

The three ADH allozymes show differences in heat stability *in vitro*: the ADH of SS homozygotes has the highest stability at high temperatures ($>40°C$) (Gibson, 1970; Gibson and Miklovich, 1971; Vigue and Johnson, 1973; Day et al., 1974a; McKay, 1981). This differential thermostability has been related to a latitudinal cline in allele frequencies observed in the Eastern part of the U.S., where the frequency of the S allele reaches a maximum of about 0.90 in the South and a minimum of about 0.50 in the North (Johnson and Schaffer, 1973; Vigue and Johnson, 1973). A comparable cline with latitude was found for the Australian continent (Oakeshott et al., 1980; Gibson et al., 1981; Anderson, 1981). This relation with latitude could be due to temperature favoring SS in the hotter areas. Further temperature-related clines of Adh alleles, though over shorter distances, have been reported from Russia (Grossman et al., 1970) and Mexico (Pipkin et al., 1973, 1975, 1976). Pipkin et al. (1973) and Malpica and Vassallo (1980) found a positive linear regression of S allele frequency on minimum temperature and on mean temperature. No correlation between Adh frequency and temperature was found when frequencies were followed in time over a 2-year period in a local population (Gionfriddo and Vigue, 1978; Gionfriddo et al., 1979). Johnson and Burrows (1976) found that allele frequencies are very stable in time and place; no decisive evidence was found for a decrease in S frequency in the winter season in local populations. McKenzie and McKechnie (1981) found a positive correlation between F frequency and temperature in wine cellars. The latitudinal cline in Australia, however, could be explained by a negative correlation between annual rainfall and Adh^F frequency (Anderson 1981).

Johnson and Powell (1974) subjected adult flies in a laboratory experiment to heat and cold shocks, and reported a higher S frequency for the survivors of heat shock and a lower frequency after cold shock compared to controls in some of the populations tested. Van Delden and Kamping (1980) confirmed the greater heat stability of SS under in vitro conditions. The ADH activity in flies exposed to 35°C for as long as 24 hr, however, did not decrease. It was further found that FS flies survive more often than both homozygotes at 35°C. It was concluded that the in vivo processes do not necessarily follow the predictions based on the results of assays in vitro. The higher survival of heterozygotes is in agreement with the results of Bijlsma-Meeles and Van Delden (1974), who found a lower extinction rate of polymorphic populations at high temperature compared to monomorphic ones. McKechnie et al. (1981) also found that heterozygotes survived better at high temperature, but only on alcohol-supplemented media.

It should be pointed out that in addition to the regular Adh alleles, Adh alleles have been found in natural populations which have ADH that is more thermostabile. These alleles, which have regular electrophoretic mobilities, are rare, however (Sampsell, 1977; Gibson et al., 1980; McKay, 1981).

RELATIONSHIPS WITH OTHER LOCI

Genes do not function independently, but have structural and functional relationships with other genes. Nevertheless, the *Adh* locus appears to take a paramount position for survival under relevant environmental conditions—the presence of alcohols—and resembles the melanism polymorphism in *Biston* and the sickle-cell polymorphism in humans in its drastic effects. Still, it is highly unlikely that there are no functional links with other loci, e.g., loci that convert the products of ADH action. Such functional relationships may lead to interaction between loci with respect to fitness (Bijlsma, 1978).

The occurrence of linkage disequilibrium with other loci may point to such fitness interactions. Surveys of linkage disequilibrium for alleles at different allozyme loci in natural populations have in general revealed little linkage disequilibrium (Langley, 1977). However, in some cases considerable linkage disequilibrium is found for inversions and allozyme loci linked to them (Hedrick *et al.*, 1978). The *Adh* locus does not take an exceptional position in this respect; in cases where a linkage disequilibrium was found between the *Adh* locus and other linked allozyme loci, the occurrence of inversions cannot be excluded (Grossman and Koreneva, 1970).

Relatively little research has been directed to the direct relation between the *Adh* locus and other loci metabolically linked to it. In this respect an interaction with the *Aldox* locus, which produces aldehyde oxidase for the conversion of the toxic acetaldehyde formed by ADH action on ethanol, is obvious. It was mentioned that, contrary to expectation, *Aldox*-negative homozygotes have about the same tolerance to ethanol as *Aldox*-positive strains (David *et al.*, 1978), suggesting no strong relationship. Another potentially interacting locus is the octanol dehydrogenase (*Odh*) locus (Courtright *et al.*, 1966). ODH converts a variety of primary, long-chain alcohols, of which the smaller ones can also be converted by ADH. It was found (Van Delden and Kamping, 1976, and unpublished) that egg-to-adult survival of the *Adh* genotypes on ethanol-supplemented food depends on the allelic constitution at the *Odh* locus. The nature of this phenomenon has not been made clear; it may have to do with the presumptive role of ODH in ecdysis (Madhavan *et al.*, 1973). Cavener and Clegg (1981) have estimated two-locus joint viabilities from allele frequency changes for the *Adh* and $\alpha Gpdh$ loci in control and ethanol environments, which indicated a fitness interaction between the two loci. They suggest a possible role of αGPDH in ethanol metabolism in connection with the NADH/NAD ratio.

CONCLUDING REMARKS

It was stated in the introduction that the *Adh* polymorphism forms an excellent test case for a direct approach to the problem of selection versus neutrality of allozyme polymorphisms, because of the presence of external, ecologically relevant, substrates. What can be concluded from the numerous studies devoted to this polymorphism? It appears that the *Adh* polymorphism meets the demands made for the direct detection of selection in view of the relation between the enzyme activities *in vitro* of the genotypes and their survival in the presence of alcohols. The laboratory experiments nearly unanimously confirm that FF, the genotype with the highest ADH activity, survives better than SS on alcohols. Some questions still need to be answered: e.g., why do heterozygotes, which generally have ADH activities intermediate between FF and SS, in general still have higher numbers of survivors than expected from the midparent values and often equal to those of FF? Further elucidation is also required with respect to the interaction of toxic and useful effects, the latter resulting from the use of ethanol as food, in relation with the *Adh* locus. Though the data from laboratory experiments are very conclusive, the available data on natural populations, exclusively winery populations, do not always harmonize with expectations. Winery populations failed to show higher F frequencies in a number of cases when they were compared with surrounding populations. The different findings in this respect need further consideration, as does the selective role of ethanol and other alcohols in less extreme natural conditions. Selection for ethanol resistance (partly) independent of the *Adh* locus may be involved in these cases. A further study of the mechanisms of adaptation to alcohols in relation to the *Adh* locus is therefore needed and will probably also involve the role of modifiers of ADH activity and the importance of other loci. This makes the *Adh* locus a promising research object for both the study of modifying genes and for interacting genes. More needs to be known about the biochemical properties of ADH and the pathways in which it is involved.

More uncertain than its role in the detoxification of alcohols is the unknown mechanism of the maintenance of the *Adh* polymorphism in nature under those conditions where the ethanol content of the food is low and its selective impact is negligible. In this respect differential survival of genotypes connected with differences in temperature stability of ADH *in vitro* has been suggested, but the data are far from conclusive. It appears that the study of fitness differences under these conditions requires a better understanding of the role of ADH in the physiology of *Drosophila*.

In conclusion it can be stated that in the case of the *Adh* polymorphism the neutrality hypothesis appears highly unlikely, although much more information is needed about other allozyme polymorphisms before one can decide for or against the generality of allozyme selection.

REFERENCES

Ainsley, R., and Kitto, G. B., 1975, Selection mechanisms maintaining alcohol dehydrogenase polymorphisms in *Drosophila melanogaster,* in: *Isozymes II* (C. L. Markert, ed.), pp. 733–742, Academic Press, New York.

Alahiotis, S., 1976, Genetic variation and the ecological parameter "food medium" in cage populations of *Drosophila melanogaster, Can. J. Genet. Cytol.* **18:**379–383.

Alahiotis, S., and Pelecanos, M., 1978, Induction of gene pool differentiation in *Drosophila melanogaster, Can. J. Genet. Cytol.* **20:**265–273.

Anderson, P. R., 1981, Geographic clines and climatic associations of Adh and α-Gpdh gene frequencies in *Drosophila melanogaster,* in: *Genetic Studies of Drosophila Populations* (J. B. Gibson and J. G. Oakeshott, eds.), pp. 237–250, Australian National University Press, Canberra, Australia.

Anderson, S. M., and McDonald, J. F., 1981a, A method for determining the *in vivo* stability of *Drosophila* alcohol dehydrogenase (E.C. 1.1.1.1), *Biochem. Genet.* **19:**411–419.

Anderson, S. M., and McDonald, J. F. 1981b, Effect of environmental alcohol on *in vivo* properties of *Drosophila* alcohol dehydrogenase, *Biochem. Genet.* **19:**421–430.

Anderson, S. M., McDonald, J. F., and Santos, M., 1981, Selection at the Adh locus in *Drosophila melanogaster*: Adult survivalship–mortality in response to ethanol, *Experientia* **37:**463–464.

Arqués, L. V., and Duarte, R. G. 1980, Effect of ethanol and isopropanol on the activity of alcohol dehydrogenase, viability and life-span in *Drosophila melanogaster* and *D. funebris, Experientia* **36:**828–830.

Ayala, F. J., and McDonald, J. F., 1980, Continuous variations: Possible role of regulatory genes, *Genetica* **52/53:**1–15.

Barnes, B. W., and Birley, A. J., 1975, Genetic variation for enzyme activity in a population of *Drosophila melanogaster.* II. Aspects of the inheritance of alcohol dehydrogenase activity in Adh$^{S/S}$ flies, *Heredity* **35:**115–119.

Barnes, B. W., and Birley, A. J., 1978, Genetic variation for enzyme activity in a population of *Drosophila melanogaster.* IV. Analysis of alcohol dehydrogenase activity in chromosome substitution lines, *Heredity* **40:**51–57.

Bernstein, S. C., Throckmorton, L. H., and Hubby, J. L., 1973, Still more genetic variability in natural populations, *Proc. Natl. Acad. Sci. USA* **70:**3928–3931.

Bijlsma, R., 1978, Polymorphism at the G 6PD and 6 PGD loci in *Drosophila melanogaster.* II. Evidence for interaction in fitness, *Genet. Res.* **31:**227–237.

Bijlsma-Meeles, E., 1979, Viability in *Drosophila melanogaster* in relation to age and ADH activity of eggs transferred to ethanol food, *Heredity* **42:**79–89.

Bijlsma-Meeles, E., 1982, Measurement of fitness components and population prediction for the alcohol dehydrogenase locus in *Drosphila melanogaster* (in preparation).

Bijlsma-Meeles, E., and Van Delden, W., 1974, Intra- and interpopulation selection concerning the alcohol dehydrogenase locus in *Drosophila melanogaster, Nature* **247:**369–371.

Birley, A. J., and Barnes, B. W., 1973, Genetical variation for enzyme activity in a population of *Drosophila melanogaster*. I. Extent of the variation for alcohol dehydrogenase activity, *Heredity* 31:413–416.

Birley, A. J., and Barnes, B. W., 1975, Genetic variation for enzyme activity in a population of *Drosophila melanogaster*. III. Dominance relationships for alcohol dehydrogenase activity, *Heredity* 35:121–126.

Birley, A. J., and Marson, A., 1981, Genetical variation for enzyme activity in a population of *Drosophila melanogaster*. VII. Genotype–environment interaction for alcohol dehydrogenase (ADH) activity, *Heredity* 46:427–441.

Birley, A. J., Marson, A., and Phillips, L. C., 1980, Genetical variation for enzyme activity in a population of *Drosophila melanogaster*. V. The genetical architecture, as shown by diallele analysis, of alcohol dehydrogenase (ADH) activity. *Heredity* 44:251–268.

Briscoe, D. A., Robertson, A., and Malpica, J. M., 1975, Dominance at Adh locus in response of adult *Drosophila melanogaster* to environmental alcohol, *Nature* 255:148–149.

Brown, A. H. D., 1979, Enzyme polymorphism in plant populations, *Theor. Popul. Biol.* 15:1–42.

Brown, M. J., 1934, Alcohol tolerance of *Drosophila*, *J. Hered.* 25:244–246.

Carton, Y., 1976, Attraction de *Cothonaspis sp. (Hym. Cynipidae)* par le milieu trophique de son hôte: *Drosophila melanogaster, Collog. Int. CNRS* 265:285–303.

Castelli, T., 1954, Les agents de la fermentation vinaire, *Arch. Microbiol.* 20:323–342.

Cavener, D., 1979, Preference for ethanol in *Drosophila melanogaster* associated with the alcohol dehydrogenase polymorphism, *Behav. Genet.* 9:359–365.

Cavener, D. R., and Clegg, M. T., 1978, Dynamics of correlated genetic systems. IV. Multilocus effects of ethanol stress environments, *Genetics* 90:629–644.

Cavener, D. R., and Clegg, M. T., 1981, Multigenic adaptation to ethanol in *Drosophila melanogaster, Evolution* 35:1–10.

Chakrabartty, P. K., 1978, Isozymes in *Drosophila* cell line, *Experientia* 34:438–439.

Chambers, G. K., 1981, Biochemistry of alcohol dehydrogenase variation in *Drosophila mellanogaster* in: *Genetic Studies of Drosophila Populations* (J. B. Gibson and J. G. Oakeshott, eds.), pp. 77–93, Australian National University Press, Canberra, Australia.

Chambers, G. K., McDonald, J. F., McElfresh, M., and Ayala, F. J., 1978, Alcohol oxidizing enzymes in 13 *Drosophila* species, *Biochem. Genet.* 16:757–767.

Chambers, G. K., Laver, W. G., Campbell, S., and Gibson, J. B., 1981, Structural analysis of an electrophoretically cryptic alcohol dehydrogenase variant from an Australian population *Drosophila melanogaster, Proc. Natl. Acad. Sci. USA* 78:3103–3107.

Clarke, B., 1975, The contribution of ecological genetics to evolutionary theory: Detecting the direct effects of natural selection on particular polymorphic loci, *Genetics* 79:101–113.

Clarke, B., and Allendorf, F. W., 1979, Frequency-dependent selection due to kinetic differences between allozymes, *Nature* 279:732–734.

Clarke, B., Camfield, R. G., Galvin, A. M., and Pits, C. R., 1979, Environmental factors affecting the quantity of alcohol dehydrogenase in *Drosophila melanogaster, Nature* 280:517–518.

Courtright, J. B., Imberski, R. B., and Ursprung, H., 1966, The genetic control of alcohol dehydrogenase and octanol dehydrogenase isoenzymes in *Drosophila, Genetics* 54:1251–1260.

Coyne, J. A., and Felton, A. A., 1977, Genic heterogeneity at two alcohol dehydrogenase loci in *Drosophila pseudoobscura* and *Drosophila persimilis, Genetics* 87:285–304.

Daggard, G. E., 1981, Alcohol dehydrogenase, aldehyde oxidase and alcohol utilization in *Drosophila melanogaster, D. immigrans* and *D. busckii,* in: *Genetic studies of Dro-*

sophila Populations (J. B. Gibson and J. G. Oakeshott, eds.), pp. 59–75, Australian National University Press, Canberra, Australia.

Daly, K., and Clarke, B., 1981, Selection associated with the alcohol dehydrogenase locus in _Drosophila melanogaster_: Differential survival of adults maintained on low concentrations of ethanol, _Heredity_ **46**:219–226.

David, J., 1973, Toxicité de faibles concentrations d'alcool ethylique pour une espèce tropicale de _Drosophila_: _Drosophila bromeliae_ Sturtevant, _C. R. Acad. Sci. Paris_ **277**:2235–2238.

David, J., 1977a, Métabolisme de l'alcool chez _Drosophila melanogaster_. I. Rôle de la déshydrogénase alcoolique dans la détoxification de ce produit, _Bull. Soc. Zool. Fr._ **102**:298.

David, J., 1977b, Signification d'un polymorphisme enzymatique: La déshydrogénase alcoolique chez _Drosophila melanogaster_, _Ann. Biol._ **16**:451–472.

David, J., and Bocquet, C., 1974, L'adaptation génétique à l'éthanol: Un paramètre important dans l'évolution des races géographiques de _Drosophila melanogaster_, _C. R. Acad. Sci. Paris_ **279**:1385–1388.

David, J. R., and Bocquet, C., 1975, Similarities and differences in latitudinal adaptation of two _Drosophila_ sibling species, _Nature_ **257**:588–590.

David, J., and Bocquet, C., 1976, Compared toxicities of different alcohols for two _Drosophila_ sibling species: _Drosophila melanogaster_ and _Drosophila simulans_, _Comp. Biochem. Physiol._ **54C**:71–74.

David, J. R., and Bocquet, C., 1977, Genetic tolerance to ethanol in _Drosophila melanogaster_: Increase by selection and analysis of correlated responses, _Genetica_ **47**:43–48.

David, J., Fouillet, P., and Arens, M. F., 1974, Comparaison de la sensibilité à l'alcool éthylique des six espèces de _Drosophila_ du sous-groupe _melanogaster_, _Arch. Zool. Exp. Gen._ **115**:401–410.

David, J. R., Bocquet, C., Arens, M. F., and Fouillet, P., 1976, Biological role of alcohol dehydrogenase in the tolerance of _Drosophila melanogaster_ to aliphatic alcohols: Utilization of an ADH-null mutant, _Biochem. Genet._ **14**:989–997.

David, J., Bocquet, C., Fouillet, P., and Arens, M. F., 1977, Tolérance génétique à l'alcool chez _Drosophila_: Comparaison des effects de la sélection chez _Drosophila melanogaster_ et _Drosophila simulans_, _C. R. Acad. Sci. Paris_ **285**:405–408.

David, J., Bocquet, C., v. Herrewege, J., Fouillet, P., and Arens, M., 1978, Alchol metabolism in _Drosophila melanogaster_: Uselessness of the most active aldehyde oxidase produced by the Aldox locus, _Biochem. Genet._ **16**:203–211.

David, J. R., v. Herrewege, J., Monclus, M., and Prevosti, A., 1979, High ethanol tolerance in two distantly related _Drosophila_ species; A probable case of recent convergent adaptation, _Comp. Biochem. Physiol._ **63C**:53–56.

Day, T. H., and Buckley, P. A., 1980, Alcohol dehydrogenase polymorphism in the seaweed fly, _Coelopa frigida_, _Biochem. Genet._ **18**:727–742.

Day, T. H., and Needham, L., 1974, Properties of alcohol dehydrogenase isozymes in a strain of _Drosophila melanogaster_ homozygous for the Adh-Slow allele, _Biochem. Genet._ **11**:167–175.

Day, T. H., Hillier, P. C., and Clarke, B., 1974a, Properties of genetically polymorphic isozymes of alcohol dehydrogenase in _Drosophila melanogaster_, _Biochem. Genet._ **11**:141–153.

Day, T. H., Hillier, P. C., Clarke, B., 1974b, The relative quantities and catalytic activities of enzymes produced by alleles at the alcohol dehydrogenase locus in _Drosophila melanogaster_, _Biochem. Genet._ **11**:155–165.

Day, T. H., Dobson, T., Hillier, P. C., Parkin, D. T., and Clarke, B., 1980, Different rates of development, associated with the alcohol dehydrogenase locus in the seaweed fly, *Coelopa frigida, Heredity* **44:**321–326.

De Jong, G., 1976, A model of competition for food. I. Frequency-dependent viabilities, *Am. Nat.* **110:**1013–1027.

De Jong, G., and Scharloo, W., 1976, Environmental determination of selective significance or neutrality of amylase variants in *Drosophila melanogaster, Genetics* **84:**77–94.

Deltombe-Lietaert, M. C., Delcour, J., Lenelle-Montfort, N., and Elens, A., 1979, Ethanol metabolism in *Drosophila melanogaster, Experientia* **35:**579–581.

Dickinson, W. J., and Sullivan, D. T., 1975, *Gene–Enzyme Systems in Drosophila*, Springer, Berlin, West Germany.

Dolan, R., and Robertson, A., 1975, The effect of conditioning the medium in *Drosophila*, in relation to frequency-dependent selection, *Heredity* **35:**311–316.

Dunn, G. R., Wilson, T. G., and Jacobson, K. B., 1969, Age-dependent changes in alcohol dehydrogenase in *Drosophila, J. Exp. Zool.* **171:**185–190.

Ewens, W. J., and Feldman, M. W., 1976, The theoretical assessment of selective neutrality, in: *Population Genetics and Ecology* (S. Karlin and E. Nevo, eds.), pp. 303–337, Academic Press, New York.

Fletcher, T. S., Ayala, F. J., Thatcher, D. R., and Chambers, G. K., 1978, Structural analysis of the ADH[S] electromorph of *Drosophila melanogaster, Proc. Natl. Acad. Sci. USA* **75:**5609–5612.

Fontdevila, A., Santors, M., and Gonzalez, R., 1980, Genotype–isopropanol interaction in the Adh locus of *Drosophila buzzatii, Experientia* **36:**398–400.

Franklin, R., 1981, An analysis of temporal variation at isozyme loci in *Drosophila melanogaster* in: *Genetic Studies of Drosophila Populations* (J. B. Gibson and J. G. Oakeshott, eds.), pp. 217–236, Australian National University Press, Canberra, Australia.

Fuyama, Y., 1976, Behavior genetics of olfactory responses in *Drosophila*. I. Olfactometry and strain differences in *Drosophila melanogaster, Behav. Genet.* **6:**407–420.

Gelfant, L. J., and MacDonald, J. F., 1980, Relationship between ADH activity and behavioral response to environmental alcohol in *Drosophila, Behav. Genet.* **10:**237–249.

Gibson, J., 1970, Enzyme flexibility in *Drosophila melanogaster, Nature* **227:**959–960.

Gibson, J. B., 1972, Differences in the number of molecules produced by two allelic electrophoretic enzyme variants in *Drosophila melanogaster, Experientia* **28:**975–976.

Gibson, J. B., and Miklovich, R., 1971, Modes of variation in alcohol dehydrogenase in *Drosophila melanogaster, Experientia* **27:**99–100.

Gibson, J. B., Lewis, N., Adena, M. A., and Wilson, S. R., 1979, Selection for ethanol tolerance in two populations of *Drosophila melanogaster* segregating alcohol dehydrogenase allozymes, *Aust. J. Biol. Sci.* **32:**387–398.

Gibson, J. B., Chambers, G. K., Wilks, A. V., and Oakeshott, J. G., 1980, An electrophoretically cryptic alcohol dehydrogenase variant in *Drosophila melanogaster*. I. Activity ratios, thermostability, genetic localization and comparison with two other thermostable variants, *Aust. J. Biol. Sci.* **33:**479–489.

Gibson, J. B., Wilks, A. V., and Chambers, G. K., 1981, Population variation in functional properties of alcohol dehydrogenase in *Drosophila melanogaster*, in: *Genetic Studies of Drosphila populations* (J. B. Gibson and J. G. Oakeshott, eds.), pp. 251–267, Australian National University Press, Canberra, Australia.

Gionfriddo, M. A., and Vigue, C. L., 1978, *Drosophila* alcohol dehydrogenase frequencies and temperature, *Genet. Res.* **31:**97–101.

Gionfriddo, M. A., Vigue, C. L., and Weisgram, P. A., 1979, Seasonal variation in the frequencies of the alcohol dehydrogenase isoalleles of *Drosophila*: Correlation with environmental factors, *Genet. Res.* **34**:317–319.

Grell, E. H., Jacobson, K., and Murphy, J. B., 1965, Alcohol dehydrogenase in *Drosophila melanogaster*: Isozymes and genetic variants, *Science* **149**:80–82.

Grell, E. H., Jacobson, K. B., and Murphy, J. B., 1968, Alterations of genetic material for analysis of alcohol dehydrogenase isozymes of *Drosophila melanogaster*, *Ann. N.Y. Acad. Sci.* **151**:441–455.

Grossman, A., 1980, Analysis of genetic variation affecting the relative activities of fast and slow ADH dimers in *Drosophila melanogaster*, *Biochem. Genet.* **18**:765–780.

Grossman, A. J., and Koreneva, L. G., 1970, Correlation between alleles of alcohol dehydrogenase and α-glycerophosphate dehydrogenase in natural populations of *Drosophila melanogaster* (in Russian), *Genetika* **6**(8):95–101.

Grossman, A. J., Koreneva, L. G., and Ulitskaya, L. E., 1970, Variation of the alcohol dehydrogenase locus in natural populations of *Drosophila melanogaster* (in Russian), *Genetika* **6**(2):91–96.

Hamrick, J. L., Linhart, Y. B., and Mitton, J. B., 1979, Relationships between life history characteristics and electrophoretically detectable genetic variation in plants, *Annu. Rev. Ecol. Syst.* **10**:173–200.

Harris, H., 1966, Enzyme polymorphisms in man, *Proc. R. Soc. Ser. B* **164**:298–310.

Harris, H., 1976, Molecular evolution: The neutralist–selectionist controversy, *Fed. Proc. Fed. Am. Soc. Exp. Biol.* **35**:2079–2082.

Hedrick, P., Jain, S., and Holden, L., 1978, Multilocus systems in evolution, *Evol. Biol.* **11**:101–184.

Heed, W. B., 1978, Ecology and genetics of sonoran desert *Drosophila*, in: *Ecological Genetics: The Interface* (P. F. Brussard, ed.), pp. 109–126, Springer, New York.

Hewitt, N. E., Pipkin, S. B., Williams, N., and Chakrabartty, P. K., 1974, Variation in ADH activity in class I and class II strains of *Drosophila*, *J. Hered.* **65**:141–148.

Hickey, D. A., and McLean, M. D., 1980, Selection for ethanol tolerance and Adh allozymes in natural populations of *Drosophila melanogaster*, *Genet. Res.* **36**:11–15.

Hisey, B. N., Thompson, J. N., and Woodruff, R. C., 1979, Position effect influencing alcohol dehydrogenase activity in *Drosophila melanogaster*, *Experientia* **35**:591–592.

Hoorn, A. J. W., and Scharloo, W., 1979, Selection on enzyme variants in *Drosophila*, *Aquilo Ser. Zool.* **20**:41–48.

Horikawa, M., Ling, L. N. L., and Fox, A. S., 1967, Effects of substrates on gene-controlled enzyme activities in cultured embryonic cells of *Drosophila*, *Genetics* **55**:569–583.

Huang, S. L., Singh, M., and Kojima, K., 1971, A study of frequency-dependent selection observed in the esterase-6 locus of *Drosophila melanogaster* using a conditioned media method, *Genetics* **68**:97–104.

Hulme, A. C., 1970, *The Biochemistry of Fruits and their Products*, Volume 1, Academic Press, New York.

Hulme, A. C., 1971, *The Biochemistry of Fruits and their Products*, Volume 2, Academic Press, New York.

Jacobson, K. B., 1968, Alcohol dehydrogenases of *Drosophila*. Interconversion of isoenzymes, *Science* **159**:324–325.

Jacobson, K. B., and Pfuderer, P., 1970, Interconversion of isoenzymes of *Drosophila* alcohol dehydrogenase. II. Physical characterization of the enzyme and its subunits, *J. Biol. Chem.* **245**:3938–3944.

Jacobson, K. B., Murphy, J. B., and Hartman, F. C., 1970, Isoenzymes of *Drosophila* alcohol dehydrogenase. I. Isolation and interconversion of different forms, *J. Biol. Chem.* **245**:1075–1080.

Jacobson, K. B., Murphy, J. B., Knopp, J. A., and Ortiz, J. R., 1972, Multiple forms of *Drosophila* alcohol dehydrogenase. III. Conversion of one form to another by NAD or aceton, *Arch. Biochem. Biophys.* **149**:22–35.

Johnson, F. M., and Burrows, P. M., 1976, Isozyme variability in species of the genus *Drosophila*. VIII. The alcohol dehydrogenase polymorphism in North Carolina populations of *Drosophila melanogaster*, *Biochem. Genet.* **14**:47–58.

Johnson, F. M., and Denniston, C., 1964, Genetic variation of alcohol dehydrogenase in *Drosophila melanogaster*, *Nature* **204**:906–907.

Johnson, F. M., and Powell, A., 1974, The alcohol dehydrogenases of *Drosophila melanogaster*: Frequency changes associated with heat and cold shock, *Proc. Natl. Acad. Sci. USA* **71**:1783–1784.

Johnson, F. M., and Schaffer, H. E., 1973, Isozyme variability in species of the genus *Drosophila*. VII. Genotype–environment relationships in populations of *Drosophila melanogaster* from the Eastern United States, *Biochem. Genet.* **10**:149–163.

Johnson, G. B., 1977, Hidden heterogeneity among electrophoretic alleles, in: *Measuring Selection in Natural Populations* (F. B. Christiansen and T. M. Fenchel, eds.), pp. 223–244, Springer, Berlin, West Germany.

Johnson, G. B., 1978, Structural flexibility of isozyme variants: Genetic variants in *Drosophila* disguised by cofactor and subunit binding, *Proc. Natl. Acad. Sci. USA* **75**:395–399.

Juan, E., and González-Duarte, R., 1980, Purification and enzyme stability of alcohol dehydrogenase from *Drosophila simulans, Drosophila virilis* and *Drosophila melanogaster* Adh^S, *Biochem. J.* **189**:105–110.

Kamping, A., and Van Delden, W., 1978, The alcohol dehydrogenase polymorphism in populations of *Drosophila melanogaster*. II. Relation between ADH activity and adult mortality, *Biochem. Genet.* **16**:541–551.

Kilias, G., Alahiotis, S. N., and Onoufriou, A., 1979, The alcohol dehydrogenase locus affects meiotic crossing-over in *Drosophila melanogaster*, *Genetica* **50**:173–177.

Kimura, M., 1968, Genetic variability maintained in a finite population due to mutational production of neutral and nearly neutral isoalleles, *Genet. Res.* **11**:247–269.

Kimura, M., 1977, Causes of evolution and polymorphism at the molecular level, in: *Molecular Evolution and Polymorphism* (M. Kimura, ed.), pp. 1–28, National Institute of Genetics, Mishima, Japan.

Kimura, M., and Ohta, T., 1971, Protein polymorphism as a phase of molecular evolution, *Nature* **229**:467–469.

King, J. L., and Jukes, T. H., 1969, Non-Darwinian evolution, *Science* **164**:788–798.

Knopp, J. A., and Jacobson, K. B., 1972, Multiple forms of *Drosophila* alcohol dehydrogenase. IV. Protein fluorescense studies, *Arch. Biochem. Biophys.* **149**:36–41.

Knoppien, P., Pot, W., and v. Delden, W., 1980, Effects of rearing conditions and age on the difference in mating success between alcohol dehydrogenase genotypes of *Drosophila melanogaster*, *Genetica* **51**:197–202.

Kojima, K., and Schaffer, H. E., 1967, Survival process of linked mutant genes, *Evolution* **21**:518–531.

Kojima, K., and Tobari, Y. N., 1969, The pattern of variability changes associated with genotype frequency at the alcohol dehydrogenase locus in a population of *Drosophila melanogaster*, *Genetics* **61**:201–209.

Korotchkin, L. I., Korotchkina, L. S., and Serov, O. L., 1972, Histochemical study of alcohol dehydrogenase in Malpighian tubules of *Drosophila melanogaster* larvae, *Folia Histochem. Cytochem.* **10**:287–292.

Kreitman, M., 1980, Assessment of variability within electromorphs of alcohol dehydrogenase in *Drosophila melanogaster*, *Genetics* **95**:467–475.

Langley, C. H., 1977, Non random associations between allozymes in natural populations of *Drosophila melanogaster* in: *Measuring Selection in Natural Populations* (F. B. Christiansen and T. M. Fenchel, eds.), pp. 265–273, Springer, Berlin, West Germany.

Laurie-Ahlberg, C. C., Maroni, G., Bewley, G. C., Lucchesi, J. C., and Weir, B. S., 1980, Quantitative genetic variation of enzyme activities in natural populations of *Drosophila melanogaster*, *Proc. Natl. Acad. Sci. USA* **77**:1073–1077.

Leibenguth, F., 1977, Selection against homozygotes in the ADH polymorphic system of *Drosophila melanogaster* associated with the lethal factor 1(2)Stm, *Biochem. Genet.* **15**:93–100.

Leibenguth, F., and Steinmetz, H., 1976, Association of the Adh locus with a lethal factor (1(2)Stm) in *Drosophila melanogaster*, *Biochem. Genet.* **14**:299–308.

Leibenguth, F., Rammo, E., and Dubiczky, R., 1979, A comparative study of embryonic gene expression in *Drosophila* and *Ephestia*, *Wilhelm Roux' Arch. Dev. Biol.* **187**:81–88.

Lewis, N., and Gibson, J., 1978, Variation in amount of enzyme protein in natural populations, *Biochem. Genet.* **16**:159–170.

Lewontin, R. C., 1974, *The Genetic Basis of Evolutionary Change*, Columbia University Press, New York.

Lewontin, R. C., and Hubby, J. L., 1966, A molecular approach to the study of genic heterozygosity in natural populations. II. Amount of variation and degree of heterozygosity in natural populations of *Drosophila pseudoobscura*, *Genetics* **54**:595–609.

Libion-Mannaert, M., Delcour, J., Deltombe-Lietart, M. C., Lenelle-Montfort, N., and Elens, A., 1976, Ethanol as a "food" for *Drosophila melanogaster*: Influence of the ebony gene, *Experientia* **32**:22–23.

MacIntyre, R. J., 1972, Studies of enzyme evolution by subunit hybridization, in: *Proceedings of the Sixth Berkeley Symposium on Mathematical Statistics and Probability* (L. M. LeCam, J. Neyman, and E. C. Scott, eds.), Volume V, pp. 129–154, University of California Press, Berkeley.

Madhavan, K., Conscience-Egli, M., Sieber, F., and Ursprung, H., 1973, Farnesol metabolism in *Drosophila melanogaster*: Ontogeny and tissue distribution of octanol dehydrogenase and aldehyde oxidase, *J. Insect Physiol.* **19**:235–241.

Malpica, J. M., and Vassallo, J. M., 1980, A test for the selective origin of environmentally correlated allozyme patterns, *Nature* **286**:407–408.

Marks, R. W., Brittnacher, J. G., McDonald, J. F., Prout, T., and Ayala, F. J., 1980, Wineries, *Drosophila*, alcohol and ADH, *Oecologia* **47**:141–144.

Maroni, C., 1978, Genetic control of alcohol dehydrogenase levels in *Drosophila*, *Biochem. Genet.* **16**:509–523.

Maynard Smith, J., and Haigh, J., 1974, The hitch-hiking effect of a favourable gene, *Genet. Res.* **23**:23–35.

McDonald, J. F., and Avise, J. C., 1976, Evidence for the adaptive signifiance of enzyme activity levels: Interspecific variation in α-GPDH and ADH in *Drosophila*, *Biochem. Genet.* **14**:347–355.

McDonald, J. F., and Ayala, F. J., 1978, Genetic and biochemical basis of enzyme variation in natural populations. I. Alcohol dehydrogenase in *Drosophila melanogaster*, *Genetics* **89**:371–388.

McDonald, J. F., Chambers, G. K., David, J., and Ayala, F. J., 1977, Adaptive response due to changes in gene regulation: A study with *Drosophila*, *Proc. Natl. Acad. Sci. USA* **74**:4562–4566.

McDonald, J. F., Anderson, S. M., and Santos, M., 1980, Biochemical differences between products of the Adh locus in *Drosophila, Genetics* **95**:1013–1022.

McKay, J., 1981, Variation in activity and thermostability of alcohol dehydrogenase in *Drosophila melanogaster, Genet. Res.* **37**:227–237.

McKechnie, S. W., Kohane, M., and Phillips, S. C., 1981, A search for interacting polymorphic enzyme loci in *Drosophila melanogaster,* in: *Genetic Studies of Drosophila Populations* (J. B. Gibson and J. G. Oakeshott, eds.), pp. 121–138, Australian National University Press, Canberra, Australia.

McKenzie, J. A., 1975, Gene flow and selection in a natural population of *Drosophila melanogaster, Genetics* **80**:349–361.

McKenzie, J. A., 1980, An ecological study of the alcohol dehydrogenase (Adh) polymorphism of *Drosophila melanogaster, Aust. J. Zool.* **28**:709–716.

McKenzie, J. A., and Fegent, J. C., 1980, Mating patterns of virgin and inseminated *Drosophila melanogaster* of different alcohol dehydrogenase (ADH) genotypes, *Experientia* **36**:1160–1161.

McKenzie, J. A., and McKechnie, S. W., 1978, Ethanol tolerance and the Adh polymorphism in a natural population of *Drosophila melanogaster, Nature* **272**:75–76.

McKenzie, J. A., and McKechnie, S. W., 1979, A comparative study of resource utilization in natural populations of *Drosophila melanogaster* and *Drosophila simulans, Oecologia* **40**:299–309.

McKenzie, J. A., and McKechnie, S. W., 1981, The alcohol dehydrogenase polymorphism in a vineyard cellar population of *Drosophila melanogaster,* in: *Genetic Studies of Drosphila Populations* (J. B. Gibson and J. G. Oakeshott, eds.), pp. 201–215, Australian National University Press, Canberra, Australia.

McKenzie, J. A., and Parsons, P. A., 1972, Alcohol tolerance: An ecological parameter in the relative success of *Drosophila melanogaster* and *Drosophila simulans, Oecologia* **10**:373–388.

McKenzie, J. A., and Parsons, P. A., 1974, Microdifferentiation in a natural population of *Drosophila melanogaster* to alcohol in the environment, *Genetics* **77**:385–394.

Milkman, R., 1976, Further evidence of thermostability variation within electrophoretic mobility classes of enzymes, *Biochem. Genet.* **14**:383–387.

Morgan, P., 1975, Selection acting directly on an enzyme polymorphism, *Heredity* **34**:124–127.

Morgan, P., 1976, Frequency-dependent selection at two enzyme loci in *Drosophila melanogaster, Nature* **263**:765–766.

Nevo, E., 1978, Genetic variation in natural populations: Patterns and theory, *Theor. Popul. Biol.* **13**:121–177.

Oakeshott, J. G., 1976a, Selection at the alcohol dehydrogenase locus in *Drosophila melanogaster* imposed by environmental ethanol, *Genet. Res.* **26**:265–274.

Oakeshott, J. G., 1976b, Biochemical differences between alcohol dehydrogenases of *Drosophila melanogaster, Aust. J. Biol. Sci.* **29**:365–373.

Oakeshott, J. G., 1977, Variation in the direction of selection applied by pentanol to the alcohol dehydrogenase locus in *Drosophila melanogaster, Austr. J. Biol. Sci.* **30**:259–267.

Oakeshott, J. G., 1979, Selection affecting enzyme polymorphisms in laboratory populations of *Drosophila melanogaster, Oecologia* **143**:341–354.

Oakeshott, J. G., and Gibson, J. B., 1981, Is there selection by environmental ethanol on the alcohol dehydrogenase locus in *Drosophila melanogaster*? in: *Genetic Studies of Drosophila Populations* (J. B. Gibson and J. G. Oakeshott, eds.), pp. 103–120, Australian National University Press, Canberra, Australia.

Oakeshott, J. G., Gibson, J. B., Anderson, P. R., and Champ, A., 1980, Opposing modes of selection on the alcohol dehydrogenase locus in *Drosophila melanogaster, Aust. J. Biol. Sci.* **33**:105–114.

O'Brien, S. J., and MacIntyre, R., 1978, Genetics and biochemistry of enzymes and specific proteins of *Drosophila*, in: *The Genetics and Biology of Drosophila* (M. Ashburner and T. R. F. Wright, eds.), Volume 2a, pp. 395–551, Academic Press, London.

O'Donnell, J., Gerace, L., Leister, F., and Sofer, W., 1975, Chemical selection of mutants that affect alcohol dehydrogenase in *Drosophila*. II. Use of 1-pentyne-3-ol, *Genetics* **79**:73–83.

O'Donnell, J., Mandel, H. C., Krauss, M., and Sofer, W., 1977, Genetic and cytogenetic analysis of the Adh region in *Drosophila melanogaster, Genetics* **86**:553–566.

Papel, J., Henderson, M., v. Herrewege, J., David, J., and Sofer, W., 1979, *Drosophila* alcohol dehydrogenase activity *in vitro* and *in vivo*: Effects of acetone feeding, *Biochem. Genet.* **17**:553–563.

Parsons, P. A., 1977, Larval reaction to alcohol as an indicator of resource utilization differences between *Drosophila melanogaster* and *Drosophila simulans, Oecologia* **30**:141–146.

Parsons, P. A., 1980a, Responses of *Drosophila* to environmental ethanol from ecologically optimal and extreme habitats, *Experientia* **36**:1070–1071.

Parsons, P. A., 1980b, Adaptive strategies in natural populations of *Drosophila, Theor. Appl. Genet.* **57**:257–266.

Parsons, P. A., 1980c, Ethanol utilization: Threshold differences among six closely related species of *Drosophila, Aust. J. Zool.* **28**:535–542.

Parsons, P. A., and King, S. B., 1977, Ethanol: Larval discrimination between two *Drosophila* sibling species, *Experientia* **33**:898–899.

Parsons, P. A.., and Spence, G. E., 1981, Ethanol utilization: Threshold differences among six *Drosophila* species, *Am. Nat.* **117**:568–571.

Parsons, P. A., and Stanley, S. M., 1981, Comparative effects of environmental ethanol on *Drosophila melanogaster* and *Drosophila simulans* adults including geographic differences in *Drosophila melanogaster*, in: *Genetic Studies of Drosophila Populations* (J. B. Gibson and J. G. Oakeshott, eds.), pp. 47–57, Australian National University Press, Canberra, Australia.

Pieragostini, E., Sangiorgi, S., and Cavicchi, S., 1979, ADH system and genetic background: Interaction with wing length in *Drosophila melanogaster, Genetica* **50**:201–206.

Pipkin, S. B., and Hewitt, N. E., 1972a, Variation of alcohol dehydrogenase levels in *Drosophila* species hybrids, *J. Hered.* **63**:267–270.

Pipkin, S. B., and Hewitt, N. E., 1972b, Effect of gene dosage on level of alcohol dehydrogenase in *Drosophila, J. Hered.* **63**:335–336.

Pipkin, S. B., Rhodes, C., and Williams, N., 1973, Influence of temperature on *Drosophila* alcohol dehydrogenase polymorphism, *J. Hered.* **64**:181–185.

Pipkin, S. B., Potter, J. H., Lubega, S., and Springer, E., 1975, Further studies on alcohol dehydrogenase polymorphism in Mexican strains of *Drosophila melanogaster*, in: *Isozymes IV* (C. L. Markert, ed.), pp. 547–561, Academic Press, New York.

Pipkin, S. B., Franklin-Springer, E., Law, S., and Lubega, S., 1976, New studies of the alcohol dehydrogenase cline in *Drosophila melanogaster* from Mexico, *J. Hered.* **67**:258–266.

Pot, W., v. Delden, W., and Kruijt, J. P., 1980, Genotypic differences in mating success and the maintenance of the alcohol dehydrogenase polymorphism in *Drosophila melanogaster*: No evidence for overdominance or rare genotype mating advantage, *Behav. Genet.* **10**:43–58.

Powell, J. R., 1975, Protein variation in natural populations of animals, *Evol. Biol.* **8**:79–119.

Prout, T., 1971a, The relation between fitness components and population prediction in *Drosophila*. I. The estimation of fitness components, *Genetics* **68**:127–149.

Prout, T., 1971b, The relation between fitness components and population prediction in *Drosophila*. II. Population prediction, *Genetics* **68**:151–167.

Rasmuson, B., Nilson, L. R., Rasmuson, M., and Zeppezauer, E., 1966, Effects of heterozygosity on alcohol dehydrogenase (ADH) activity in *Drosophila melanogaster*, *Hereditas* **56**:313–316.

Reddy, A. R., Pelliccia, J. G., and Sofer, W., 1980, Adh-negative mutants: Detection of an altered tryptic peptide in a mutant enzyme of *Drosophila*, *Biochem. Genet.* **18**:339–351.

Retzios, A. D., and Thatcher, D. R., 1979, Chemical basis of the electrophoretic variation observed at the alcohol dehydrogenase locus of *Drosophila melanogaster*, *Biochimie* **61**:701–704.

Richmond, R. C., and Gerking, J. L., 1979, Oviposition site preference in *Drosophila*, *Behav. Genet.* **9**:233–241.

Sampsell, B., 1977, Isolation and genetic characterization of alcohol dehydrogenase thermostability variants occurring in natural populations of *Drosophila melanogaster*, *Biochem. Genet.* **15**:971–988.

Sampsell, B. M., and Milkman, R., 1978, Van der Waals bonds as unit factors in allozyme thermostability variation, *Biochem. Genet.* **16**:1139–1141.

Schwartz, M. F., and Jörnvall, H., 1976, Structural analysis of mutant and wild-type alcohol dehydrogenase from *Drosophila melanogaster*, *Eur. J. Biochem.* **68**:159–168.

Schwartz, M., and Sofer, W., 1976a. Alcohol dehydrogenase-negative mutants in *Drosophila*: Defects at the structural locus? *Genetics* **83**:125–136.

Schwartz, M., and Sofer, W., 1976b, Diet-induced alterations in distribution of multiple forms of alcohol dehydrogenase in *Drosophila*, *Nature* **263**:129–131.

Schwartz, M., Gerace, L., O'Donnell, J., and Sofer, W., 1975, *Drosophila* alcohol dehydrogenase: Origin of the multiple forms, in: *Isozymes I* (C. L. Markert, ed.), pp. 725–751, Academic Press, New York.

Schwartz, M., O'Donnell, M. J., and Sofer, W., 1979, Origin of multiple forms of alcohol dehydrogenase from *Drosophila melanogaster*, *Arch. Biochem. Biophys.* **194**:365–378.

Singh, R. S., Lewontin, R. C., and Felton, A. A., 1976, Genetic heterogeneity within electrophoretic "alleles" of xanthine dehydrogenase in *Drosophila pseudoobscura*, *Genetics* **84**:609–629.

Sofer, W. H., and Hatkoff, M. A., 1972, Chemical selection of alcohol dehydrogenase negative mutants in *Drosophila*, *Genetics* **72**:545–549.

Sofer, W., and Ursprung, H. J., 1968, *Drosophila* alcohol dehydrogenase, *Biol. Chem.* **243**:3110–3115.

Soliman, M. H., and Knight, M. L., 1981, Behavioural responses to ethanol and octanol by alcohol dehydrogenase genotypes of *Drosophila melanogaster*, in: *Genetic Studies of Drosophila Populations* (J. B. Gibson and J. G. Oakeshott, eds.), pp. 95–102, Australian National University Press, Canberra, Australia.

Starmer, W. T., Heed, W. B., and Rockwood-Sluss, E. S., 1977, Extension of longevity in *Drosophila mojavensis* by environmental ethanol: Differences between subraces, *Proc. Natl. Acad. Sci. USA* **74**:387–391.

Thatcher, D. R., 1977, Enzyme instability and proteolysis during purification of an alcohol dehydrogenase from *Drosophila melanogaster*, *Biochem. J.* **163**:317–323.

Thatcher, D. R., 1980, The complete amino acid sequence of three alcohol dehydrogenase alleloenzymes (Adh[N-11], Adh[S] and Adh[UF]) from the fruitfly *Drosophila melanogaster*, *Biochem. J.* **187**:875–886.

Thatcher, D. R., and Camfield, R., 1977, Chemical basis of the electrophoretic variation between two naturally occurring alcohol dehydrogenase alloenzymes from *Drosophila melanogaster, Biochem. Soc. Trans.* **5**:271–272.

Thatcher, D. R., and Retzios, A., 1980, Mutations affecting the structure of the alcohol dehydrogenase from *Drosophila melanogaster, Proteins and Related Subjects* **28**:157–160.

Thompson, J. N., and Kaiser, T. N., 1977, Selection acting upon slow-migrating Adh alleles differing in enzyme activity, *Heredity* **38**:191–195.

Thompson, J. N., Ashburner, M., and Woodruff, R. C., 1977, Presumptive control mutation for alcohol dehydrogenase in *Drosophila melanogaster, Nature* **270**:363.

Thörig, G. E. W., Schoone, A. A., and Scharloo, W., 1975, Variation between electrophoretically identical alleles at the alcohol dehydrogenase locus in *Drosophila melanogaster, Biochem. Genet.* **13**:721–731.

Ursprung, H., and Carlin, L., 1968, *Drosophila* alcohol dehydrogenase: *In vitro* changes of isozyme patterns, *Ann. N.Y. Acad. Sci.* **151**:456–475.

Ursprung, H., and Leone, J., 1965, Alcohol dehydrogenase: A polymorphism in *Drosophila melanogaster, J. Exp. Zool.* **160**:147–154.

Ursprung, H., Sofer, W. H., and Burroughs, N., 1970, Ontogeny and tissue distribution of alcohol dehydrogenase in *Drosophila melanogaster, Wilhelm Roux' Arch.* **164**:201–208.

Van Delden, W., and Kamping, A., 1976, Polymorphism at the Adh and Odh loci in *Drosophila melanogaster,* in: *Proc. 5th Eur. Drosophila Res. Conf.,* Louvain-la-Neuve, Belgium.

Van Delden, W., and Kamping, A., 1979a, Selection against Adh null mutants in *Drosophila melanogaster,* in: *Proc. 6th Eur Drosophila Res. Conf.,* Kupari, Yugoslavia.

Van Delden, W., and Kamping, A., 1979b, The alcohol dehydrogenase polymorphism in populations of *Drosophila melanogaster.* III. Differences in developmental times, *Genet. Res.* **33**:15–27.

Van Delden, W., and Kamping, A., 1980, The alcohol dehydrogenase polymorphism of *Drosophila melanogaster.* IV. Survival at high temperature, *Genetica* **51**:179–185.

Van Delden, W., and Kamping, A., 1981, Selection against an Adh null allele, *Drosophila Information Service* **56**:149–150.

Van Delden, W., and Kamping, A., 1982, Adaptation to alcohols in relation to the alcohol dehydrogenase locus in *Drosophila melanogaster* (in press).

Van Delden, W., Kamping, A., and van Dijk, H., 1975, Selection at the alcohol dehydrogenase locus in *Drosophila melanogaster, Experientia* **31**:418–419.

Van Delden, W., Boerema, A. C., and Kamping, A., 1978, The alcohol dehydrogenase polymorphism in populations of *Drosophila melanogaster.* I. Selection in different environments, *Genetics* **90**:161–191.

Van Dijk, H., 1981, The relationship between ADH activity and body weight in *Drosophila melanogaster, Drosophila Information Service* **56**:150–151.

Van Herrewege, J., and David, J., 1974, Utilisation de l'alcool éthylique dans le métabolisme énergétique d'un insecte: influence sur la durée de survie des adultes de *Drosophila melanogaster, C. R. Acad. Sci. Paris* **279**:335–338.

Van Herrewege, J., and David, J. R., 1978, Feeding an insect through its respiration: Assimilation of alcohol vapors by *Drosophila melanogaster* adults, *Experientia* **34**:163–164.

Van Herrewege, J., and David, J. R., 1980, Alcohol tolerance and alcohol utilization in *Drosophila*: Partial independence of two adaptive traits, *Heredity* **44**:229–235.

Van Herrewege, J., David, J. R., and Grantham, R., 1980, Dietary utilization of aliphatic alcohols by *Drosophila, Experientia* **36**:846–847.

Vigue, C. L., and Johnson, F. M., 1973, Isozyme variability in species of the genus *Drosophila*. VI. Frequency-property-environment relationships of allelic alcohol dehydrogenases in *Drosophila melanogaster*, *Biochem. Genet.* **9**:213–227.

Vigue, C., and Sofer, W., 1974, Adh[n5]; A temperature-sensitive mutant at the Adh locus in *Drosophila*, *Biochem. Genet.* **11**:387–396.

Vigue, C., and Sofer, W., 1976, Chemical selection of mutants that affect ADH activity in *Drosophila*. III. Effects of ethanol, *Biochem. Genet.* **14**:127–135.

Voelker, R. A., Langley, C. H., Leigh Brown, A. J., Onishi, S., Dickson, B., Montgomery, E., and Smith, S. C., 1980, Enzyme null alleles in natural populations of *Drosophila melanogaster*: Frequencies in a North Carolina population, *Proc. Natl. Acad. Sci. USA* **77**:1091–1095.

Ward, R. D., 1974, Alcohol dehydrogenase in *Drosophila melanogaster*. Activity variation in natural populations, *Biochem. Genet.* **12**:449–458.

Ward, R. D., 1975, Alcohol dehydrogenase activity in *Drosophila melanogaster*: A quantitative character, *Genet. Res.* **26**:81–93.

Ward, R. D., and Hebert, P. D. N., 1972, Variability of alcohol dehydrogenase activity in a natural population of *Drosophila melanogaster*, *Nature* **236**:243–244.

Wilks, A. V., Gibson, J. B., Oakeshott, J. G., and Chambers, G. K., 1980, An electrophoretically cryptic alcohol dehydrogenase variant in *Drosophila melanogaster*. II. Post-electrophoresis heat treatment screening of natural populations, *Aust. J. Biol. Sci.* **33**:575–585.

Wilson, P. G., and McDonald, J. F., 1981, A comparative study of enzyme activity variation between α glycerophosphate and alcohol dehydrogenases in *Drosophila melanogaster*, *Genetica* **55**:75–79.

Woodruff, R. C., and Ashburner, M., 1979a, The genetics of a small autosomal region of *Drosophila melanogaster* containing the structural gene for alcohol dehydrogenase. I. Characterization of deficiencies and mapping of Adh and visible mutations, *Genetics* **92**:117–132.

Woodruff, R. C., and Ashburner, M., 1979b, The genetics of a small autosomal region of *Drosophila melanogaster* containing the structural gene for alcohol dehydrogenase. II. Lethal mutations in the region. *Genetics* **92**:133–149.

Wright, D. A., and Shaw, C. R., 1970, Time of expression of genes controlling specific enzymes in *Drosophila* embryos, *Biochem. Genet.* **4**:385–394.

Yamazaki, T., 1971, Measurement of fitness at the esterase-5 locus in *Drosophila pseudoobscura*, *Genetics* **67**:579–603.

Yoshimaru, H., and Mukai, T., 1979, Lack of experimental evidence for frequency-dependent selection at the alcohol dehydrogenase locus in *Drosophila melanogaster*, *Proc. Natl. Acad. Sci. USA* **79**:876–878.

5

Developmental Changes in the Orientation of the Anuran Jaw Suspension

A Preliminary Exploration into the Evolution of Anuran Metamorphosis

RICHARD J. WASSERSUG

and

KARIN HOFF

Department of Anatomy
Dalhousie University
Halifax, Nova Scotia, Canada B3H 4H7

INTRODUCTION

Many major taxa have evolved complex life cycles with distinct larval and adult forms separated by a rapid transformation, or metamorphosis. Holometabolous insects and anurans are perhaps the best known examples. Much of the attention given the process of metamorphosis by developmental biologists is stimulated by the spectacular amount of morphologic change that occurs in these organisms [see Gilbert and Frieden (1981) for a recent review of metamorphosis from a developmental biology perspective]. In frogs this transformation seems canalized in that for species with a free-living larva, metamorphosis is obligatory; i.e., there are no known anurans that can breed as tadpoles. Wassersug (1975) and Wilbur (1980) have independently suggested that the anuran larva is spe-

223

cifically adapted for using temporary resources in fluctuating environments. Such habitats may be very rich in nutrients, but only briefly.
Organisms such as tadpoles and many insect larvae that live in these
habitats, may grow very rapidly, but their accelerated growth rate requires
a massive commitment to the organ systems responsible for ingestion
and digestion. This commitment is at the expense of the differentiation
of many other organ systems, such as those associated with defense,
locomotion, and, most importantly, reproduction. In order to migrate to
new resources and to breed, an extensive transformation of body form
is necessary. How did the extreme metamorphosis of the Anura evolve?
Can we examine the evolutionary history of this ontogenetic process
proper?

Entomologists have long perceived a general pattern for the evolution
of the metamorphic process (Sharov, 1957, 1966; Kukalova-Peck, 1978).
The generalized or primitive condition is seen in hemimetabolous insects,
which show gradual progress through instar stages. The "advanced metamorphosis" occurs in holometabolous insects, which have a much more
abrupt transition from larva to adult. Data in Brown (1977) show that in
both intensity (defined as the morphologic distance between larva and
adult) and abruptness (number of instars showing morphologic transformation from the previous stage), metamorphosis is greater in holometabolous than hemimetabolous insects. A similar situation may hold for the
evolution of metamorphosis in the Anura. Nopsca (1930), in fact, asserted
that the morphologic divergence observed between larval and adult amphibians has increased in the course of evolution in a way analogous to
that observed in insects. In the following study we explore the idea that
evolution has increased both the intensity and abruptness of anuran metamorphosis.

Aspects of this idea have been explored by Pusey (1943), Orton
(1944), Stephenson (1950), Szarski (1957), and de Jongh (1968). Their
work implies that primitive anurans should have less difference between
larvae and adults than more derived anurans. This hypothesis has not
been tested.

The purpose of this paper is to determine if one can distinguish
between primitive and advanced metamorphosis within the Anura. Other
amphibians can serve as a reference out-group. The primitive condition
is taken as one in which there is little difference between larvae and
adults. Indeed, compared to anurans, living urodeles change little with
metamorphosis (Orton, 1953; Regel, 1963). Fossil urodeles (Carroll and
Holmes, 1980; Ivakhnenko, 1978) and labyrinthodonts (Boy, 1974), likewise had larvae that differed little from their adults compared to fossil or
living anurans.

In this study we quantify the change associated with metamorphosis of the jaw suspension for 38 frog species. The species used were selected to represent as many genera as possible; particular effort was taken to examine those genera considered archaic by systematic herpetologists. We have accepted a taxonomy that recognizes "archaic" (Archaeobatrachia) and "advanced" (Neobatrachia) grades within the Anura (Lynch, 1973; Duellman, 1975). These grades as used here can be defined exclusively on adult characters. The discoglossoid (Ascaphidae, Leiopelmatidae, Discoglossidae) and pipoid (Pipidae and Rhinophrynidae) superfamilies make up the Archaeobatrachia (Lynch, 1973). The majority of living anuran families and species are in the Neobatrachia. Some version of this Archaeobatrachia–Neobatrachia distinction is recognized by most herpetologists, although there is admittedly disagreement about the terminology for, and membership in, anuran suprafamilial taxa.* It is important to emphasize here that our study is not directed toward reworking anuran classification; rather, our goal is to explore the evolution of an ontogeny—to suggest that the anuran metamorphic process itself has evolved.

For this proposed comparison of metamorphosis we must examine, in developmental series, features that are unique to tadpoles, and that change with metamorphosis. Many such features can be seen in the tadpole suspension-feeding and respiratory mechanism. In tadpoles, feeding and respiratory currents are pulled into the mouth by depression of the medial portion of the ceratohyal (Fig. 1) (de Jongh, 1968; Kenny, 1969; Gradwell, 1972). This depression is accomplished by contraction of muscles that run from the jaw suspension (the palatoquadrate) to the ceratohyal. The ceratohyal rotates about a ceratohyal–palatoquadrate articulation. Elongation of the anterior end of the palatoquadrate allows the ceratohyal to move up and down rather than back and forth. Severtzov (1969) credited this vertical motion of the ceratohyal as the outstanding feature that distinguished anuran larvae from other Amphibia. The horizontal orientation of the palatoquadrate is fundamental to this mechanism. This type of jaw suspension can be understood as an adaptation

---

* In the past decade the Pelobatidae have been classified alternatively as primitive (Duellman, 1975; Dowling and Duellman, 1978), advanced (Porter, 1972; Sokol, 1977b), and transitional (Lynch, 1973; Goin et al., 1978; Laurent, 1979). Although we take the traditional view that pipoid frogs are archaic, Laurent (1979) separates them into an intermediate grade and Sokol (1977b) considers them advanced (i.e., a specialized offshoot of a ranoidean rather than discoglossoidean lineage). Morescalchi (1980) has presented some karyological data in support of Sokol's taxonomy. Most recent workers consider the Microhylidae as advanced (Sokol, 1975; Duellman 1975; Dowling and Duellman, 1978; Morescalchi, 1980; Wassersug, 1980), although several workers follow Starrett (1973) in consigning them to a more primitive grade.

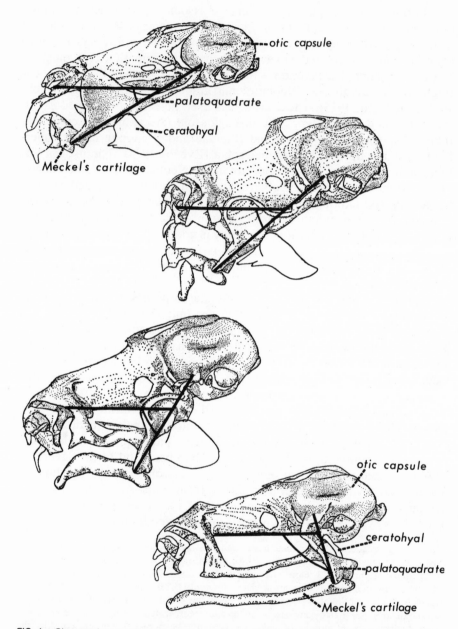

FIG. 1.   Changes in suspensorial angle in *Rana temporaria* through metamorphosis. The developmental progression is from the tadpole at the top to the frog at the bottom. This figure indicates how this angle was measured in the present study. Redrawn from Pusey (1938), with the author's permission.

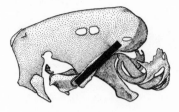

Ascaphus truei

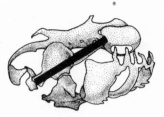

Bombina orientalis

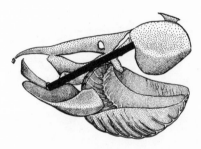

Rhinophrynus dorsalis

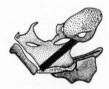

Hymenochirus boettgeri

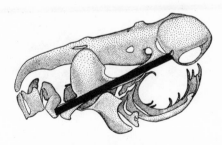

Alytes obstetricans

Xenopus laevis

FIG. 2. The chondrocrania of representative archaeobatrachian anuran larvae used in this study. Heavy lines mark the long axis of the palatoquadrates. Note that the palatoquadrates are elongated and anteriorly directed in all species, despite the great diversity of larval chondrocrania. See also Fig. 3.

Gastrophryne carolinensis

Pleurodema borelli

Heleophryne natalensis

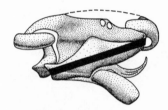

Anotheca spinosa

Scaphiopus bombifrons

Bufo americanus

FIG. 3.   The chondrocrania of representative neobatrachian anuran larvae plus one microhylid, *Gastrophryne carolinensis,* and one pelobatid, *Scaphiophus bombifrons,* used in this study. Heavy lines mark the long axis of the palatoquadrates. Note the elongated and anteriorly directed palatoquadrate bars. Larval suspensorial angles are generally smaller for these forms than those in Fig. 2. See also Table I.

for increasing the volume of the ceratohyal pump and the branchial (filter) baskets in the buccopharynx of the tadpole (Wassersug and Hoff, 1979). The horizontal suspensorium in tadpoles is a key feature in an elaborate functional complex and, as such, is likely to have a single evolutionary origin.

The uniqueness of the tadpole suspensorium has been appreciated for over 100 years (Goette, 1875; Gaupp, 1893, 1894, 1898; Parker, 1871, 1876, 1881) and the larval chondrocrania of over two dozen species are now described and illustrated. Larval chondrocrania from most of the genera of concern to this study have been described [see De Beer (1937) and de Jongh (1968) for reviews of this literature], and all show a relatively horizontally oriented palatoquadrate (Figs. 2 and 3) when compared with the palatoquadrate of salamanders (Carroll and Holmes, 1980) or adult frogs.

The palatoquadrate in anurans reorients extensively during metamorphosis (de Jongh, 1968; Edgeworth, 1935; van Eeden, 1951; Gaupp, 1898; Kotthaus, 1933; Litzelmann, 1923; Okutomi, 1937; Parker, 1876, 1881; Pusey, 1938; Kraemer, 1974; Reinbach, 1939; Ridewood, 1897, 1898a,b,c; Sedra, 1950; Sedra and Michael, 1957, 1958; Stephenson, 1951; Takisawa and Sunaga, 1951; Takisawa et al., 1952a,b; Weisz, 1945a,b). It shortens relative to the neurocranium, an assumes a more vertical or posteriorly directed position (Fig. 1); the farther back the articulation of the palatoquadrate with Meckel's cartilage, the larger the adult gape. A large gape may be a specialization in the adult for eating large prey (directly or indirectly associated with the tongue protrusion mechanism) (Nopsca, 1930) and a large buccal floor area may be important for respiration (de Jongh and Gans, 1969a,b) or resonance during vocalization.

In this study we show that there is an inverse relationship between the palatoquadrate orientation, relative to the base of the braincase, in larval anurans and orientation in the adult. In addition, the anuran families recognized by others as archaic have the least change in palatoquadrate orientation with metamorphosis. This leads to some speculation on the feeding adaptations of archaic frog larvae and the morphology of the ancestral anuran.

MATERIALS AND METHODS

Data were taken from larvae that were cleared and stained by a modification of the Simons and Van Horn (1971) technique (Wassersug,

1976b), and postmetamorphic individuals that were either cleared and stained, partially dissected, skeletonized, or x-rayed.

The orientation of the palatoquadrate was measured in lateral view as the angle between the floor of the neurocranium and the long axis of the palatoquadrate, usually defined as a line between the articulations of the palatoquadrate with Meckel's cartilage and with the otic capsule* (see Fig. 1). Palatoquadrate orientation was measured with a protractor in large specimens, and with an ocular goniometer on a Zeiss dissecting microscope for smaller specimens (including those x-rayed). One series of five *Ambystoma tigrinum* was measured for comparison as an outgroup.

Interspecific comparison of the metamorphic process requires that the larvae and adults of the different species compared be at the same stage of development. Larvae were staged according to Gosner (1960) and only individuals in stages 36 and 37 were used. The Gosner staging table uses hind limb development to identify stages. Thus, a tacit assumption of our study is that the growth rate and differentiation of the tadpole hindlimbs is not affected by metamorphosis; in other words, although a tadpole has hindlimbs, these limbs function only for the adult and are not a larval adaptation. Observations of swimming tadpoles at stages 36 and 37 (Wassersug and Sperry, 1977; Wassersug and Feder, 1979; Feder and Wassersug, 1979) support this assertion.

A second assumption is that stages 36 and 37 are "mature" premetamorphic anurans. The idea of a "mature" tadpole morphology has been discussed and verified for other features of the tadpole feeding apparatus by Nichols (1937) and Wassersug (1976a,c). We measured as many specimens as possible for several species at stages 36 and 37 and could discern no differences in the orientation of the palatoquadrate between these two stages. In many cases we only had one or two specimens;

---

* Because the palatoquadrate is anchored to the chondrocranium anteriorly (by a *commissura quadrato-cranialis anterior*) and is thus akinetic, this angle is not altered by jaw movements. The angle we measured should not be confused with the squamosal angle measured by Griffiths (1954) and shown to have little phylogenetic significance by Starrett (1967). In certain species an otic process does not form until metamorphosis and for these larvae the closest approximation of the palatoquadrate to the otic capsule was taken as its cranial termination. Most workers of this century have considered the common larval and adult otic processes homologous. But as recently as 1951 Eeden wrote "except for the pars articularis palatoquadrati, the articulation with Meckel's cartilage, there is no single process of the anuran palatoquadrate concerning the morphology of which there is general agreement" (Eeden, 1951). Close to 100 pages have been published in the last 100 years on the homologies of various processes of the tadpole palatoquadrate. For the sake of our study of metamorphosis we take a robust view of the palatoquadrate as a simple single bar.

however, the largest range of variation within a single species (*Pseudacris triseriata*) was only 3° ($N = 13$).

Comparable concerns of maturity and character stability apply to adults. Only postmetamorphic individuals that had ossification of the squamosal, the dermal bone overlying the adult palatoquadrate, were considered adult.

RESULTS

While there is much inter- and intrafamilial variation in suspensorial angles, small angles in larvae tend to be associated with large angles in adults (Table I). This pattern is most evident when one looks at family means. Because of small samples within families and the nonrandom nature of these samples, statistical analyses of the data should be interpreted with caution. Spearman's rank correlation test does show a weak negative correlation (38 anuran species; $r = -0.22$, $0.05 < p < 0.1$) be-

TABLE I. Orientation of the Suspensorium in Larval and Adult Amphibians[a]

	Angle of suspensorium		
	Larva	Adult	Change with metamorphosis
Urodeles			
Ambystomatidae			
Ambystoma tigrinum	59	76	17
Anura			
Rhinophrynidae			
Rhinophrynus dorsalis	28	81	53
Ascaphidae			
Ascaphus truei	35	99	64
Pipidae			
Hymenochirus boettgeri	37	89	52
Pipa parva	35	95	60
Xenopus laevis	23	117	94
Mean	32	100	69
Microhylidae			
Gastrophryne carolinensis	23	107	84
Glyphoglossus molossus	31	67	36
Hypopachus barberi	24	100	76
Kaloula picta	28	107	79
Microhyla heymonsi	27	102	75
Mean	27	96	70

(*continued*)

TABLE I. (*Continued*)

	Angle of suspensorium		
	Larva	Adult	Change with metamorphosis
Discoglossidae			
Alytes cisternasii	29	107	78
Alytes obstetricans	29	100	71
Bombina orientalis	32	109	77
Discoglossus pictus	35	110	75
Mean	31	107	75
Bufonidae			
Bufo americanus	27	104	77
Pelobatidae			
Pelobates cultripes	24	97	73
Scaphiopus bombifrons	21	110	89
Scaphiopus holbrooki	25	96	71
Mean	23	101	78
Hylidae			
Anotheca spinosa	26	128	102
Hyla rubra	28	100	72
Hyla versicolor/chrysoscelis	20	100	80
Osteopilus septentrionalis	26	113	87
Pachymedusa dachnicolor	19	95	76
Pseudacris triseriata	22	103	81
Mean	24	107	83
Dendrobatidae			
Colostethus nubicola	22	110	88
Ranidae			
Amolops jerboa	23	96	73
Rana everetti	22	113	91
Rana magna	27	121	94
Rana pipiens	25	110	85
Ooeidozyga laevis	27	127	100
Staurois latopalmatus	23	108	85
Mean	25	113	89
Rhacophoridae			
Rhacophorus leucomystax	23	112	89
Leptodactylidae			
Heleophryne purcelli	25	122	97
Leptodactylus wagneri	24	135	111
Limnodynastes tasmaniensis	23	103	80
Pleurodema borelli	24	106	76
Mean	25	117	90
Centrolenidae			
Centrolenella fleischmanni	21	114	93
Hyperoliidae			
Hyperolius viridiflavus fernique	25	118	93

[a] Families ranked according to the mean difference between larval and adult measurements. Slight discrepancies between column means and row differences reflect rounding error.

tween the angle of the suspensorium in adults and in larvae, i.e., the smaller this angle in the larva, the larger the angle in the adult.

If the evolution of metamorphosis parallels the evolution of the anurans, then the most archaic frogs, as indicated by adult morphology, should show the least specialization in terms of the amount of change associated with metamorphosis, i.e., the most archaic anurans should resemble salamanders and other groups of amphibians in this regard. The data in Table I and Fig. 4 are ranked according to the mean family difference between larval and adult angles of the suspension. The difference between the larval and the adult angle is smallest for the salamander *Ambystoma tigrinum* (17°), and data taken from illustrations in Regel (1963) also show very little change (27°) in suspensorial angle during development for the hynobiid salamander *Hynobius keyserlingii*. The anuran families with the least difference are the three archaeobatrachian

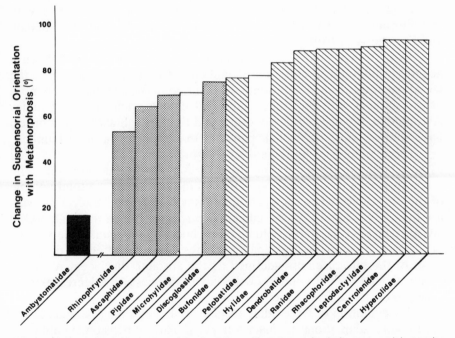

FIG. 4.   Mean changes in suspensorial orientation with metamorphosis for one urodele species and 14 anuran families (see Table I). The ambystomatid salamander *Ambystoma* (black) shows far less change than any anuran family. Among the anurans, the Archaeobatrachians (stippled) show the least change, while the Neobatrachians (diagonal lines) show the greatest change. Because of controversies on the taxonomic grade of the Microhylidae and Pelobatidae, they are assigned to neither grade here. They are, however, intermediate between the Archaeobatrachia and the Neobatrachia (see text).

families Rhinophrynidae, Ascaphidae, and Pipidae. The ones with the
greatest difference between larvae and adults are the neobatrachian families Hyperoliidae, Centrolenidae, and Leptodactylidae (Fig. 4). The mean
amount of change with metamorphosis for the archaeobatrachian species
examined (excluding the controversial Pelobatidae and Microhylidae) was
69°. The mean amount of change of the neobatrachian species was 87°.
These means are significantly different at the $p < 0.01$ level ($T = 63.5$,
Wilcoxon's two-sample rank test). In general, families traditionally recognized as archaic or advanced have a metamorphic pattern that reflects
their evolutionary grade.

DISCUSSION

Our data suggest that archaic frog families share, as a primitive feature, relatively little change at metamorphosis compared to the more
advanced anurans. *Ascaphus* and *Rhinophrynus* change the least with
metamorphosis and may be the two most primitive living anurans (Maxson
and Daugherty, 1980).

As first suggested by Pusey (1938), there appears to have been a
general trend in the evolution of extant anurans toward elongation of the
larval palatoquadrate and reorientation of that palatoquadrate to be more
nearly parallel with the trabecular cartilages. Tadpoles of the families
Ascaphidae, Pipidae, Discoglossidae, and Rhinophrynidae (in that order)
have shorter, more vertical suspensions than tadpoles of other families.

The evolution of horizontal suspensoria in tadpoles can be viewed
as an adaptation for increasing the efficiency of the buccal pump. The
pump displaces the greatest possible volume relative to the total volume
of the oral cavity when the palatoquadrates that support the ceratohyals
are themselves parallel to the tabecular cartilages. When the angle of the
jaw suspension is zero, the volume of the buccal cavity equals the volume
of the buccal pump. As this angle becomes larger the effective displacement of the buccal pump—defined as the absolute displacement of the
pump divided by the volume of the cavity—necessarily decreases as a
function of the cosine of the suspensorial angle.

The evolution of large adult suspensorial angles may be understood
as possible adaptations to macrophagy, tongue protrusion, ventilatory
behaviors, and vocalization. Variation among species is great, but one
clue concerning the functional significance of the adult angle is the fact
that the two most specialized fossorial frogs in our sample, *Glyphoglossus*
and *Rhinophrynus*, which eat ants, termites, and other ground insects,

have the smallest angles. Frogs with small angles occur in the Leptodactylidae and Pelobatidae; they typically have truncated snouts and fossorial habits. There are, however, some clear exceptions to this pattern in that some frogs, like *Pachymedusa* and *Amolops*, which are not fossorial, also have small adult angles.

Microhylids and bufonids have less change in suspensorial orientation than do other non-archaeobatrachians. The sample size for the Bufonidae is too small to warrant discussion of this family. Larval features not involving the jaw suspension have been used by some workers (Starrett, 1973; Inger, 1967; Savage, 1973) to assign the Microhylidae to the archeaobatrachian grade despite much contrary evidence (Blommer-Schlosser, 1975; Laurent, 1979; Wassersug, 1980; Sokol, 1975). It is worth pointing out, however, that it is the characteristically small angles of the adults, specifically *Glyphoglossus molossus*, and not the suspensorial design of microhylid larvae that give the microhylids the appearance of a "primitive" metamorphosis. Excluding the bizarre *Glyphoglossus*, the remaining microhylids have an average change in suspensorial angle of 78°. This is identical to that of the pelobatids and intermediate between the Archaeobatrachia and Neobatrachia *sensu stricto* (Table I).

If most anurans have evolved an efficient larval buccal pump by specializations of the suspensorium and hyoid, how have archaeobatrachian larvae with a more generalized suspensorium remained competitive? Starting with a more generalized suspensorium, archeobatrachian larvae have evolved a variety of novel solutions to maintain the mechanical efficiency of their feeding/respiratory mechanism. Discoglossoid larvae (e.g., *Ascaphus*, *Alytes*, and *Bombina* in Fig. 2) and to some extent pelobatoid larvae (e.g., *Scaphiopus* in Fig. 3) have rotated the trabecular horns downward toward the palatoquadrate. This rotation provides an efficient buccal pump without the extreme upward rotation of the palatoquadrate that characterizes most advanced larvae. But a "cost" of this design is a ventrally directed mouth, which more or less confines these animals to bottom feeding. The pipoid solution has been forward displacement of the whole apparatus. This is seen in the extreme in *Xenopus*, but also to a lesser extent in *Pipa* [see the figures in Sokol (1975, 1977b)] and *Rhinophrynus* (Fig. 2). *Hymenochirus* (Fig. 2), which is an obligate carnivore, has a unique solution. It has basically redesigned most of its head around an otherwise generalized jaw suspension (Wassersug and Hoff, 1979). The suspensorium is massive and extensively braced to the neurocranium (a condition developed independently in *Anotheca*, which is also a specialized carnivore; Fig. 3) and the ceratohyals move in an oblique, rather than a vertical plane.

A Secondary Hypothesis on the Abruptness of Metamorphosis

In going from an aquatic, suspension-feeding way of life to a terrestrial, carnivorous way of life, it is assumed that the anuran larvae pass
through a series of stages in which they are not well adapted to either
land or water. De Jongh (1968) states this belief as follows: "the tadpole
stages represent an ontogenetic adaptive peak; the larvae are highly specialized and well adapted to their environment and mode of life. After
metamorphosis a frog reaches a similar adaptive peak, although in a
different mode of life. The metamorphic stages are less completely
adapted in many respects." Larvae with four legs exposed, but still retaining a tail, are exceptionally vulnerable to predation by snakes. Stom-
ach contents from wild-caught natricine snakes are disproportionately
high in transforming larvae even though the snakes examined ate both
larvae and frogs (Arnold and Wassersug, 1978). In laboratory experiments
using garter snakes (*Thamnophis*) as predators and chorus frogs (*Pseudacris*) as prey, snakes in the water caught many more transforming
individuals than they did premetamorphic larvae; on land these snakes
caught significantly more transforming individuals than fully transformed
frogs (Wassersug and Sperry, 1977). The forelimbs of transforming larvae
appeared to hinder smooth sinusoidal locomotion in water, and their tails
retard saltatory locomotion on land. In either environment the transforming larvae have little success in avoiding the snakes.

Williams (1966, p. 89) wrote: "When its [an organism's] normal development takes it through a succession of stages of different mortality
rates it can be expected to hurry through the stages of high mortality and
to proceed slowly (by comparison) through those that are less dangerous."
Szarski (1957) specifically argued that natural selection has favored those
anurans that have shortened the "stage during which the individual is
adapted neither to land nor to water surroundings, and is therefore most
exposed to danger." It should follow from the above that the more divergent a tadpole is from its adult, the less adapted the transformational
forms are to either the adult or the larval environment and the more
abrupt should be the metamorphosis. Stephenson (1950) phrased this by
saying "the greater the degree of larval specialization of the tadpole, the
more striking is the anuran metamorphosis."

It is unfortunately very difficult to test the hypothesis that evolution
has modified the abruptness of metamorphosis in the Anura. Such a test
requires comparing complete developmental series from a variety of different species. Data are available for the absolute time spent in metamorphosis for four archaeobatrachian Anura: *Ascaphus truei*, 60 days
(Metter, 1967, and personal communication); *Hymenochirus boettgeri*,

8 days (Sokol, 1962); *Xenopus laevis*, 14 days; and *Alytes obstetricans*, 8 days [references cited in Wassersug and Sperry (1977)]. This contrasts with a range of 2–15 days ($\bar{x} = 7$) spent in metamorphosis for nine neobatrachian species (Wassersug and Sperry, 1977). These data offer at best only weak support for the hypothesis, but it must be noted that the different species were not raised under the same conditions, and even if they had been, the ideal conditions for growth of one species would not necessarily be the same for any other species. In fact, since thermal tolerances of amphibians change through metamorphosis (Cupp, 1980; Sherman, 1980), it becomes nearly impossible to identify appropriate conditions for raising larvae for the purpose of making interspecific comparisons of the absolute time they take to metamorphose. Dettlaff and Dettlaff (1961) discussed the problem of comparing developmental patterns on amphibians and proposed examining ratios of the time for the initiation of key embryonic events. They, however, did not consider metamorphosis and conceded that available data are insufficient to quantitively characterize temporal differences in developmental schedules among species. We have attempted to get at this problem by looking at relative, rather than absolute, rates of metamorphosis.

Where we have complete developmental series of a species (21 total, 10 Archaeobatrachia), we have plotted the angle of the jaw suspension against standard Gosner developmental stages. Examples of these plots are shown in Fig. 5. Such plots indicate that the Archaeobatrachia have a more gradual metamorphosis than do the Neobatrachia. While these plots support the hypothesis that evolution of metamorphosis has gone from less abrupt to more abrupt, such a conclusion from these data is potentially circular. The problem is that the staging table uses the position of the corner of the mouth with respect to the eye to assign stage during metamorphosis, and this character is itself a reflection of the orientation of the underlying suspensorium. In point of fact, this exercise demonstrated for us how difficult it can be to stage archaeobatrachian larvae, just because they have a more generalized (i.e., vertical) suspension. An excellent illustration of this problem is *Ascaphus truei*, in which, because of the orientation of the suspensorium, a larva identified as stage 40 or earlier based on leg morphology may also key out as stage 42 or later based on the position of the corner of the mouth. Those who have raised *Xenopus laevis* will undoubtedly concur that metamorphic transformation appears more gradual in this species than in other common laboratory anurans, such as *Rana pipiens*, but such an impression is difficult to quantify.

Other features of anurans may be ultimately useful in discerning a primitive versus advanced metamorphosis. Biochemical systems that

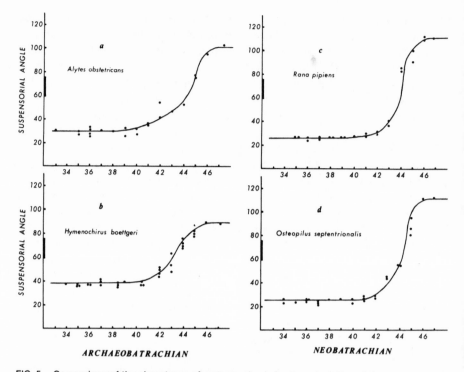

FIG. 5. Comparison of the abruptness of metamorphosis for the orientation of the suspensorium in representative anurans. Horizontal axis indicates developmental stage (Gosner, 1960). A stage 47, not in the Gosner staging table, is included; individuals are assigned to this stage if they have grown since completing metamorphosis. Curves fitted by eye. Black bar on the vertical axis indicates the total range for suspensorial angle in larvae and adults of the salamander *Ambystoma tigrinum*. The figures imply that archaeobatrachian anurans have a less abrupt metamorphosis than the neobatrachian anurans.

have a characteristic larval and adult state could be used to study the evolution of the metamorphic process. Some data are already available to demonstrate differential patterns of metamorphosis in archaeobatrachian versus neobatrachian hemoglobins. In the neobatrachian *Hyla arborea* the shift from larval to adult hemoglobin "starts later and develops more rapidly than in most other species" (Cardellini and Sala, 1978); in the archeobatrachian *Bombina variegata* the hemoglobin shift "starts earlier and develops slower than in most species" (Cardellini and Sala, 1979). Broyles (1981) similarly reports that the ontogenetic hemoglobin shift "begins earlier and ends later in *Xenopus*" than in *Rana catesbeiana*.

Other relevant data can be found in the neurophysiology literature. Grobstein and Comer (1977) report that *Xenopus* exhibits substantially more postmetamorphic eye migration than does *Rana*. In the context of a study of metamorphosis, it may be more appropriate to say that *Xenopus* has a more gradual and prolonged metamorphosis of its visual system than does *Rana*. However, *Xenopus* has unusual chondrocrania in both the larva and adult (see Table I and Fig. 2) and without more comparative data, these observations must be interpreted cautiously.

The Ancestral Anuran

If the evolution of the Anura involves the evolution of metamorphosis, then it should be possible to make some prediction about the life cycle of the ancestral anuran. This ancestor presumably had little change with metamorphosis and that change was rather gradual. Did this animal look more like a tadpole or a frog? The orientation of the palatoquadrate in nonanuran Amphibia, such as *Ambystoma* at any stage (Table I), is neither like that of a typical tadpole (approximately 25°) nor of a typical frog (approximately 105°). Rather the palatoquadrate of a typical larval or adult salamander is oriented at 65°–70° [see many figures in Regel (1963); Carroll and Holmes (1980)]. This is its orientation in anurans in the middle of metamorphosis (i.e., Gosner stages 44–45; see Fig. 5). Anurans at this stage are clumsy, with all four limbs exposed and a short tail. While these creatures seem ill suited for this world, a protoanuran of this generalized body form may have been quite successful in the Triassic. Indeed, most of the predators on contemporary transforming anurans, such as snakes and birds, did not exist at that time.

The nearest fossil form to the ancestral anurans is *Triadobatrachus*, which has both larval and adult characteristics (Griffiths, 1963; Hecht, 1963; Estes and Reig, 1973).The skull is particularly froglike, but there are many more vertebrae than in adult frogs, including six posterior to the sacrum. Estes and Reig (1973) concluded that the *Triadobatrachus* material they studied was in a "young stage." While *Triadobatrachus* is a more elongate animal than living anurans, and may not be a direct ancestor to the Anura, in having a mixture of larval and adult features it is consistent with our idea of what an ancestral anuran may have looked like. It must be emphasized that we do not conceive of the ancestral anuran as a strict mosaic of contemporary larval and adult anuran features, but we do conceive of that organism as being morphologically intermediate between contemporary larval and adult anurans.

Metamorphosing anurans are generalized in more of their morphology than just their jaw suspension. At stage 44 the ceratohyals are oriented obliquely from anteromedial to posterolateral and are wider medially. At this stage the whole hyoid skeleton looks more like the salamander's than like that of a tadpole or frog [Compare figures in Stephenson (1951), Gaupp (1904, 1906), de Jongh (1968), Ridewood (1897, 1898a,b,c) of transforming anurans with figures in Ozeti and Wake (1969) and Dowling and Duellman (1978) of salamanders]. It is also during this brief period, in the middle of metamorphosis, that anurans look most like each other. Orton (1944) came close to recognizing this when she said, "In the course of the evolution of the Salienta there has thus been not only an increased morphological divergence, between adult and larvae . . . but also an increasing departure of each stage from its respective ancestral condition." Modern anurans in the middle of metamorphosis may closely resemble the primitive anuran in body form.

SUMMARY

We have examined the hypothesis that evolution has increased the difference between larvae and adults within the Anura. For this study we have measured the angle of the jaw suspension in both the larvae and adults of 38 frog and one salmander species. For anurans there is an inverse relationship between the size of this angle in the larva and its size in the adult. Frog families traditionally recognized as archaic tend to have the least difference in this feature between larvae and adults, while families recognized as advanced have the greatest difference. There is, however, much interfamilial variation, which clouds the picture.

The small suspensorial angle of most tadpoles allows for efficient suspension feeding and respiration associated with a unique buccal pumping mechanism. Larvae of archaic frogs with a larger suspensorial angle have evolved a variety of otherwise novel cranial features that allow them to remain competitive with tadpoles of advanced families.

A secondary hypothesis is presented that the more a tadpole differs from its adult, the more abrupt should be its metamorphosis. A rigorous test of this hypothesis requires interspecific comparison of the absolute time spent in metamorphosis. Some data are available which support this idea; however, these data are technically not comparable. A less rigorous test was performed by examining changes in the suspensorial angle as a function of developmental stage. This analysis indicated that the archaeobatrachian frogs do have a less abrupt metamorphosis than neobatrachian frogs, but such an interpretation must be made with caution because the staging of tadpoles itself depends on suspensorial angle.

Other features, such as the ontogenetic shifts in hemoglobin structure and in the nervous system, could be used to explore the evolution of anuran metamorphosis. In general, available data suggest that evolution has increased both the magnitude of the difference between larval and adult anurans and the abruptness of the metamorphic process. This pattern for the evolution of anuran metamorphosis is parallel to, but more subtle than, the shift from hemimetabolous to holometabolous development in insects.

The ancestral anuran probably had a gradual metamorphosis and changed little as it transformed from larva to adult. The anuran skeletal system is most similar to that of other amphibians when an anuran is in the middle of metamorphosis. The ancestral anuran probably looked more like an anuran in metamorphosis than like either contemporary tadpoles or frogs.

ACKNOWLEDGMENTS

Most of the specimens for this study were provided by an army of friends and colleagues around the world. To those acknowledged by name in Wassersug and Rosenberg (1979) we add Lynn Branch, Eneas Salati, and Paulo Vanzolini. Some specimens were borrowed from the Field Museum of Natural History; the American Museum of Natural History; the Smithsonian Institution; the Museum of Zoology, University of Michigan; the Museum of Natural History, University of Kansas; and the Museum of Vertebrate Zoology, University of California. We are grateful to the curatorial staffs of all these institutions for making these materials available to us.

Figures 3 and 4 were drawn by Claire Kryczka; Shirley Aumiller assisted with photography. Debra Randall and Andrew Lum typed many drafts of the manuscript. We thank Karen Rosenberg for her assistance throughout this study and Martin Feder for both his patience and stimulating perspective on natural history. P. Alberch, S. D. Busack, D. Cannatella, J. L. Edwards, D. J. Futuyma, J. Hanken, M. K. Hecht, K. M. Hiiemae, J. A. Hopson, W. R. Heyer, R. F. Inger, S. Salthe, O. M. Sokol, B. Shaffer, K. Rosenberg, D. B. Wake, and G. R. Zug offered critical comments on earlier manuscript drafts. This research was supported by the National Science Foundation.

REFERENCES

Arnold, S. J., and Wassersug, R. J., 1978, Differential predation on metamorphic anurans by garter snakes (*Thamnophis*): Social behavior as a possible defense, *Ecology* **59**:1014–1022.

Blommers-Schlosser, R., 1975, Observations on the larval development of some Malagasy frogs, with notes on their ecology and biology (Anura: Dyscophinae, Scaphiophryninae and Cophylinae), *Beaufortia* **24**:7–26.

Boy, J. A., 1974, Die Larven der rhacitomen Amphibien (Amphibia: Temnospondyli; Karbon-Trias) *Palaeontol. Z.* **48**:236–268.

Brown, V. K., 1977, Metamorphosis: A morphometric description, *Int. J. Insect Morphol. Embryol.* **6**:221–223.

Broyles, R. H., 1981, Changes in blood during amphibian metamorphosis, in: *Metamorphosis: A Problem in Developmental Biology* (L. I. Gilbert and E. Frieden, eds.), pp. 461–490, Plenum Press, New York.

Cardellini, P. and Sala, M., 1978, Metamorphic variations in the hemoglobins of *Hyla arborea* L., *Comp. Biochem. Physiol.* **61(B)**:21–24.

Cardellini, P., and Sala, M., 1979, Metamorphic variations in the hemoglobins of *Bombina variegata* (L.), *Comp. Biochem. Physiol.* **64(B)**:113–116.

Carroll, R. L., and Holmes, R., 1980, The skull and jaw musculature as guide to the ancestry of salamanders. *Zool. J. Linn. Soc.* **68**:1–40.

Cupp, P. V., Jr., 1980, Thermal tolerance of five salientian amphibians during development and metamorphosis, *Herpetologica* **36**:234–243.

de Beer, G., 1937, *The Development of the Vertebrate Skull*, Clarendon Press, Oxford.

de Jongh, H. J., 1968, Functional morphology of the jaw apparatus of larval and metamorphosing *Rana temporaria* L., *Neth. J. Zool.* **18**:1–103.

de Jongh, H. J., and Gans, C., 1969a, Bullfrog (*Rana catesbeiana*) ventilation: How does the frog breathe, *Science* **163**:1223–25.

de Jongh, H. J., and Gans, C., 1969b, On the mechanism of respiration in the bullfrog, *Rana catesbeiana*: A reassessment, *J. Morphol.* **127**:259–90.

Dettlaff, T. A., and Dettlaff, A. A., 1961, On relative dimensionless characteristics of the development duration in embryology, *Arch. Biol.* **72**:1–16.

Dowling, H. G., and Duellman, W. E., 1978, *Systematic Herpetology: A Synopsis of Families and Higher Categories*, HISS Publications, New York.

Duellman, W. E., 1975, On the classification of frogs, *Occas. Pap. Mus. Nat. Hist. Univ. Kansas* **42**:1–14.

Edgeworth, F. H., 1930, On the masticatory and hyoid muscles of larvae of *Xenopus laevis*, *J. Anat.* **64**:184–188.

Edgeworth, F. H., 1935, *The Cranial Muscles of Vertebrates*, University Press, Cambridge.

Eeden, van, J. A., 1951, The development of the chondrocranium of *Ascaphus truei* Stejneger with special reference to the relations of the palatoquadrate to the neurocranium, *Acta Zool.* **1932**:41–176.

Estes, R., and Reig, O. A., 1973, The early fossil record of frogs: A review of the evidence, in: *Evolutionary Biology of the Anurans: Contemporary Research on Major Problems* (J. L. Vial, ed.), pp. 11–63, University of Missouri Press, Columbia, Missouri.

Feder, M. E., and Wassersug, R. J., 1979, Activity and energy metabolism in larvae of the toad *Bufo woodhousei*, *Am. Zool.* **19**:863.

Gaupp, E., 1893, Beiträge zur Morphologie des Schädels. I. Primordial-cranium und Kieferbogen von *Rana fusca*, *Morphologische Arbeiten* **2**:275–481.

Gaupp, E., 1894, Beitrage zur Morphologie des Schädels. II. Das Hyobranchialskelet der Anuren und seine Umwandlung, *Morphologische Arbeiten* **3**:399–437.

Gaupp, E., 1898, Ontogenese und Phylogenese des schall-leitenden Apparates bei den Wirbeltieren, *Ergeb. Anat. Entwicklungs-gesch.* **8**:990–1149.

Gaupp, E., 1904, Das Hyobranchialskelet der Wirbeltiere, *Ergebn. Anat. Entwicklungs-gesch.* **14**:808–1048.

Gaupp, E., 1906, Die Entwicklung der Kopfskelettes, in: *Hertwig's Handbuch Entwicke-lungslehre der Wirbeltiere* Volume 3, pp. 573–874.

Gilbert, L. I., and Frieden, E. (eds.), 1981, *Metamorphosis: A Problem in Developmental Biology*, 2nd ed., Plenum Press, New York.

Goette, A., 1875, *Atlas zur Entwicklungsgesschichte der Unke*, Verlag von Leopold Voss, Leipzig.

Goin, C. J., Goin, O. B., and Zug, G. R., 1978, *Introduction to Herpetology*, 3rd ed., W. H. Freeman and Co., San Francisco.

Gosner, K. L., 1960, A simplified table for staging anuran embryos and larvae with notes on identification, *Herpetologica* **16**:183–190.

Gradwell, N., 1972, Gill irrigation in *Rana catesbeiana*. Part II. On the musculoskeletal mechanism, *Can. J. Zool.* **50**:501–521.

Griffiths, I., 1954, On the otic element in Amphibia, Salientia, *Proc. Zool. Soc. Lond.* **124**:35–50.

Griffiths, I., 1963, The phylogeny of the Salientia, *Biol. Rev.* **38**:241–292.

Griffiths, I., and de Carvalho, A. L., 1965, On the validity of employing larval characters as major phyletic indices in Amphibia, Salientia, *Rev. Bras. Biol.* **25**:113–121.

Grobstein, P., and Comer, C., 1977, Post-metamorphic eye migration in *Rana* and *Xenopus*, *Nature* **269**:54–56.

Hecht, M., 1963, A reevaluation of the early history of the frogs. Part II. *Syst. Zool.* **12**:20–35.

Inger, R., 1967, The development of a phylogeny of frogs, *Evolution* **21**:369–384.

Ivakhnenko, M. F., 1978, Urodeles from the Triassic and Jurassic of Soviet Central Asia, *Paleontol. Zh.* **12**:362–368.

Kenney, J. S., 1969, Feeding mechanism in anuran larvae, *J. Zool. Lond.* **157**:225–246.

Kluge, A. G., and Farris, J., 1969, Quantitative phyletics and the evolution of anurans, *Syst. Zool.* **18**:1–32.

Kotthaus, A., 1933, Die Entwicklung des Primordial-craniums von *Xenopus laevis* bis zur Metamorphose, *Z. Wiss. Zool.* **144**:510–572.

Kraemer, M., 1974, La Morphogenese du chondrocrane de *Discoglossus pictus* Otth (Amphibiens, anoure), *Bull. Biol.* **108**:211–228.

Kukalova-Peck, J., 1978, Origin and evolution of insect wings and their relationship to metamorphosis, as documented by the fossil record, *J. Morphol.* **156**:53–126.

Laurent, R. F., 1979, Esquisse d'une phylogenese des anoures, *Bull. Soc. Zool. Fr.* **104**:397–422.

Litzelmann, E., 1923, Entwicklungsgeschichtliche und vergleichend-anatomische Unter-suchungen über den Visceralapparat der Amphibien, *Z. Anat. Entwicklungs-gesch.* **67**:457–493.

Lynch, J. D., 1973, The transition from archaic to advanced frogs, in: *Evolutionary Biology of the Anurans: Contemporary Research on Major Problems* (J. L. Vial, ed.), pp. 133–182, University of Missouri Press, Columbia, Missouri.

Maxson, L. R., and Daugherty, C. H., 1980, Evolutionary relationships of the mono-typic toad family Rhinophrynidae: A biochemical perspective, *Herpetologica* **36**:275–280.

Mertens, R., 1960, Die Larven der Amphibien und ihre evolutive Bedeutung, *Zool. Anz.* **164**:337–358.

Metter, D. E., 1967, Variation in the ribbed frog *Ascaphus truei* Stejneger, *Copeia* **1967**:634–649.

Morescalchi, A., 1980, Evolution and Karyology of the amphibians, *Boll. Zool.* **47** (Suppl.):113–126.

Nichols, R. J., 1937, Taxonomic studies on the mouth parts of larval Anura, *Univ. Illinois Bull.* **34**:1–73.

Nopsca, F., 1930, Notes on Stegocephalia and Amphibia, *Proc. Zool. Soc. Lond.* **64**:979–995.

Okutomi, K., 1937, Die Entwicklung des Chondrocraniums von *Polypedates buergeri schlegelii, Z. Anat. Entwicklungsgesch.* **107**:28–64.

Orton, G. L., 1944, Studies on the systematic and phylogenetic significance of certain larval characters in the Amphibia Salientia, Ph.D. Thesis, University of Michigan, Ann Arbor, Michigan (unpublished).

Orton, G. L., 1953, Systematics of vertebrate larvae, *Syst. Zool.* **1953**:63–75.

Orton, G. L., 1957, The bearing of larval evolution on some problems in frog classification, *Syst. Zool.* **6**:79–86.

Ozeti, N., and Wake, D. B., 1969, The morphology and evolution of the tongue and associated structures in salamanders and newts (Family Salmandridae), *Copeia* **1969**:91–123.

Parker, W. K., 1871, On the structure and development of the skull of the common frog (*Rana temporaria* L.), *Phil. Trans. R. Soc. Lond.* **161**:137–211.

Parker, W. K., 1876, On the structure and development of the skull in the Batrachia. Part II, *Phil. Trans. R. Soc. Lond.* **166**:601–669.

Parker, W. K., 1881, On the structure and development of the skull in the Batrachia. Part III, *Phil. Trans. R. Soc. Lond. (Part I)* **172**:1–266.

Porter, K. R., 1972, *Herpetology*, Saunders, Philadelphia, Pennsylvania.

Pusey, H. K., 1938, Structural changes in the anuran mandibular arch during metamorphosis, with reference to *Rana temporaria, Q. J. Microsc. Sci.* **80**:479–553.

Pusey, H. K., 1943, On the head of the liopelmid frog, *Ascaphus truei*. I. The chondrocranium, jaws, arches and muscles of a partly-grown larva, *Q. J. Microsc. Sci.* **94**:105–185.

Regel, E. D., 1963, Development in *Hynobius keyserlingii, Trans. Zool. Inst. Akad. Nauk SSSR* **1963**:33 (cited in Schmalhausen, I. I., 1968, *The Origin of Terrestrial Vertebrates*, Academic Press, New York).

Reinbach, W., 1939, Untersuchungen uber die Entwicklung des Kopfskeletts von *Calyptochephalus gayi, Jena. Z. Naturwiss.* **72**:211–362.

Ridewood, W. G., 1897, On the structure and development of the hyobranchial skeleton of the parsley-frog (*Pelodytes punctatus*), *Proc. Zool. Soc. Lond.* **1897**:577–595.

Ridewood, W. G., 1898a, On the structure and development of the hyobranchial skeleton and larynx in *Xenopus* and *Pipa*; with remarks on the affinities of the Aglossa, *J. Linn. Soc.* **26**:53–128.

Ridewood, W. G., 1898b, On the development of the hyobranchial skeleton of the midwife-toad (*Alytes obstetricans*), *Proc. Zool. Soc. Lond.* **1898**:4–12.

Ridewood, W. G., 1898c, On the larval hyobranchial skeleton of the anurous batrachians with special reference to the axial parts, *J. Linn. Soc.* **26**:474–487.

Savage, J. M., 1973, The geographic distribution of frogs: patterns and predictions, in: *Evolutionary Biology of the Anurans: Contemporary Research on Major Problems* (J. L. Vial, ed.), pp. 351–445, University of Missouri Press, Columbia, Missouri.

Sedra, S. N., 1950, The metamorphosis of the jaws and their muscles in the toad, *Bufo regularis* Reuss, correlated with the changes in the animal's feeding habits, *Proc. Zool. Soc. Lond.* **120**:405–449.

Sedra, S. N. and Michael, M. I., 1957, The development of the skull, visceral arches, larynx and visceral muscles of the South African clawed toad, *Xenopus laevis* (Daudin) during the process of metamorphosis (from stage 55 to stage 66). *Verh. K. Ned. Akad. Wet. Afd. Natuurk.* **51**:5–80.

Severtzov, A. S., 1969, Food seizing mechanism of anuran larvae, *Dokl. Akad. Nauk SSSR* **187**:211–214 (Transl.).

Sharov, A. G., 1957, Types of insect metamorphosis and their relationship, *Rev. Entomol. SSSR* **36**:569–576.

Sharov, A. G., 1966, *Basic Arthropodan Stock*, Pergamon Press, New York.

Sherman, E., 1980, Ontogenetic change in thermal tolerance of the toad *Bufo woodhousei fowleri*, *Comp. Biochem. Physiol.* **65A**:227–230.

Simons, E. V., and Van Horn, J. R., 1971, A new procedure for whole-mount alcian blue staining of the cartilagenous skeleton of the chicken embryos, adapted to the clearing procedure in potassium hydroxide, *Acta Morphol. Neerl. Scand.* **8**:281–292.

Sokol, O. M., 1962, The tadpole of *Hymenochirus boettgeri*, *Copeia* **1962**:272–284.

Sokol, O. M., 1975, The phylogeny of anuran larvae: a new look, *Copeia* **1975**:1–23.

Sokol, O., 1977a, A subordinal classification of frogs (Amphibia: Anura), *J. Zool. Lond.* **182**:505–508.

Sokol, O., 1977b, The free swimming *Pipa* larvae, with a review of pipid larvae and pipid phylogeny (Anura: Pipidae), *J. Morphol.* **154**:357–426.

Starrett, P. H., 1967, The phylogenetic significance of the jaw musculature in anuran amphibians, Doctoral Dissertation, University of Michigan.

Starrett, P. H., 1973, Evolutionary patterns in larval morphology, in: *Evolutionary Biology of the Anurans: Contemporary Research on Major Problems* (J. L. Vial, ed.), pp. 252–271, University of Missouri Press, Columbia, Missouri.

Stephenson, N. G., 1950, Observations on the development of the amphicoelous frogs, *Leioplema* and *Ascaphus*, *J. Linn. Soc.* **42**:18–28.

Stephenson, N. G., 1951, On the development of the chondrocranium and visceral arches of *Leiopelma archeyi*, *Trans. Zool. Soc. Lond.* **27**:203–251.

Szarski, H. J., 1957, The origin of the larva and metamorphosis in Amphibia, *Am. Nat.* **91**:283–301.

Takisawa, A., and Sunaga, Y., 1951, Über die Entwicklung des M. depressor manidibulae bei Anuren im Laufe der Metamorphose, *Folia Anat. Jpn.* **23**:273–293.

Takisawa, A., Ohara, Y., and Kano, K., 1952a, Die Kaumuskulatur der Anuren (*Bufo vulgaris japonicus*)*Während der Metamorphose* **24**:1–28.

Takisawa, A., Ohara, Y., and Sunaga, Y., 1952b, Über die Umgestalttung der Mm. intermandibulares der Anuren während der Metamorphose, *Folia Anat. Jpn.* **24**:215–241.

Tihen, J. A., 1965, Evolutionary trends in frogs, *Am. Zool.* **5**:309–391.

Wassersug, R. J., 1972, The ultraplanktonic entrapment mechanisms of anuran larvae, *J. Morphol.* **137**:279–288.

Wassersug, R. J., 1975, The adaptive significance of the tadpole stage with comments on the maintenance of complex life cycles in anurans, *Am. Zool.* **15**:405–417.

Wassersug, R. J., 1976a, Oral morphology of anuran larvae: Terminology and general description, *Occas. Pap. Mus. Nat. Hist. Univ. Kansas* **48**:1–23.

Wassersug, R. J., 1976b, A procedure for differential staining of cartilage and bone in whole, formalin fixed vertebrates, *Stain Technol.* **51**:131–134.

Wassersug, R. J., 1976c, Internal oral features in *Hyla regilla* (Anura: Hylidae) larvae: An ontogenetic study, *Occas. Pap. Mus. Nat. Hist. Univ. Kansas* **49**:1–24.

Wassersug, R. J., 1980, Internal oral features of larvae from eight anuran families: Functional, systematic, evolutionary, and ecological considerations, *Univ. Kansas Mus. Nat. Hist. Misc. Publ.* **68**:1–146.

Wassersug, R. J., and Feder, M. E., 1979, Respiratory behaviors of *Xenopus laevis* larvae, *Am. Zool.* **19**:863.

Wassersug, R. J., and Hoff, K., 1979, A comparative study of the buccal pumping mechanism of tadopoles, *Biol. J. Linn. Soc.* **12:**225–259.

Wassersug, R. J., and Rosenberg, K., 1979, Surface anatomy of branchial food traps of tadpoles: A comparative study, *J. Morphol.* **159:**393–425.

Wassersug, R. J., and Sperry, D. G., 1977, The relation of locomotion to differential predation on *Pseudacris triseriata* (Anura: Hylidae), *Ecology* **58:**830–839.

Weisz, P. B., 1945a, The development and morphology of the larva of the South African Clawed Toad, *Xenopus laevis.* I. The third-form tadpole, *J. Morphol.* **77:**163–191.

Weisz, P. B., 1945b, The development and morphology of the larva of the South African Clawed Toad, *Xenopus laevis, J. Morphol.* **77:**193–217.

Wilbur, H. M., 1980, Complex life cycles, *Annu. Rev. Ecol. Syst.* **11:**67–93.

Williams, G. C., 1966, *Adaptation and Natural Selection*, Princeton University Press, Princeton, New Jersey.

6

Regulatory Genes and Adaptation

Past, Present, and Future

ROSS J. MACINTYRE

Section of Genetics and Development
Cornell University
Ithaca, New York 14853

INTRODUCTION

Ever since the publication of *The Origin of Species,* evolutionary biologists have fixed their attention on how natural selection brings about adaptation. The study of adaptation has most often involved both comparative and functional morphology or physiology. As a result, the relevance of special body parts or metabolic systems to the survival of particular organisms has been shown many times in the last 120 years. The genetic basis for most of the adaptations which have been described, however, has not been elucidated, often because the organisms themselves are not amenable to genetic analysis, or because the traits are very complex, i.e., the genetic basis can only be described as polygenic. Thus, the effects of single genes cannot be parsed out, and the evolutionary dynamics of those genes within and between populations cannot be easily examined.

In addition, the fairly recent discipline of molecular biology has not contributed in any substantial way to our understanding of the evolution of adaptation, although here also, elegant *descriptions* of adaptation at the molecular level are now available.

This failure is not because the organisms studied by molecular biologists are refractory to genetic analysis. Indeed, since genetics is a major component of molecular biology, virtually every organism used has a well-understood genetic system. Not only that, the level at which molecular biologists work is at or very close to the level of the gene itself. As a result, there is often a simple relationship between the phenotype and the gene. Simple Mendelian analyses are the rule, not the exception, as is the case with complex morphologic or physiologic characters.

In this chapter, I will attempt to outline why molecular biology has told us so little about the origin of adaptations. First, I will briefly review the reasons structural gene products, which have been intensively studied, are not likely to be the molecular "stuff" of adaptation. I will also summarize the results of several studies which indicate regulatory gene variation may be abundant in natural populations and will discuss how our present ignorance of the details of the regulation of structural gene action in eukaryotes poses new dilemmas to the evolutionary biologist interested in the evolution of adaptation.

THE CASE FOR STUDYING THE EVOLUTION OF REGULATORY GENES

Historical Considerations

Only after a distinction between regulatory and structural genes was established experimentally by Jacob and Monod (1961) could any meaningful discussion on the relative importance of the two kinds of genes in evolution begin. Wallace (1963) first noted that globin chains, then only partially sequenced, displayed very little interspecific variation despite the presence of obvious morphologic differences between the species themselves. At that time there was extensive but circumstantial evidence that the gene pools of most species were very heterogeneous. This prompted Wallace to propose that it was not genes coding for proteins that were polymorphic, but rather, it was the genes controlling the amount and the appearance of those proteins during development that were responsible for intra- and interpopulational heterosis. Wallace proposed that there would be strong selection for heterozygosity in regulatory elements if most of them controlled more than one structural gene.

With the advent of gel electrophoresis and as more amino acid sequences were published, the notion that structural genes were invariant was quickly dispelled. In fact, as is now well known, most soluble enzymes are polymorphic in outbreeding species, and only a very few proteins do not vary interspecifically. Nevertheless, in the early to mid seventies, the idea that it is mainly regulatory genes that respond to selection and that results in adaptation was revived, most forcefully by Wilson and his colleagues (Wilson, 1975; King and Wilson, 1975; Bush *et al.*, 1977). In my opinion, no single compelling observation was responsible for stirring this revival of interest in the evolution of regulatory genes; rather, it was due to several observations coming from studies on structural gene variation. These observations involve the rates of structural gene evolution, selection coefficients associated with enzyme polymorphisms, and structural gene differences between species.

Evolutionary Rates of Change of Structural Gene Products

From the initial comparisons of the primary sequences of homologous proteins from species with well-documented fossil records came the concept of the unit evolutionary period (UEP) of a protein, i.e., the time required for a 1% sequence divergence to occur in the protein from two independent lineages (Dickerson, 1971). At first, when a large number of interspecific comparisons were made there appeared to be very little variance in the independent estimates of the UEP of any one protein. These comparisons generally involved species with very different histories of adaptive radiation; thus, the monotonic evolution of the particular proteins appeared to be uncoupled from those morphologic and physiologic phenotypes, many of which had evolved rapidly and radically in different lineages. As more sequence data became available (e.g., Romero-Herrera *et al.*, 1973) along with more sophisticated tree-building methods (Goodman and Moore, 1977) and better statistical tests (Langley and Fitch, 1974), it became clear that few, if any, proteins have evolved at a constant rate. Nevertheless, there are very few large differences in the rates of change which have been measured (Fitch, 1975).

Goodman and Moore's (1977) analysis of the globins showed that after gene duplication there were very rapid rates of change in the primary sequences of at least one of the duplicate gene products. This rate then slowed down in most derived lineages. They propose that any large and presumably adaptive changes in the function of a new gene product take place very soon after the gene duplication event. If this is true, it follows that studies on newly duplicated structural genes may provide pertinent information on the evolution of adaptation at the molecular level. The studies of Whitt (Whitt *et al.*, 1975), Goodman (Goodman *et al.*, 1975), and Gottlieb (Gottlieb and Greve, 1981) and their colleagues are certainly relevant in this regard. In addition, there is evidence that regulatory changes in newly duplicated structural gene products may occur very rapidly (Ferris and Whitt, 1979; Markert *et al.*, 1975).

Even though gene and genome duplication has been a major evolutionary force [see review by MacIntyre (1976)], most structural gene products examined to date are probably in the "fine tuning" stage of their evolutionary history. That is, following the gene duplication, and after the major changes in the structure and function of the protein have taken place, each new amino acid substitution will alter the protein only very slightly in a functional sense (Fitch, 1972). Then there will be strong stabilizing selection to maintain the basic function of the proteins. This has been demonstrated experimentally in several instances. For example, the homologous subunits of tryptophan synthetase from *Escherichia coli* and *Salmonella typhimurium* differ by 40 amino acid substitutions out of 268 positions. Schneider *et al.* (1981), using a host strain of *E. coli* that carried a deletion for the *trpA* gene (which codes for the α subunit of tryptophan synthetase), were able to select for *trpA*⁺ recombinants between two plasmids, one of which carried a mutant *trpA* gene cloned from *E. coli*, the other a mutant *trpA* gene from *S. typhimurium*. Each of the *trpA*⁺ recombinant genes coded for a hybrid α subunit, that is, a polypeptide with an NH_2 terminal amino acid sequence from *E. coli* and a COOH terminus from *S. typhimurium*. The relative extent of each species' contribution to the primary sequence of the α subunit depended upon the position of the crossover between the two plasmids. They analyzed six of the hybrid α subunits both for their ability to catalyze the conversion of indole to tryptophan and to associate with the β subunits (from the *trpB* gene). The latter reaction makes possible the conversion of indole glycerol phosphate to tryptophan. All six α subunits were very similar to subunits from *E. coli* or *S. typhimurium* with regard to these essential catalytic activities. The data are presented in Table I. Note that the hybrid subunits tested ranged from one in which *E. coli* contributed only six of the 40 amino acid substitutions differentiating the subunits from the two species to one in which 15 of the different amino acids were coded by the *E. coli* part of the recombinant gene. It appears, then, that the 40 substitutions do not affect basic catalytic properties of the subunit in any important way. As the data in Table I show, however, they do affect at least one other property of the subunit, its heat stability. It may be that these differences result from amino acid substitutions that have subtly adapted the molecule in each species to particular cellular conditions and microhabitats, but this has not been demonstrated experimentally.

MacIntyre and Dean (1978) present another instance in which members of a group of homologous proteins from ten *Drosophila* species retain their basic catalytic properties despite what must be substantial numbers of amino acid differences. They made several so-called heterospecific dimers of acid phosphatase by subunit dissociation and reassociation.

TABLE I. Comparison of Enzymatic Activities and
Thermal Stabilities of Tryptophan Synthetase from *E.
coli*, *S. typhimurium*, and Six Strains with
Recombinant α Subunits[a]

Source of a subunit[b]	Activity ratio[c]	Thermal stability[d]
E. coli	2.2	95
6–34a	2.5	25
6–34b	2.2	25
8–32	2.8	60
12–28	3.0	25
14–26	2.6	15
15–25	3.1	35
S. typhimurium	2.0	45

[a] After Schneider *et al.* (1981).
[b] The first number in the designations of the hybrid α subunits indicates how many of the 40 amino acid residues which are different in the two species are from the *E. coli* sequence. The second number refers to the number of *S. typhimurium* specific residues in the hybrid subunit.
[c] Ratio (indole → tryptophan)/(indole glycerol phosphate → tryptophan).
[d] Percent activity remaining in extract after 26 min at 52°C.

The enzymes they isolated were evolutionary novelties in that they were composed of two subunits, but each from a different species. Fifteen such enzymes were analyzed for the amount of acid phosphatase activity per unit acid phosphastase protein and were found to be identical to each other and to five homospecific enzymes [see Fig. 5 in MacIntyre and Dean (1978)]. On the other hand, both by immunologic criteria and by tests involving subunit association, the acid phosphatases examined were very different. The relevant data are shown in Table II. In almost all of the tests, the isolated subunits differ sharply in their affinities for one another. Perhaps more instructive, however, are the immunologic distances between the homologous acid phosphatases. These immunologic distances are directly related to the differences in the antisera concentrations necessary to complex equivalent amounts of antigen in homologous versus heterologous tests. Distances of 200–300 mean that antiserum must be concentrated 100–1000 times to get an equivalent reaction in a heterologous test. This is virtually at the limit of measurable cross-reaction, and, indeed, in two cases there was no cross-reaction. From observations on other proteins where sequence data are available, a lack of cross-reaction implies that there is at least a 30–40% difference in the

TABLE II. Subunit Association Ratios and Immunologic Distances between Fifteen
Heterospecific *Drosophila* Acid Phosphatases[a]

Heterospecific enzyme	Ratio of homospecific to heterospecific enzymes formed during subunit association[b]	Immunologic distances
Melanogaster–virilis	0.45	284: anti-*D. melanogaster*; no cross-reaction with anti-*D. virilis*
Simulans–virilis	0.61	No cross-reaction with anti-*D. virilis*
Melanogaster–mercatorum	0.71	245: anti-*D. melanogaster*
Simulans–mercatorum	1.36	Not tested
Melanogaster–mulleri	0.67	230: anti-*D. melanogaster*
Simulans–mulleri	0.90	Not tested
Nebulosa–virilis	1.08	172: anti-*D. nebulosa*, 201: anti-*D. virilis*
Nebulosa–mercatorum	1.67	187: anti-*D. nebulosa*
Nebulosa–mulleri	2.07	179: anti-*D. nebulosa*
Paulistorum–virilis	0.79	209: anti-*D. paulistorum*, 221: anti-*D. virilis*
Paulistorum–mercatorum	1.01	215: anti-*D. paulistorum*
Paulistorum–mulleri	1.69	209: anti-*D. paulistorum*
Willistoni–virilis	0.92	Not tested
Willistoni–mercatorum	1.32	Not tested
Willistoni–mulleri	1.12	Not tested

[a] From MacIntyre and Dean (1978).
[b] Ratio is 1.00 if subunit association is random.

amino acid sequences of the two proteins (Champion *et al.*, 1974). It is remarkable that despite extensive substitutions in the amino acids involved in the antigenic determinants and in the subunit binding areas of these acid phosphatases, their enzymatic activities, when determined both qualitatively and quantitatively, remain unchanged. Clearly, this must be a property which is under strong stabilizing selection pressure.

Since selection has acted to stabilize the essential functions of these and presumably most other structural gene products, the amino acid substitutions that have occurred are likely to be either neutral in selective value or to be associated with relatively small selection coefficients. The fairly constant evolutionary rate of change of most structural gene products can be explained either by stochastic (Ewens, 1977) or by selective processes (Hartl and Dykhuizen, 1979). Nevertheless, the stabilizing selective forces acting on a protein make it unlikely that its structural gene will become involved in any rapid or evolutionary process.

Selection Coefficients and Enzyme Polymorphisms

It follows that, if proteins are under intense stabilizing selection pressure, enzyme polymorphisms may well be selectively neutral or, at best, have relatively small selection coefficients associated with them. Indeed, it has been difficult to get *direct* evidence for a selective difference between allozymes, even in those instances where other evidence implies there is selection maintaining the enzyme polymorphism. Thus, it is clearly not enough just to show a strong genotype–environment correlation exists with regard to a particular polymorphism or that there is a parallel pattern of geographic variation in two sympatric species (Clarke, 1975). Such evidence for selection is still not compelling, because of the problem of linkage. That is, alleles at the locus in question may, for historical reasons, be in linkage disequilibrium with another gene or gene complex which is being strongly selected by some environmental parameter (Laurie-Ahlberg and Merrell, 1979). It is important, then, to show with biochemical tests that the allelic differences in the gene product are associated with the environmental factor, and, if possible, the genotypes should be subjected to experimental manipulations which allow an outcome, in terms of allele frequency changes, to be judged against a specific prediction (Koehn, 1978). As mentioned above, this has not been a trivial task. Only in a few instances has it even been partially accomplished. Koehn and his co-workers have investigated the population dynamics and the biochemical characteristics of a leucine aminopeptidase from the mollusc *Mytilis edulis* [see review by Koehn (1978); Young *et al.* (1979)] Levels of this enzyme are positively affected by increases in salinity, and Koehn proposes that it plays a role in regulating the intracellular pool of free amino acids, a critical factor in osmoregulation. The frequency of one allele in particular, LAP^{94}, is strongly correlated with salinity, and its product is catalytically the most active of the LAP allozymes. The role of selection with regard to the other allozymes in this stable polymorphism (it is shared by a related species, *Modiolus demissus,* in the same geographic area) is not yet understood.

In two other systems, LDH in the killifish *Fundulus heteroclitus* and α-GPDH in *D. melanogater,* clines in allele frequencies are strongly correlated with temperature, and kinetic parameters of the allozymes vary significantly and predictably with that same variable (Place and Powers, 1979; Miller *et al.,* 1975). Table III presents some of the abstracted data on α-GPDH from *D. melanogaster.* In this insect, the enzyme is critical for flight by virtue of its participation in the α-glycerophosphate cycle in the thoracic flight muscle (O'Brien and MacIntyre, 1978). Note that flies

TABLE III. Biochemical and Population Genetic Data on α-GPDH from *Drosophila melanogaster*

Temperature, °C	Frequency of α-GPDH^F [a]	K_m,[b] μM DHAP		Turns per second[c]	
		α-GPDH^{FF}	α-GPDH^{SS}	α-GPDH^{FF}	α-GPDH^{SS}
10	0.80	780	470	NE	NE
20	0.87	550	680	0.8	1.3
30	0.92	400	880	1.7	1.1

[a] From Johnson and Schaffer (1973); estimated from regression line.
[b] From Miller *et al.* (1975).
[c] From Glen Collier, personal communication. Data are taken from experiments conducted on an insect flight mill. NE, not estimated.

carrying the allozyme of α-GPDH with the lowest K_m at 30°C are also the best fliers at that temperature. The allele coding for the α-GPDH^F allozyme is also more frequent in populations from warmer climates. The measurements of K_m and flight ability were made on inbred lines, however, and should be repeated on strains which have randomized genetic backgrounds.

Despite the fact that it has been studied very intensively, the role of selection in the maintenance of the alcohol dehydrogenase (Adh) allozyme polymorphism in *Drosophila melanogaster* is still ambiguous. There are two electromorphs, Adh^F and Adh^S, but within each there are also thermostability variants (Sampsell, 1977; Sampsell and Milkman, 1978). A definite correlation exits between the electrophorphs and larval preference for ethanol (Cavener, 1979) and, in some cases, the adults [Briscoe *et al.* (1975), but see also McKenzie and McKechnie (1978)]. Thus, Adh^F increases in frequency in experimental populations raised on food with high concentrations of different short-chain alcohols (Cavener and Clegg, 1978, 1981). The Adh^F-containing flies generally have more Adh protein than do flies homozygous for Adh^S alleles (Thompson and Kaiser, 1977; McDonald *et al.*, 1980). It is not clear whether this is due to a property of the Adh^F structural gene product (i.e., the polypeptide is more resistant to intracellular degradation) or to some aspect involving the regulation of Adh. In this regard a few Adh^F strains do have uncharacteristically low levels of Adh activity (Day *et al.*, 1974; Sampsell, 1981). At any rate, careful kinetic analyses of the electromorphs by McDonald *et al.* (1980) indicate that at high alcohol concentrations flies with the Adh^F allozyme would degrade toxic alcohols more rapidly than would flies with the Adh^S allozyme. On the other hand, they propose that flies with the Adh^S allozyme, which has a lower K_m, would have an advantage in environments with very low levels of those same alcohols. It may be, then, that the selective basis for the polymorphism of the Adh electromorphs resides partly in a structural gene product difference, namely the K_m difference, and partly to a difference in the regulation of the alleles of the structural gene, namely the difference between the per fly amounts of Adh^F and Adh^S allozymes.

Temperature also appears to affect the Adh electromorph polymorphism in *D. melanogaster* (Vigue and Johnson, 1973; Van Delden and Kamping, 1980). In general, flies with Adh^S "alleles" survive better at higher temperatures. Sampsell, however, has preliminary information (1981, and personal communication) that the flies carrying the thermal resistant variants of both fast *and* slow electromorphs may, in fact, survive better at higher ambient temperatures. Since thermal stability variants are generally due to structural gene differences, this may turn out to be a particularly clear case of a direct action of selection on an enzyme polymorphism.

Let me repeat that there are few examples where *both* population and biochemical studies have indicated that there is a selective basis for an enzyme polymorphism, given the vast number of polymorphisms that have been described. Even in these rare examples, it is my impression that the contributions of the allozyme alleles to overall fitness are relatively slight, for surely if enzyme polymorphisms were under strong selection pressure, we should not lack for cases in which the role of individual loci in determining fitness was understood. These judgments support the proposition that changes in most enzyme loci are not major components of organismic evolution.

Genetic Differences between Species

Another observation which indicates structural genes are not primarily involved in adaptation at the morphologic or physiologic level comes from electrophoretic comparisons between species which recently diverged from a common ancestor. Evolutionary theory maintains that isolated populations of a species can become adapted in various ways to their local environments via the formation of novel coadapted gene complexes. Reproductive isolating mechanisms, which may evolve during the isolation of two local populations and/or when two such populations reestablish contact, serve to preserve the genetic bases of such new adaptations. Clearly, if structural gene differences underlie these adaptations, then two new species should be genetically more distant than two interbreeding populations of a single species when differences in homologous proteins are used to assay genetic identities. There are now many reported instances where two or more species are genetically as similar as geographic populations from a single species. In a number of studies, the species that were compared have been morphologically or cytogenetically quite different. The Hawaiian *Drosophila* provide some of the clearest examples; e.g., the species pairs *D. heteroneura* vs *D. sylvestris* and *D. ochrobasis* vs *D. setimentosum* both have genetic identities greater than 0.90. The former, however, have quite different head shapes and behavioral patterns (Sene and Carson, 1977; Carddock and Johnson, 1979) and the latter are quite distinct karyotypically (Carson *et al.*, 1975; Nair *et al.*, 1977). Turner *et al.* (1979) found that the races of the butterfly species *Heliconius melpomene* and *H. erato*, where different alleles have been fixed for wing color patterns, exhibit genetic identities of 0.93 or greater. Since the races of each species are Mullerian mimics, the color patterns of the wings are good examples of the selectively established local adaptations alluded to above. In this case, genetic tests clearly show that none of the 17 structural genes sampled by gel electrophoresis is involved in the morphologic difference between the races. Templeton (1979) has provided similar evidence from studies on two new "species," namely parthenogenetic strains derived from a sexually reproducing population of *Drosophila mecatorum*. The two parthenogenetic strains have the same allele fixed at each of 17 allozyme loci, yet each strain successfully "adapted" to parthenogenesis in very different ways. The differences appear both in morphology and in tests involving relative fitness. Humans and chimpanzees, of course, provide still another example of two morphologically distinct species [even by criteria not dependent upon anthropocentric judgments—see Cherry *et al.* (1978)] which are nevertheless very similar in allele frequencies at structural gene loci (King and Wilson, 1975; Bruce and Ayala, 1979). Finally, Avise and Ayala (1975, 1976; Avise, 1977) showed that in several fish genera the rate of accumulation of structural gene differences in diverging taxa does not depend upon the number of speciation events, but is quite dramatically a function of elapsed time since divergence. The available evidence says, then, that speciation and structural gene evolution are essentially unrelated phenomena. Thus, there is a strong implication that structural gene differences are not involved in the genetic basis for the adaptations that speciation, as an evolutionary process, preserves.

To sum up: the rather constant rates of change of proteins, the paucity of cases in which selection coefficients have been associated with enzyme polymorphisms, and the absence of an association between speciation and structural gene product differences all make it clear that we must look elsewhere if we are to understand the genetic basis of adaptation. If we assume, for the moment at least, that the distinction between structural and regulatory genes is a valid one, then obviously we should turn our attention to the latter class of genetic units.

GENE REGULATION IN HIGHER EUKARYOTES: THE CURRENT STATE OF AFFAIRS

The control of gene action is well understood in prokaryotes. It is in the higher eukaryotes, however, that morphologic adaptations are most apparent, and we certainly understand the historical development of higher eukaryotic phyla and their ecologic relationships in much greater detail than we do the prokaryotes. For these reasons, I will limit my discussion of gene control to the eukaryotes, but unfortunately, this is a complex and a poorly understood subject at the present time. I will indicate below how this complexity and our present ignorance of eukaryotic gene control at the molecular level pose some definite problems for anyone interested in the evolution of adaptation.

Let us consider the complexity underlying the control of a single eukaryotic structural gene. First, the presence of a nuclear membrane uncouples the two basic processes involved in gene expression, transcription and translation. Probably as a result of this uncoupling, transcripts are much more stable in eukaryotes and are subject to modification and processing. This adds more levels at which regulation can occur. Furthermore, the DNA itself in eukaryotes is generally complexed with histones and other proteins, which adds still another dimension to gene control at the transcriptional level, a dimension which is absent in prokaryotes. It is possible, however, to diagram the places at which regulation can occur in the path from gene to phenotype; in this case the phenotype is a biologically active structural gene product. Such a diagram is presented in Fig. 1. It is designed to show that regulation is actually a system involving at least six major processes. It is also deceptively simple, for each of these steps can be subdivided into sometimes quite different processes. For example, and I will elaborate on this below, the beginning of transcription involves *at least* a change in the state of the chromatin around the structural gene, as well as intiation by RNA polymerase. Primary transcripts must be modified and may frequently be spliced, with the concomitant release of intron sequences. Translation involves the interaction of a variety of factors at each of its three basic steps, which are initiation, elongation, and termination. There are, in addition, many possible post-translational modifications, of which the best understood are glycosylation, phosphorylation, adenylation, and spe-

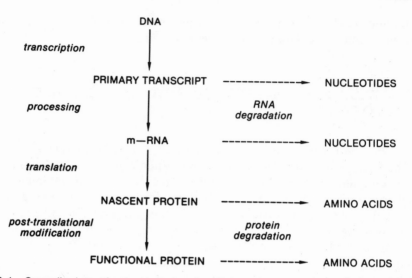

FIG. 1.   Generalized steps involved in the transfer of information from DNA to a functional protein.

cific proteolysis [for reviews of most of these processes, see O'Malley *et al.* (1977), Walker (1977), Prescott and Goldstein (1980), Goldstein and Prescott (1980), and Wold (1981)].

The complexity of regulation can also be seen in a particularly well-understood gene-enzyme system, β-glucuronidase in the mouse. The different factors that affect the expression of this lysosomal enzyme are shown in Fig. 2 and Table IV, which are adapted from Paigen (1979), and Swank *et al.* (1978). The structural gene *Gus-s* is part of a group of tightly

TABLE IV.   Genes Involved in the Regulation of Murine β-Glucuronidase

Gene	Chromosome	Function
Gus-s	5	Structural gene for β-glucuronidase
Gus-t	5	Affects the developmental profile of β-glucuronidase
Gus-u	5	Affects systemic rate of β-glucuronidase synthesis
Gus-r	5	Affects level of β-glucuronidase induction by testosterone
Eg	8	Structural gene for egasyn, a protein which binds to β-glucuronidase in the endoplasmic reticulum
tfm	X	Structural gene for testosterone receptor protein
dw	16	Controls general production of pituitary hormones
lit	6	Controls production of growth hormone in the pituitary gland
pe	13	Affects the rate of β-glucuronidase synthesis in the kidney
bg	13	Affects the amount of β-glucuronidase secreted into the urine by lysosomal exocytosis

linked sites, the others of which behave like *cis*-acting regulatory elements. Thus the *Gus-r* site is responsible for interacting with the complex containing testosterone and its receptor protein. This interaction results in a 100-fold increase in the rate of β-glucuronidase synthesis in the kidney cell. Pituitary growth hormone is another systemic inducer of β-glucuronidase synthesis and may interact at the *Gus-u* site. *Gus-t* is within the *Gus* gene complex and affects the rate of β-glucuronidase synthesis in different tissues during development. Any unlinked locus, like *tfm*, that

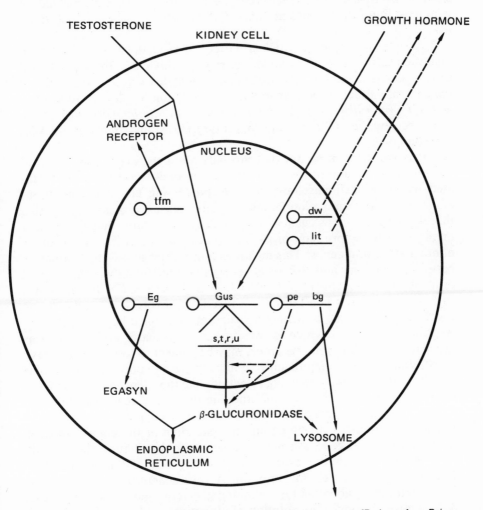

FIG. 2.   Genetic control of β-glucuronidase in the kidney cell of the mouse. [Redrawn from Paigen (1979).]

affects the production of the hormone receptor protein, or, like *dw* and *lit,* affects the hormone itself, will also act to regulate the synthesis of the structural gene product. (In this regard, the *pe* gene affects the synthesis of β-glucuronidase, but it is not known at what level this occurs.) The genes *eg* and *bg* affect the levels or subcellular localization of β-glucuronidase, but do so after translation. They may be quite specific with regard to the *Gus* structural gene product (e.g., egasyn) or affect a number of other lysosomal enzymes (e.g., *bg*). This is only part of the story, since Swank *et al.* (1978) report elevated levels of β-glucuronidase in at least four other mouse mutants, *Pallid, Pale ear, Maroon,* and *Ruby eye.*

Another example in which a variety of regulatory genes has been identified is in *Drosophila melanogaster.* In this case, three structural gene products, xanthine dehydrogenase, aldehyde oxidase, and pyridoxal oxidase, are affected, presumably after translation, by at least four other genes, *maroon-like, aldox-2, cinnamon,* and *low-xanthine dehydrogenase* [for a review, see O'Brien and MacIntyre (1978) and Bentley and Williamson (1979).

It is worth pointing out that virtually all of the major processes involved in expression of a structural gene product (see Fig. 1) are mediated by proteins themselves, such as processing enzymes, initiation, elongation, and termination factors, protein kinases, proteases, etc. Also those proteins, such as egasyn or the androgen receptor, that are directly involved in the expression of β-glucuronidase are, in fact, the products of other structural genes. This means that in a strict sense, the distinction between regulatory and structural genes is meaningful only in the context of a specific and defined system such as β-glucuronidase. I shall return to this point below when I discuss why a direct involvement of a putative regulatory gene in the control of a structural gene must be demonstrated. A protein, however, which is involved in some basic process such as translation, or a hormone receptor, may be subject to too many selective constraints to play a significant role in adaptive evolution. Such proteins may be under intense stabilizing selection, like those enzymes I discussed above when making the case for studying regulatory gene evolution.

Since structural gene expression involves such a complex system of regulatory processes, reaching an understanding of the control of even a single structural gene will not be a easy task. In addition, we are still very ignorant about the exact details of any one of those processes. Consider, for example, transcription. A very general picture of the eukaryotic transcriptional unit is beginning to emerge and is depicted in Fig. 3. When we consider the evolution of *cis*-acting regulatory sites, it is the TATA or Hogness/Goldberg box and the so-called "upstream regulation

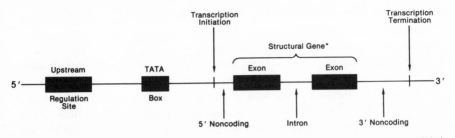

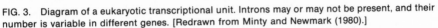

FIG. 3. Diagram of a eukaryotic transcriptional unit. Introns may or may not be present, and their number is variable in different genes. [Redrawn from Minty and Newmark (1980).]

site" with which we are probably dealing. The former is concerned with the initiation of transcription by RNA polymerase II, while the latter site presumably interacts with *trans*-acting regulatory gene products. In each instance, however, the actual mechanisms involved and the requirements in terms of base sequences are poorly understood.

Consider first the TATA box, or what is considered to be the analog of the bacterial promoter. The modal sequence is TATAT/AAT/A [McKeown and Firtel (1981)] and is generally located approximately 30 nucleotides upstream from the place at which transcription begins. Recently, a variety of *in vivo* experimental systems have been described where the transcription of cloned genes containing a wild-type 5' sequence can be compared with the transcription of genes deleted for parts or all of the 5' flanking sequence. It turns out from these studies that the TATA box is necessary but probably not sufficient for precise transcription initiation. Sequences outside of the TATA box that are important in determining whether transcription will occur normally have been identified in SV40 early genes (Benoist and Chambon, 1981), rabbit β-globin (Dierks *et al.*, 1981), and ovalbumin in an *in vitro* system (Tsai *et al.*, 1981). In the latter study, some deletions brought other TATA box sequences close to the structural gene. These, however, did not act as promoters, which again indicates sequences outside the TATA box are important in transcription initiation. Both Guarente and Ptashne (1981) and Faye *et al.* (1981) have examined the effects of deletions of 5' flanking sequences of the yeast iso-1-cytochrome C gene on transcription in homologous host cells. Again, regions as far upstream as 130 base pairs from the TATA box turn out to be important in determining accurate transcription initiation. The exact number and kind of sequences important for transcription in the 5' flanking regions of structural genes, then, are still very much open questions.

The nature of sequences common to "upstream regulatory sites" is also largely unknown. Indeed, the relationship between sequences around the TATA box that influence the initiation of transcription and the "upstream regulatory site" may not always be clearly defined. An obvious strategy when looking for upstream sites is to search for similarities in 5' flanking sequences of structural genes under coordinate control, while keeping in mind that any duplicated structural genes may contain homologous 5' sequences as a result of the duplication event (Robinson and Davidson, 1981). In other words, sequence similarities due solely to historical events such as gene duplication or gene conversion must be differentiated from those due to functional considerations. The heat shock system of *Drosophila melanogaster* is a system in which common, *cis*-acting regulatory sequences should be present. If *Drosophila* larvae, excised tissues, or cultured cells are exposed to a high temperature, e.g., 37°C, for 40 min, an almost unique set of mRNAs is transcribed, while the messages present in the cells at normal temperatures, i.e., 25°C, disappear from the polysomes. From these new messages, a characteristic set of at least seven polypeptides is translated with 8–10 min after the heat shock. The appearance of the mRNAs and the heat shock polypeptides can be correlated in larval salivary glands with at least five distinct puffs on the polytene chromosomes [see review by Ashburner and Bonner (1979)]. It has been possible to associate particular heat shock polypeptides with the various puffs; thus, the puff at 63BC produces a transcript responsible for the polypeptide of 84,000 mol. wt.; 67B for 22,000-, 23,000-, 26,000-, and 27,000-dalton polypeptides; 87A and 87C for the 70,000-dalton polypeptide; and 95D for the polypeptide of mol. wt. 68,000. The abundance of the heat shock mRNAs has made it possible to clone most of the genes in the heat shock system and thus to search for sequences outside of the coding regions that might control the concerted response of these genes. Initial work centered on the heat shock genes at 87A and 87C. At each locus, there are multiple copies of this so-called z gene; two copies at 87A and from three to five at 87C, depending upon the strain. Figure 4 provides a schematic diagram of the two gene clusters and their transcriptional orientation. In 87C, two of the z genes are separated by about 40,000 base pairs. This region is composed of the highly repeated α, β sequences (α is 400 base pairs long, β is 1100). Also present are a few copies of a γ sequence (900 bases long), which is of interest because it is partially homologous to the z_{nc} (z noncoding) sequence which is found at the 5' ends of the coding parts of all the z genes (z_c elements) (Lis *et al.*, 1981). The similarity between γ and z_{nc} is significant since, after heat shock, RNAs with sequences complementary to α,β DNA become abundant, as do, of course, mRNAs complementary to z_c DNA. The α,β RNA is probably not translated, but its coordinate induction with z_c mRNA indicates that both the γ and z_{nc} sequences are important with regard to

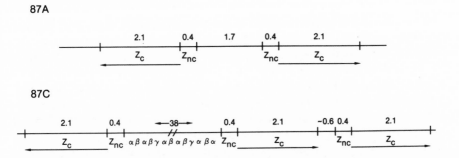

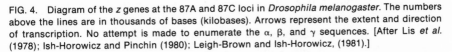

FIG. 4. Diagram of the z genes at the 87A and 87C loci in *Drosophila melanogaster.* The numbers above the lines are in thousands of bases (kilobases). Arrows represent the extent and direction of transcription. No attempt is made to enumerate the α, β, and γ sequences. [After Lis *et al.* (1978); Ish-Horowicz and Pinchin (1980); Leigh-Brown and Ish-Horowicz, (1981).]

the transcription of sequences *cis* to them. There is indeed homology in the z_{nc} sequences from different *z* genes, as shown by Ingolia *et al.* (1980). The homology can be detected as far as 250 base pairs upstream from the TATA box, and is, in fact, so extensive that no "spacers," which might be expected to diverge in sequence, could be detected. Another question, of course, concerns homology between z_{nc} and 5′ flinking sequences from heat shock genes not located at 87A and 87C. Neither Lis *et al.* (1981) nor Holmgren *et al.* (1981) could detect any sequences homologous to z_{nc} at these other loci. Lis *et al.* probed the genome with a section of z_{nc} 170 bases long by in situ hybridization to polytene chromosomes and by Southern blot analysis of genomic DNA, while Holmgren *et al.* compared actual sequence data from z_{nc} with the 5′ ends of three other heat shock genes, 63BC, 67B, and 95D. The latter investigators found two smaller homologous regions between the TATA box and the transcription initiation site but no apparent strong homology in the next 120 bases upstream from the TATA box. It may well be, however, that the upstream regulatory site is still farther upstream. In this regard, Jack *et al.* (1981) have found a complex of three polypeptides which specifically bind to DNA from two heat shock genes from different chromosomal loci. The binding occurs at a sequence some 1000 base pairs upstream from the point at which transcription starts. There is also some recent evidence that at least some of the control of the heat shock genes occurs post-transcriptionally (Scott and Pardue, 1981). In summary, despite their promise, the heat shock genes of *Drosophila* have to date as yet told us very little about the nature of "upstream regulatory sites."

Some recent evidence indicates that not only may upstream regulatory sites be located at surprisingly large distances from structural genes, but that there may be a multiplicity of such sites. The multiplicity may explain how there can be different rates of synthesis of a gene product in different tissues, a phenotype which has been found to exhibit both intra- and interspecific variation (see below for examples). A paradigm for this multiplicity is the mouse α-amylase system [see also Marie *et al.* (1981)]. From the work of Hagenbuchle *et al.* (1981) and Young *et al.* (1981) it is clear that salivary gland and liver α-amylase mRNAs have the same translated sequence and share the first 48 bases immediately adjacent to the initiator codon. They differ, however, in the rest of their nontranslated 5′ ends. The nontranslated 5′ portion of the major liver mRNA is 206 bases long but is only 95 bases long in the salivary gland message. Figure 5 presents a diagram of the structural gene and its 5′ end. Note that there are two distinct promoters, one for each tissue-specific mRNA, and one is located 4500 base pairs away from the structural genes and the other almost 7500 base pairs away. It is still not known if there are tissue-specific "upstream regulatory sites" associated with each promoter or, indeed, if initiation occurs at only one promoter in each tissue. In the salivary gland, however, the liver promoter is included in an intron which must be eliminated in some way in the salivary gland message. It is tempting to attribute the 100-fold difference in the relative abundance of the two mRNAs in the two tissues (Flavell, 1980) to differences in the strengths of the two promoters, which could well depend on upstream sequence differences. On the other hand, differences in mRNA turnover rates or even differences in the rates of processing of the primary transcripts [see also Nevins and Wilson (1981)] could be responsible.

Finally, to point out once again how little we know about transcriptional control in eukaryotes, we must consider how the putative "upstream regulatory sites" might act to enhance or shut down transcription.

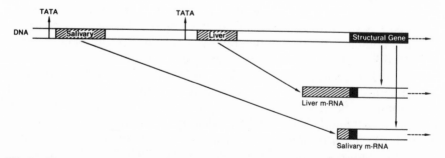

FIG. 5. Diagram of the mouse α-amylase gene and its promoters. The numbers above the line indicate kilobases. Untranslated sequences are shaded, and the positions of the TATA boxes are indicated by arrows. [Redrawn from Young *et al.* (1981).]

Paigen (1979) proposes that, in the induction of β-glucuronidase in the mouse kidney cell, the complex between testosterone and androgen receptor acts at the *gus-r* site to "activate" chromatin prior to transcription. This activation may involve an unpacking of the chromatin, perhaps with the release of nucleosomes. There is, of course, rather extensive evidence that transcriptionally active genes have an "exposed" chromatin structure, i.e., experimentally, they are sensitive to DNase-1 digestion (Wu *et al.*, 1979). The DNase-1 sensitivity can extend several hundred base pairs upstream from the 5' ends of the active genes. There is also some evidence that very short DNase-1 hypersensitive sites are present near the 5' ends of unexpressed heat shock genes (Keene *et al.*, 1981), perhaps to allow access of *trans*-acting regulatory proteins to the 5' ends of these genes [see also Samal *et al.* (1981)].

Recently, the effect of DNA methylation on gene control has been extensively analyzed [for a review see Razin and Riggs (1980)], since methylated cytosines affect the interactions of proteins and DNA. It seems that most transcriptionally active genes are undermethylated (Weintraub *et al.*, 1981) and, in one case, blocking the methylation of a transcriptionally inactive gene with an analog of cytosine resulted in its expression in later cell generations (Groudine *et al.*, 1981). This activated gene also became DNase-1 sensitive. Thus, if the upstream regulatory site, when activated, "prepares" packed chromatin for transcription, does it do it by first demethylating the cytosines, or does this come after the reduction in the number of nucleosomes? Once again, we are ignorant of the molecular details of the process. Also, recall that transcription is but one of many processes involved in the expression of a structural gene, and regulation may occur in any or all of them. As with transcription, there are still enormous gaps in our understanding of each of the other basic processes.

A WORKING DEFINITION OF A REGULATORY GENE

Considering the complexity of regulation and how little we know of its details, it may be foolish to even attempt to define a regulatory gene. Nevertheless, it is important to have at least a working definition if we are to study the evolution of regulatory genes and their role in adaptation. Thus, let us define a regulatory gene as any gene that directly affects the amount, the tissue distribution, or the developmental profile of another gene product. This working definition tells us, if nothing else, what kinds of phenotypes might result from genetic changes at regulatory loci, namely

(a) differences in the levels of a structural gene product in some or all of the tissues of an organism (quantity variants), (b) differences in the presence or absence of the structural gene product in different tissues of the organism (tissue variants), and (c) differences in the time of appearance during development of the structural gene product (temporal variants). It should be clear that these need not be mutually exclusive categories of regulatory gene variants. In fact, in most cases (see below) there is substantial overlap in the phenotypes, e.g., a particular strain may have an elevated level of a structural gene product because that gene becomes active earlier in development.

REGULATORY VARIANTS IN NATURAL POPULATIONS

There is abundant interspecific and intraspecific variation affecting protein levels, much of which can be analyzed genetically and/or biochemically. Thus, Wilson *et al.* (1977; Table 2) have listed 18 cases where in the same tissue from different mammalian species a particular protein exhibited a tenfold or greater difference in specific activity. McDonald and Avise (1976) found that *Drosophila* species are extremely variable in their levels of alcohol dehydrogenase (Adh). It has also been possible to select for increased levels of Adh in *D. melanogaster* with ethanol as a selective agent (David and Bocquet, 1977). McDonald *et al.* (1977) showed that the response did not involve the selection of mutants at the structural gene locus. In a similar study, Devonshire (1977) selected an insecticide-resistant strain of an aphid which had a 60-fold increase in the specific activity of a carboxyl esterase, an enzyme which can detoxify the insecticide. In the above cases, the enzymes from both control and selected lines were identical when several biochemical parameters, including K_m, were examined. Indeed, it may not always be necessary to use selection to find intraspecific variants affecting enzyme levels. In *D. melanogaster,* for example, McDonald and Ayala (1978) constructed a number of strains homozygous for a particular Adh allele—the structural gene is on the second chromosome—but which had different third chromosomes, and found several significant interstrain differences in Adh levels. In a more extensive study, Laurie-Ahlberg *et al.* (1980) screened a number of lines whose genomes had been held constant except for one of the three major chromosomes. They measured the specific activities of seven enzymes whose structural gene locations were known. There was highly significant interstrain variation in the levels of five enzymes that could only be accounted for by factors on chromosomes that did not carry the structural gene. The specific activity differences were as high as twofold in some cases [see also Wilson and McDonald (1981)]. It is obviously possible to carry out a genetic analysis on such a large phenotypic difference.

Elegant descriptions of tissue-specific enzyme patterns in closly related *Drosophila* species have been reported by Dickinson (1981). He examined by gel electrophoresis five enzymes in 13 tissues from 27 Hawaiian *Drosophila* species. Out of the 65 combinations, there were 19 cases where in an interspecific comparison, the enzyme was present in one species, but essentially absent in the other. There were also many comparisons where the amount of the enzyme in the same tissue from different species was clearly different in a quantitative sense. A similar example can be seen in Table V, which is taken from a study by Ahearn and Kuhn (1981) on aldehyde oxidase in various larval tissues from ten species of Hawaiian *Drosophila*. There are both qualitative and quantitative differences in the tissue distribution of the enzyme in the ten species; in fact, each species has its own unique quantitative pattern in the imaginal discs. The last four species in the table appear to have generally lower overall levels of aldehyde oxidase activity, and three of the four are from the same taxonomic subgroup. However, *D. formella* is from that same subgroup, yet has a very different aldehyde oxidase pattern in the larval tissues. The impression is, then, that tissue-specific enzyme patterns are evolving rapidly in these *Drosophila* species. Similar studies on fish species by Whitt and his co-workers (Whitt *et al.*, 1977; Phillip *et al.*, 1979), however, indicate there may be greater evolutionary conservatism within the vertebrates with regard to the tissue distribution of enzymes. Only one enzyme, Adh, was common to the studies on the Hawaiian *Drosophila* species and the fish species examined by Phillip *et al.* (1979), however. Also, the majority of the 33 structural gene loci scored in the latter study code for enzymes critical in intermediary metabolism and may be less prone to vary than the enzymes used in the *Drosophila* investigation.

Finally, it is now apparent that many tissue-specific enzyme pattern differences may really be due to the effects of temporal variants at regulatory loci. This could well be true of many of the cases cited by McDonald and Avise (1976) and by Dickinson (1981). A dramatic example of an interspecific difference in the developmental profile of a structural gene product comes from Nair *et al.* (1977). Two Hawaiian *Drosophila* species, *D. setimentosum* and *D. ochrobasis,* share an allozyme polymorphism at an esterase locus with two of the four alleles common to both species. In *D. ochrobasis,* however, the gene is not expressed in adult tissues, only in the larvae. Adult interspecific hybrids express only

TABLE V. Distribution of Aldehyde Oxidase Activity in Imaginal Discs from Larvae of Ten Species of Hawaiian *Drosophila*[a]

Imaginal disc	Heteroneura	Silvestris	Formella	Adiastola	Setosimentum	Pilimana	Grimshawi	Bostrycha	Disjuncta	Prostopalpis
Labial	+++	+++	+++	++	+	++	+	+	+	+
Eye-antennal	+++	+	+++	+	+	+	-	+	-	+
First leg	+++	+	++	+	+	++	-	+	+	-
Wing	++	++	+	++	++	++	-	+	-	-
Second leg	+++	+	++	+	++	++	-	+	+	-
Haltere	+++	+	+	++	++	+	-	+	-	-
Third leg	++	+	++	+	+	+	-	+	+	-
Genital	+	NS	NS	++	NS	++	NS	-	-	NS

[a] +++, ++, +: dark, moderate, very light staining intensity. -: no detectable adehyde oxidase activity. NS: not scored. After Ahearn and Kuhn (1981).

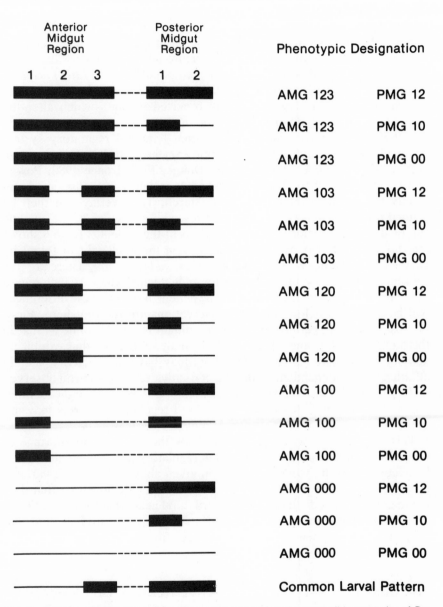

FIG. 6. Patterns of α-amylase activity in the adult midgut found among wild-type strains of *Drosophila melanogaster*. Dark bars indicate the presence of α-amylase activity. [After Doane (1980).]

the allele from *D. setimentosum,* which implies the putative regulatory effect is *cis*-acting.

Temporal gene variation can also be found within as well as between species. The recombinant inbred mouse strains used by Paigen and co-workers to find temporal genes for β-glucuronidase, β-galactosidase, and other enzymes were derived from outbred populations of *Mus musculus* [for a review see Paigen (1979)]. Dickinson (1975), Bewley (1981), and Hoorn and Scharloo (1981) have all reported variants which alter the developmental program of a *Drosophila* enzyme.

Perhaps the most extensively studied temporal regulatory system which ultimately affects a tissue-specific enzyme pattern is the α-amylase system of *Drosophila* (Abraham and Doane, 1978; Doane 1980; Powell and Lichtenfels, 1979). The larval or adult midgut can be dissected and simply laid on a medium containing starch. The overlay can then be stained for amylase activity, and the presence or absence of the enzyme can be directly detected in various regions of the midgut. Figure 6 shows the different patterns of amylase activity in the adult midgut found by Doane (1980) among wild-type strains of *D. melanogaster*. There are five subregions in the anterior and posterior midgut which can vary with regard to the presence of amylase activity. The genetic basis for these differences will be discussed below, but two observations are relevant here. First, in most cases, the pattern in the larval midgut is not the same as that in the adult from the same strain, and, second, subregions that exhibit no activity in the midguts of young adults may have high activity later in adulthood. Age-dependent switching of α-amylase has also been observed in *D. pseudoobscura* (Powell and Licthenfels, 1979). This indicates that a complex system of temporally expressed regulatory controls may underlie that ontogenetic expression of amylase in *Drosophila*.

This brief review was intended to show that there is no shortage of naturally occurring putative variants in regulatory genes; regulatory, that is, according to the working definition given above. Once such variants are discovered, it becomes very important to then demonstrate that they are in genetic units distinct from the structural gene, that they specifically affect the regulation of the particular structural gene product, and that they affect some phenotype that influences fitness.

THE GENETIC ANALYSIS OF PUTATIVE REGULATORY VARIANTS

It is important, but not sufficient, to show biochemically that the structural gene product from the putative regulatory variant strain is the same as that from a normal or wild-type strain. In lieu of primary sequence determinations, these comparisons most often involve K_m determinations, inhibitor sensitivities, and thermal stability measurements (McDonald and Ayala, 1978; Edwards *et al.*, 1977). Since one expects essentially negative results when comparing the protein from a regulatory variant with the homolog from a standard strain, it is difficult to know when the number of tests is adequate. But the real reason genetic tests are essential is that structural gene alterations can mimic phenotypes due to regulatory gene mutations. Thus, a duplication of the structural gene locus can produce a nice mimic of a quantity variant. It is also easy to imagine that a mutant enzyme, because of its own particular amino acid sequence, might be more liable to proteolytic destruction in some tissue or at some particular time during development, thus appearing to be a tissue or temporal variant. Such an amino acid sequence difference may not alter the structure of the enzyme enough to affect the K_m or even its thermal stability.

Even apart from the necessity of ruling out structural gene differences, it is important to analyze the genetic basis of putative regulatory variants for quite another reason. That is, the genetic basis for such variants must be simple enough so we can ultimately understand their evolutionary dynamics. If, for example, the phenotypic difference between a high and low line with regard to some enzyme activity level does not segregate cleanly in the F_2, i.e., it disappears into a morass of polygenes each with small effects, then such a difference is of little use to us. It is true that we would know that genetic factors are responsible for the difference, but we would have little hope of eventually understanding how they work individually in producing the phenotypic difference. Consequently, we could not measure their frequencies in populations with any precision, nor could we ever hope to assign selection coefficients to genotypes. There are several instances in fact where the genetic basis for a quantity variant could not be worked out because an F_2 produced no clear-cut classes (Powell and Lichtenfels, 1979; Byjlsma, 1980). Nevertheless, there are many more cases where the situation is simple enough so that a satisfactory genetic description of a quantity, tissue, or temporal variant can be obtained. Many cases even appear to be monogenic, or, if more complex, the number of genes and their relative contributions to the phenotype can be ascertained [see especially Synder (1978)]. Consider once again (see Fig. 1) the control of murine β-glucuronidase. The situation, as we have seen, is complicated, but the separation of mutants at the many loci into different recombinant inbred lines made a complete genetic analysis possible. Likewise in *Drosophila*, the ability to use balancer chromosomes, i.e., chromosomes with dominant markers and inversions that suppress crossing over, enables one to partition the genome chromosome by chromosome when performing a genetic analysis. One can then quickly judge whether a genetic basis of the particular phenotype will be simple enough to justify further intrachromosomal mapping of the responsible factors.

Assuming, then, that initial crosses indicate the genetic basis for a putative regulatory variant is simple enough to analyze, what then? Because the kind of control exerted by regulatory genes linked to the structural gene may be very different from unlinked regulatory genes, some further mapping should be done. In *Drosophila*, with the use of balancer chromosomes, it is possible to screen for linked and unlinked regulatory variants independently. The strategy is to construct strains with only the chromosome carrying the structural gene held constant (i.e., isogenic for the whole chromosome). Any difference in the level or the tissue or temporal specificity of the structural gene product would then be due to mutants at loci on other chromosomes. Alternatively, in order to detect possible linked regulatory loci, the chromosome carrying the structural gene can be varied among strains while the rest of the genome is held constant. Emerging from the studies so far is a general rule, that unlinked regulatory genes are *trans*-acting and dominant in their expression, whereas alleles of closely linked regulatory genes affect only the allele at the structural gene locus linked *cis* to them. There expression is said to be additive. Table VIA shows these interactions in diagrammatic form. Note that an unlinked structural gene duplication would violate the rule, i.e., the F_1 would indicate additivity, not dominance.

Mapping unlinked regulatory genes should present no particular problems to the geneticist. On the other hand, their dominance has important and troublesome ramifications for one interested in surveying natural populations (see below). Mapping putative regulatory variants that are linked to the structural gene, on the other hand, can be a major problem, but one made all the more important by the fact that structural gene mutants that mimic regulatory gene variants will also exhibit an additive expression pattern in the F_1. In order to show clearly that the regulatory site is different from that occupied by the structural gene, it is important to have structural gene variants in addition to the regulatory variants. The experimental strategy, then, is to recover recombinants between the two kinds of variants. In the example shown in Table VIA, one would look for the RG^{hi} SG^F or RG^{lo} SG^S recombinant chromosomes in the offspring. This straightforward strategy is complicated by several considerations, however. First, there should be some assurance that the structural gene variants are really that. Allozyme variants—which must, incidentally, show a codominant pattern of inheritance—are probably the best to use, since the *cis* effect of linked regulatory gene variants can be

TABLE VIA. Structural Gene Product Levels in Genotypes
Containing Either Unlinked or Linked Regulatory Genes[a]

Position of RG relative to SG	Genotype	Amount of SG product
Unlinked	RG^{hi}/RG^{hi}; SG/SG	2X
	RG^{lo}/RG^{lo}; SG/SG	X
	RG^{hi}/GR^{lo}; SG/SG	2X
Linked	RG^{hi} SG/RG^{hi} SG	2X
	RG^{lo} SG/RG^{lo} SG	X
	RG^{hi} SG/RG^{lo} SG	1.5X

[a] SG, a regulated structural gene. RG, any regulatory gene. hi/lo, alleles of RG affecting amount of SG.

seen with gel electrophoresis. Thus, the hi variant of RG would cause more of the Sg^F subunit to be produced in the recombinant class, thereby altering the electrophoretic pattern. The expected pattern is shown in Fig. 7 (also see Table VIB). The other recombinant class, RG^{lo} SG^S, would also be distinguishable from the other phenotypes on a quantitative basis or after outcrosses to an SG^F stock. Thermostability variants can and have been used as structural gene markers in crosses designed to separate regulatory and structural genes by crossing over [see Paigen (1979) for examples]. They are more cumbersome to work with than the allozymes, however, and in some cases might not be due to mutations in the structural gene (Bentley and Williamson, 1979).

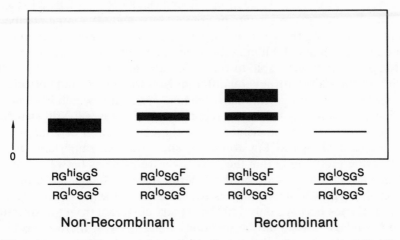

FIG. 7. Expected gel patterns of the offspring from the cross diagrammed in Table VIB.

TABLE VIB. Hypothetical Genotypes in a
Crossover Experiment Designed to Show
that Linked Regulatory and Structural
Genes Are at Separable Sites[a]

Parent 1: $RG^{hi}\ SG^S/RG^{lo}\ SG^F$
Parent 2: $RG^{lo}\ SG^S/RG^{lo}\ SG^S$
Recombinant offspring: $RG^{hi}\ SG^F/RG^{lo}\ SG^S$, $RG^{lo}\ SG^S/RG^{lo}\ SG^S$

[a] S/F: alleles of SG specifying slow or fast allozymes.

A second major consideration in mapping *cis*-acting regulatory sites
concerns the resolving power of the genetic analysis. As we have seen,
these mutations, assuming they involve promoters or "upstream regulatory sites," may be very close to the structural gene itself. In this regard,
it is worth briefly reviewing the elegant studies of Chovnick and his co-
workers on the genetic fine structure of the gene for xanthine dehydrogenase (Xdh) in *Drosophila melanogaster* [for reviews, see McCarron *et
al.* (1979) and Chovnick *et al.* (1980)]. They have shown that each of two
cis-acting regulatory mutants, ry^{i409H} (ry^{+4}) and ry^{i1005L} (ry^{+10}) map in
a region adjacent to but clearly separable genetically from the structural
gene. To do this, however, required a sample of well over one million
zygotes, and their genetic system contained, in addition to electrophoretic
variants and putative control site mutants, the following elements: (1)
lethal mutations in genes flanking the *Xdh* locus, (2) deficiencies of the
Xdh locus, (3) a chemical selection scheme which kills virtually all zygotes
except wild-type recombinants or convertants, and (4) a large number of
mutations marking the structural gene. The latter included leaky mutants
and mutants which exhibit interallelic complementation. With these tools,
so to speak, they were able to recombine the mutant ry^{+4} site, which is
analogous to RG^{hi}, with several different Xdh allozymes, and very clearly
demonstrate a *cis* effect. Similarly, the ry^{+10} mutant, which is an under-
producer of Xdh, has now been recombined with several Xdh structural
gene markers. These experiments and the construction of a map of the
Xdh locus as shown in Fig. 8 was an enormous but important task. In
considering Fig. 8, note that the cross-over frequencies between the control and structural sites are on the order of one in 100,000! Note also that
the resolving power of the Xdh system has allowed the genetic localization
of control sites perhaps only 1000 base pairs upstream from the structural
gene boundary. (It should be mentioned that, as yet, there is no direct
evidence on the length of the *Xdh* control region in terms of base pairs.

The numbers in Fig. 8 come from an estimate of gene size based on the molecular weight of the structural gene product.) This could mean that the i^{409H} and i^{1005L} sites may be in an "upstream regulation site" or even in a promoter. Recall from Fig. 5 that the promoters for murine α-amylase are well over 1000 base pairs upstream from the structural gene.

Given the genetic sophistication needed to map *cis*-acting control sites in the *Xdh* gene, it is not surprising that in most cases proposed *cis*-acting regulatory sites have not been separated genetically from the structural gene (Bewley, 1981; Dickinson, 1978, 1980; Grossman, 1981). On the other hand, it is encouraging that a few investigators have managed to show recombination between linked control sites and structural genes. For example, Dickinson (1975) found 2.3% recombination between electrophoretic sites in the aldehyde oxidase gene and a *cis*-acting temporal mutant. Abraham and Doane (1978) also showed that a gene influencing the α-amylase pattern in the adult midgut, called *map*, is about two crossover units from the α-amylase structural gene. In fact, *map* may be a complex locus, with one *cis*-acting gene *map-AMG*, determining the anterior midgut pattern, and a second *trans*-acting gene, *map-PMG*, responsible for amylase activity in the two subregions of the posterior midgut (see Fig. 6). In this regard, Powell and Lichtenfels (1979) could not genetically identify a locus analogous to *map* in *D. pseudoobscura*, even though "PMG-12"- and "PMG-00"-like phenotypes could be selected. The regulation of α-amylase in *Drosophila* appears, then, to be tantalizingly complex, and, since under some dietary conditions a high α-amylase level appears to be an adaptive trait (Hoorn and Scharloo, 1981), it is important that we understand it in detail.

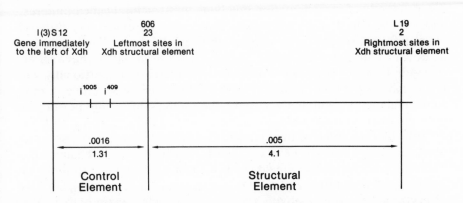

FIG. 8. A genetic map of the *Xdh* locus in *Drosophila melanogaster*. The order of i^{1005} relative to i^{409} is not known. Numbers above the line are in map units. Numbers below the line are in kilobases (estimated). [After Chovick *et al.* (1980).]

The observations on aldehyde oxidase and amylase in *Drosophila* imply that *cis*-acting elements need not always be tightly linked to the structural gene. Indeed, Paigen (1979) points out that the inability to recover large numbers of recombinant inbred lines in the mouse allows one only to separate mutant sites more than 0.1 map unit from each other, or on average, 100,000–200,000 base pairs apart. Yet in some cases recombinants between structural gene and temporal variants have been found among samples of inbred mouse lines. Thus, whereas it is important to work with an organism which can be manipulated genetically when working with regulatory genes, it may not have to be *Drosophila* or some other organism where one million or more zygotes can be screened in a reasonable number of generations.

DIRECT VERSUS INDIRECT EFFECTS OF REGULATORY GENES

I have already discussed our ignorance about the exact mechanisms of gene control, and how complex they are likely to be. It is unrealistic at the present time to undertake an in-depth study on the exact function of every quantity, tissue, or temporal variant that might affect a structural gene product. Yet at least some investigation of the role of each putative regulatory gene is necessary. The very complexity of gene product interaction implies there will be genes whose main function has virtually nothing to do with the control of a particular structural gene, yet mutations in these other genes might indirectly influence the level of the structural gene product. The evolution of these other genes might then proceed independently of any adaptation involving the structural gene of interest, and, consequently, a study of these genes would provide very little relevant information about the adaptation itself. In short, they would lead us down "blind alleys." For example, Clarke *et al.* (1979) have shown that Adh levels in *Drosophila melanogaster* increase proportionally with the square of the body weight. Body weight is very susceptible to environmental factors, and, in their study, flies of the same genotype but from different cultures exhibited as much as a fourfold difference in their levels of Adh. Indeed, some of the increase in Adh following ethanol selection of *D. melanogaster* can be accounted for by body weight changes [see Table I in Ayala and McDonald (1980)]. In another example Bulfield and Trent (1981) noted that a particular mouse strain, SM, has a higher level of erythrocyte glyceraldehyde-3-phosphate dehydrogenase activity than strain C57BL, but crosses between the two strains indicated a number of genes with small effects are responsible for the difference. It turns out that the enzymes from the two strains are more stable in the presence of NAD, and the erythrocytes in strain SM have considerably higher levels of this cofactor. The enzyme activity difference, then, is probably due to the greater protective effect of NAD on the enzyme in the cells of strain SM. Thus, while there still may be a structural gene difference—the enzymes from the two strains have different thermostabilities—the "enzyme quantity variant" appears to be due to the action of a large number of other enzymes which affect intracellular levels of NAD. Finally, consider the mouse gene *neu-1* (Peters *et al.*, 1981), which may be a structural gene for neuraminidase. This enzyme posttranslationally modifies several enzymes, including α-glucosidase, α-mannosidase, and acid phosphatase. *Neu-1* behaves like a *trans*-acting regulatory mutant, yet it is not involved specifically with the control of any one of those three structural genes.

Since the interactions that involve a particular structural gene product may range from the trivial, e.g., genes influencing body weight, to the profound, e.g., a *cis*-acting promoter, a decision must be made by each investigator as to what constitutes a "direct" effect. Thus, how should one judge a gene like *neu-1*? Should it be considered as one of the regulatory genes directly affecting, for example, acid phosphatase? Certainly, as one learns more about the regulation of any structural gene, more instances will arise where it is difficult to distinguish between a direct and an indirect effect. Perhaps we must simply consider it an occupational hazard in the study of gene regulation.

FUTURE PROBLEMS IN THE STUDY OF REGULATORY GENE VARIATION

We have already reviewed the ample evidence that regulatory gene variation is extensive in natural populations. Very few systematic surveys have been reported, however. Snyder (1978) showed that subspecies of *Peromyscus maniculatus* could be differentiated on the basis of regulatory gene variants affecting the quantitative levels of hemoglobin subunits. Powell (1979) surveyed populations of *D. pseudoobscura* for variation in the adult midgut α-amylase pattern (see Fig. 6). Thirteen distinct patterns were found and the frequencies of those patterns were unrelated to frequencies of alleles of the α-amylase structural gene. Unfortunately, the genetic basis of the α-amylase midgut patterns in this species is complex, and the frequencies of individual alleles at even one locus could not be measured.

Even in systems where individual regulatory genes can be identified, however, it may not always be possible to conduct adequate surveys, especially when the phenotype cannot be measured by gel electrophoresis. Then, too, remember that most *trans*-acting genes have dominant effects, and complementation tests can only be done with recessive mutants. If we cannot carry out complementation tests, how can we know if we are dealing with the same regulatory gene when we survey many different populations? Indeed, when complementation tests are not feasible, every regulatory variant that shows the same phenotype would have to be genetically mapped. This obviously could become a serious rate-limiting step in any population survey. It seems clear that *cis*-acting regulatory variants that are linked to a structural gene that has alleles coding for several different allozymes may be the most favorable system for population genetic analyses.

A final problem concerns adaptation. In much of the discussion so far, it has simply been assumed that some change in the quantity or in the spatial or temporal expression of a protein is more likely to be adaptive than a change in the structural gene itself. In the future, that assumption must be justified for each system studied. Thus, it may be easy to see that a polymorphism involving the control of β-galactosidase in *E. coli* would be adaptive, but it is much harder to imagine why a polymorphism involving, for example, the presence of aldehyde oxidase in an imaginal disc would be important. In short, it may be very difficult to demonstrate a correlation between a regulatory phenotype and some environmental factor and then to design experiments which can critically test predictions about regulatory gene frequency changes in appropriately stressed populations. Shifting our attention from structural genes to regulatory genes when we try to understand the evolution of adaptation will surely not make life any easier. Nevertheless, we should be encouraged by some of the reports already mentioned, e.g., the ability of Devonshire (1977) to select in aphids for increased levels of a detoxifying enzyme, or of David and Bocquet (1977) to select in *Drosophila* for increases in Adh activity with ethanol (McDonald and Ayala, 1978), or of Hoorn and Scharloo (1981), again in *Drosophila*, to select for early expression of an α-amylase allele in strains put on diets with starch as a limiting growth factor.

CONCLUSION

In this chapter, I have attempted to make a case for studying the regulation of specific structural genes rather than just the properties of the structural gene products themselves, even though the molecular basis of eukaryotic gene control is bound to be complex and, at present, poorly understood. The available evidence indicates that there are abundant regulatory gene variants in populations, but for any system it is important to (1) clearly demonstrate a separate genetic basis for any regulatory phenotype—separate, that is, from the structural gene; and (2) show that the regulatory gene is directly involved in the control of the structural gene. Finally, as evolutionary biologists we must devise efficient ways to quantify the extent of intraspecific and interspecific variation in regulatory genes, and critically examine the role of this variation in adaptation.

ACKNOWLEDGMENTS

Many of the ideas in this article resulted from a discussion of regulatory gene evolution held at the 1981 meetings of the Society for the Study of Evolution in Iowa City. I would like to thank all the participants in that discussion for their valuable comments and suggestions. The input of Douglas Cavener and Douglas Knipple, who made several important improvements of the final manuscript, is especially appreciated.

REFERENCES

Abraham, I., and Doane, W. W., 1978, Genetic regulation of tissue specific expression of amylase structural genes in *Drosophila melanogaster, Proc. Natl. Acad. Sci. USA* **75**:4446–4450.
Ahearn, J., and Kuhn, D., 1981, Aldehyde oxidase distribution in the picture-winged Hawaiian *Drosophila*: Evolutionary trends, *Evolution* **35**:635–646.
Ashburner, M., and Bonner, J., 1979, The induction of gene activity in Drosophila by heat shock, *Cell* **17**:241–254
Avise, J. C., 1977, Is evolution gradual or rectangular? Evidence from living fishes, *Proc. Natl. Acad. Sci. USA* **74**:5083–5087.
Avise, J., and Ayala, F., 1975, Genetic change and rates of cladogenesis, *Genetics* **81**:757–773.
Avise, J. C., and Ayala, F. J., 1976, Genetic differentiation in speciose versus depauperate phylads: Evidence from the California minnows, *Evolution* **30**:46–58.
Ayala, F. J., and MacDonald, J., 1980, Continuous variation: Possible role of regulatory genes, *Genetica* **52/53**:1–15.
Benoist, C., and Chambon, P., 1981, *In vivo* sequence requirements of the SV40 early promoter region, *Nature* **290**:304–310.
Bentley, M., and Williamson, J. H., 1979, The control of aldehyde oxidase and xanthine

dehydrogenase activities by the *cinnamon* gene in *Drosophila melanogaster, Can. J. Genet. Cytol.* **21**:457–471.

Bewley, G., 1981, Genetic control of the developmental program of L-glycerol-3-phosphate dehydrogenase isozymes in *Drosophila melanogaster*: Identification of a *cis* acting temporal element affecting GPDH-3 expression, *Dev. Genet.* **2**:113–130.

Briscoe, D. A., Robertson, A., and Malpica, J.-M., 1975, Dominance at the Adh locus of adult *Drosophila melanogaster* to environmental alcohol, *Nature* **255**:148–149.

Bruce, E., and Ayala, F. J., 1979, Phylogenetic relationships between man and the apes: Electrophoretic evidence, *Evolution* **33**:1040–1956.

Bulfield, G., and Trent, J., 1981, Genetic variation in erythorcyte NAD levels in the mouse and its effect on glyceraldehyde phosphate dehydrogenase activity and stability, *Biochem. Genet.* **19**:87–94.

Bush, G. L., Case, S. M., Wilson, A. C., and Patton J. L., 1977, Rapid speciation and chromosomal evolution in mammals, *Proc. Natl. Acad. Sci. USA* **74**:3942–3946.

Byjlsma, R., 1980, Polymorphism at the G6PD and 6PGD loci in *Drosophila melanogaster*, IV. Genetic factors modifying enzyme activity, *Biochem. Genet.* **18**:699–716.

Carson, H. L., Johnson, W., Nair P., and Sene, F., 1975, Allozymic and chromosomal similarity in two Drosophila species, *Proc. Natl. Acad. Sci. USA* **72**:4521–4525.

Cavener, D., 1979, Preference for ethanol in *Drosophila melanogaster* associated with the alcohol dehydrogenase polymorphism, *Behav. Genet.* **9**:359–365.

Cavener, D., and Clegg, M. T., 1978, Dynamics of correlated genetic systems. IV. Multilocus effects of ethanol stress environments, *Genetics* **90**:629–644.

Cavener, D. R., and Clegg, M. T., 1981, Multigenic response to ethanol in *Drosophila melanogaster, Evolution* **35**:1–10.

Champion, A. B., Prager, E. M., Wachter, D., and Wilson, A., 1974, Microcomplement fixation, in: *Biochemical and Immunological Taxonomy of Animals* (C. A. Wright, ed.), pp. 397–414, Academic Press, London.

Cherry, L., Case, S., and Wilson, A. C., 1978, Frog perspective on the morphological difference between humans and chimpanzees, *Science* **200**:209–211.

Chovnick, A., McCarron, M., Clark, S. H., Hilliker, A. J., and Rushlow, C. A., 1980, Structural and functional organization of a gene in *Drosophila melanogaster*, in: *Development and Neurobiology of Drosophila* (O. Siddiqui, P. Bau, Linda Hall, and Jeffrey Hall eds.), pp. 3–23, Plenum, New York.

Clarke, B., 1975, The contribution of ecological genetics to evolutionary theory: Detecting the direct effects of natural selection on particular polymorphic loci, *Genetics (Suppl.)* **79**:101–113.

Clarke, B., Camfield, R., Galvin, A., and Pitts, C., 1979, Environmental factors affecting the quantity of alcohol dehydrogenase in *Drosophila melanogaster, Nature* **280**:517–518.

Craddock, E., and Johnson, W., 1979, Genetic variation in Hawaiian Drosophila. V. Chromosomal and allozymic diversity in *Drosophila sylvestris* and its homosequential species, *Evolution* **33**:137–155.

David, J., and Bocquet, C., 1977, Genetic tolerance to ethanol in *Drosophila melanogaster*: Increase by selection and analysis of correlated responses, *Genetics* **47**:43–48.

Day, T., Hiller, P., and Clarke, B., 1974, The relative quantities and catalytic activities of enzymes produced by alleles at the alcohol dehydrogenase locus in *Drosophila melanogaster, Biochem. Genet.* **11**:155–165.

Devonshire, A., 1977, The properties of a carboxylesterase from the peach potato aphid, *Myzus persica* (Sulz), and its role in conferring insecticide resistance, *Biochem. J.* **167**:675–683.

Dickerson, R., 1971, The structure of cytochrome c and the rates of molecular evolution, *J. Mol. Evol.* **1:**26–45.

Dickinson, W. J., 1975, A genetic locus affecting the developmental expression of an enzyme in *Drosophila melanogaster, Dev. Biol.* **42:**31–140.

Dickinson, W. J. 1978, Genetic control of enzyme expression in *Drosophila*; A locus influencing tissue specificity of aldehyde oxidase, *J. Exp. Zool.* **206:**333–342.

Dickinson, W. J., 1980, Tissue specificity of enzyme expression regulated by diffusable factors: Evidence in Drosophila hybrids, *Science* **207:**995–997.

Dickinson, W. J., 1981, Evolution of patterns of gene expression in Hawaiian picture-winged Drosophila, *J. Mol. Evol.* **16:**73–94.

Dierks, P., vanOoyen, A., Mantei, N., and Weissman, C., 1981, DNA sequences preceding the rabbit β-globin gene are required for formation in mouse L cells of β-globin RNA with the correct 5′ terminus, *Proc. Natl. Acad. Sci. USA* **78:**1411–1415.

Doane, W. W., 1980, Midgut amylase activity patterns in Drosophila: Nomenclature, *Drosophila Information Service* **55:**36–39

Edwards, T. C. R., Candido, E. P. M., and Chovnick, A., 1977, Xanthine dehydrogenase from *Drosophila melanogaster*, a comparison of the kinetic parameters of the pure enzyme from two wild isoalleles differing at a putative regulatory site, *Mol. Gen. Genet.* **154:**1–6.

Ewens, W., 1977, Population genetics theory in relation to the neutralist–selectionist controversy, in: *Advances in Human Genetics*, Volume 8 (H. Harris and K. Hirschhorn, eds.), pp. 67–131, Plenum Press, New York.

Faye, G., Leung, D., Tatchell, K., Hall, B. D., and Smith, M., 1981, Deletion mapping of sequences essential for *in vivo* transcription of the iso-1 cytochrome c gene, *Proc. Natl. Acad. Sci. USA* **78:**2258–2262.

Ferris, S., and Whitt, G., 1979, Evolution of the differential regulation of duplicate genes after polyploidization, *J. Mol. Evol.* **12:**267–317.

Fitch, W. M., 1972, Does fixation of neutral mutations form a significant part of observed evolution in proteins?, in: *Evolution of Genetic Systems* (H. H. Smith, ed.), pp. 186–216, Gordon and Breach, New York.

Fitch, W. M., 1975, Molecular evolutionary clocks, in: *Molecular Evolution* (F. J. Ayala, ed.), Sinnauer, Sunderland, Massachusetts.

Flavell, R. A., 1980, The transcription of eukaryotic genes, *Nature* **285:**356–357.

Goldstein, L., and Prescott, D. M., 1980 (eds.), *Cell Biology, A Comprehensive Treatise.* Volume 3, *Genetic Expression: The Production of RNA's*, Academic Press, New York.

Goodman, M., and Moore, G., 1977, Use of Chou-Fasman amino acid conformational parameters to analyze the organization of the genetic code and to construct protein geneologies, *J. Mol. Evol.* **10:**7–47.

Goodman, M., Moore, G., and Matsuda, G., 1975, Darwinian evolution in the geneology of hemoglobin, *Nature* **253:**603–608.

Gottlieb, L., and Greve, L., 1981, Biochemical properties of duplicated isozymes of phosphoglucose isomerase in the plant, *Clarkia xantiana, Biochem. Genet.* **19:**155–172.

Grossman, A., 1981, Analysis of genetic variation affecting the relative activities of fast and slow ADH dimers in *Drosophila melanogaster* heterozygotes, *Biochem. Genet.* **18:**765–780.

Groudine, M., Eisenmann, R., and Weintraub, H., 1981, Chromatin structure of endogenous retroviral genes and activation by an inhibitor of DNA methylation, *Nature* **292:**311–317.

Guarente, L., and Ptashne, M., 1981, Fusion of *Escherichia coli lac z* to the cytochrome c gene of *Saccharomyces cerevisiae, Proc. Natl. Acad. Sci. USA* **78:**2199–2203.

Hagenbuchle, O., Tosi, M., Schibler, U., Bovey, R., Wellauer, P., and Young, R., 1981, Mouse liver and salivary gland α-amylase mRNA's differ only in 5' non-translated sequences, *Nature* **289**:643–646.

Hartl, D., and Dykhuizen, D., 1979, A selectivity driven molecular clock, *Nature* **281**:230–231.

Holmgren, R., Corces, V., Morimoto, R., Blackman, R., and Meselson, M., 1981, Sequence homologies in the 5' regions of four *Drosophila* heat shock genes, *Proc. Natl. Acad. Sci. USA* **78**:3775–3778.

Hoorn, A. J. W., and Scharloo, W., 1981, The functional significance of amylase polymorphism in *Drosophila melanogaster*. VI. Duration of development and amylase activity in larvae when starch is a limiting factor, *Genetica* **55**:195–202.

Ingolia, T., Craig, E. A., and McCarthy, B. J., 1980, Sequence of three copies of the gene for the major Drosophila heat shock induced protein and their flanking regions, *Cell* **21**:669–679.

Ish-Horowicz, D., and Pinchin, S. M., 1980, Genomic organization of the 87A7 and 87C1 heat induced loci of *Drosophila melanogaster, J. Mol. Biol.* **142**:231–245.

Jack, R. S., Gehring, W. J., and Brach, C., 1981, Protein component from Drosophila larval nuclei showing sequence specificity for a short region near a major heat shock protein gene, *Cell* **24**:321–331.

Jacob, F., and Monod, J., 1961, Genetic regulatory mechanisms in the synthesis of proteins, *J. Mol. Biol.* **3**:318–356.

Johnson, F., Schaffer, H., 1973, Isozyme variability in species of the genus *Drosophila*. VII. Genotype–environment relationships in populations of *D. melanogaster* from the eastern U.S., *Biochem. Genet.* **10**:149–163.

Keene, M. A., Corces, V., Lowenhaupt, K., and Elgin, S., 1981, DNase-1 hypersensitive sites in *Drosophila* chromatin occur at the 5' ends of regions of transcription, *Proc. Natl. Acad. Sci. USA* **78**:143–146.

King, M. J., and Wilson, A., 1975, Evolution at two levels. Molecular similarities and biological differences between humans and chimpanzees, *Science* **188**:107–116.

Koehn, R., 1979, Physiology and biochemistry of enzyme variation: The interface of ecology and population genetics, in: *The Interface of Ecology and Genetics* (P. Brussard and O. Solbrig, eds.), Springer-Verlag, Berlin.

Langley, C. H., and Fitch, W. M., 1974, An examination of the constancy of the rate of molecular evolution, *J. Mol. Evol.* **3**:161–177.

Laurie-Ahlberg, C. C., and Merrell, D., 1979, Aldehyde oxidase allozymes, inversions and DDT resistance in some laboratory populations of *Drosophila melanogaster, Evolution* **33**:342–349.

Laurie-Ahlberg, C. C., Maroni, G., Bewley, G. C., Lucchesi, J. C., and Weir, B. S., 1980, Quantative variations of enzyme activities in natural populations of *Drosophila melanogaster, Proc. Natl. Acad. Sci. USA* **77**:1073–1077.

Leigh-Brown, A. J., and Ish-Horowicz, D., 1981, Evolution of the 87A and 87C heat-shock loci in *Drosophila, Nature* **290**:677–682.

Lis, J., Prestige, L., and Hogness, D., 1978, A novel arrangement of tandemly repeated genes at a major heat shock site in *D. melanogaster, Cell* **14**:901–919.

Lis, J., Neckamayer, W., Mirault, M. E., Artavanis-Tsakonas, S., Lall, P., Martin, G., and Schedl, P., 1981, DNA sequences flanking the starts of the *hsp* 70 and αβ heat shock genes are homologous, *Dev. Biol.* **83**:291–300.

MacIntyre, R. J., 1976, Evolution and ecological value of duplicate genes, *Annu. Rev. Ecol. Syst.* **7**:421–468.

MacIntyre, R. J., and Dean, M. R., 1978, Evolution of acid phosphatase-1 in the genus Drosophila as estimated by subunit hybridization: Interspecific tests, *J. Mol. Evol.* **12:**143–171.

Marie, J., Simon, M.-P., Dreyfuss, J.-C., and Kahn, A., 1981, One gene but two messenger RNA's encode liver L and red cell L' pyruvate kinase subunites, *Nature* **292:**70–72.

Markert, C. L., Shaklee, J., and Whitt, G., 1975, Evolution of a gene, *Science* **189:**102–114.

McCarron, M., O'Donnell, J., Chovnick, A., Bhullar, B. S., Hewitt, J., and Candido, E. P. M., 1979, Organization of the *rosy* locus in *Drosophila melanogaster*: Further evidence in support of a *cis* acting control element adjacent to the xanthine dehydrogenase structural element, *Genetics* **91:**275–293.

McDonald, J. F., and Avise, J. C., 1976, Evidence for the adaptive significance of enzyme activity levels: Interspecific variation in Gpdh and Adh in Drosophila, *Biochem. Genet.* **14:**347–355.

McDonald, J. F., and Ayala, F. J., 1978, Genetic and biochemical basis of enzyme activity variation in natural populations. I. Alcohol dehydrogenase in *Drosophila melanogaster*, *Genetics* **89:**371–388.

McDonald, J. F., Chambers, G. K., David, J., and Ayala, F. J., 1977, Adaptive response due to changes in gene regulation: A study with Drosophila, *Proc. Natl. Acad. Sci. USA* **74:**4562–4566.

McDonald, J. F., Anderson, S. M., and Santos, M., 1980, Biochemical differences between products of the Adh locus in Drosophila, *Genetics* **95:**1013–1022.

McKenzie, J., and McKechnie, S., 1978, Ethanol tolerance and the Adh polymorphism in a natural population of *D. melanogaster*, *Nature* **272:**75–76.

McKeown, M., and Firtel, R., 1981, Differential expression and 5' end mapping of actin genes in *Dictyostelium*, *Cell* **24:**799–807.

Miller, S., Pearcy, R., and Berger, E., 1975, Polymorphism at the α-glycerophosphate dehydrogenase locus in *Drosophila melanogaster*. I. Properties of adult enzymes, *Biochem. Genet* **13:**175–188.

Minty, A., and Newmark, P., 1980, Gene regulation: New, old and remote controls, *Nature* **288:**210–211.

Nair, P., Carson, H. L., and Sene, F., 1977, Isozyme polymorphism due to regulatory influence, *Am. Nat.* **111:**789–791.

Nevins, J. R., and Wilson, M. C., 1981, Regulation of adenovirus-2 gene expression at the level of transcriptional termination and RNA processing, *Nature* **290:**113–118.

O'Brien, S., and MacIntyre, R., 1978, Genetics and biochemistry of enzymes and specific proteins of *Drosophila*, in: *The Genetics and Biology of Drosophila*, Volume 2a (M. Ashburner and T. R. F. Wright, eds.), pp. 396–552, Academic Press, New York.

O'Malley, B., Towle, H., and Schwartz, R., 1977, Regulation of gene expression in eucaryotes, *Annu. Rev. Genet.* **11:**239–275.

Paigen, K., 1979, Acid hydrolases as models of genetic control, *Annu. Rev. Genet.* **13:**417–466.

Peters, J., Shallow D., Andrews, S., and Evans, L., 1981, A gene (*Neu-1*) on chromosome 17 of the mouse affects acid α-glucosidase and codes for neuraminidase, *Genet. Res. Camb.* **38:**47–56.

Phillip, D. P., Childers, W. F., and Whitt, G. S., 1979, Evolution of patterns of differential gene expression: A comparison of temporal and spatial pattern of isozyme locus expression in two closely related fish species (northern largemouth bass, *Micropterus salmoides salmoides* and smallmouth bass, *Micropterus dolomieui*), *J. Exp. Zool.* **210:**473–488.

Place, A. R., and Powers, D. A., 1979, Genetic variation and relative catalytic efficiencies: Lactate dehydrogenase B allozymes of *Fundulus heteroclitus, Proc. Natl. Acad. Sci. USA* **76**:2354–2358.

Powell, J., 1979, Population genetics of Drosophila amylase. II. Geographic patterns in *D. pseudoobscura, Genetics* **92**:613–622.

Powell, J., and Lichtenfels, J., 1979, Population genetics of *Drosophila* amylase. I. Genetic control of tissue specific expression in *D. pseudoobscura, Genetics* **92**:603–612.

Prescott, D. M., and Goldstein, L., (eds.), 1980, *Cell Biology, A Comprehensive Treatise,* Volume 4, *Genetic Expression: Translation and the Behavior of Proteins,* Academic Press, New York.

Razin, A., and Riggs, A. D., 1980, DNA methylation and gene function, *Science* **210**:604–610.

Robinson, R., and Davidson, N., 1981, An analysis of a Drosophila tRNA gene cluster: Two tRNA^{leu} genes contain intervening sequences, *Cell* **23**:251–259.

Romero-Herrera, A. E., Lehmann, H., Joysey, K., and Friday, A. E., 1973, Molecular evolution of myoglobin and the fossil record: A phylogenetic synthesis, *Nature* **246**:389–395.

Samal, B., Worcel, A., Louis, C., and Schedl, P., 1981, Chromatin structure of the histone genes of *D. melanogaster, Cell* **23**:401–409.

Sampsell, B., 1977, Isolation and genetic characterization of alcohol dehydrogenase thermostability variants occurring in natural populations of *Drosophila melanogaster, Biochem. Genet.* **15**:971–988.

Sampsell, B.,1981, Survival differences between Drosophila with different Adh thermostability variants, *Drosophila Information Service* **56**:114–115.

Sampsell, B., and Milkman, R., 1978, Van der Waals bonds as unit factors in allozyme thermostability variation, *Biochem. Genet.* **16**:1139–1141.

Schneider, W., Nichols, B. P., and Yanofsky C., 1981, Procedure for production of hybrid genes and proteins and its use in assessing significance of amino acid differences in homologous tryptophan synthetase polypeptides, *Proc. Natl. Acad. Sci. USA* **78**:2169–2173.

Scott, M., and Pardue, M. L., 1981, Translational control in lysates of *Drosophila melanogaster* cells, *Proc. Natl. Acad. Sci USA* **78**:3353–3357.

Sene, F. M., and Carson, H. L., 1977, Genetic variation in Hawaiian Drosophila. IV. Allozymic similarity between *D. silvestris* and *D. heteroneura* from the island of Hawaii, *Genetics* **86**:187–198.

Snyder, L., 1978, Genetics of hemoglobin in the deer mouse, *Peromyscus maniculatus,* II. Multiple alleles at regulatory loci, *Genetics* **89**:531–550.

Swank, R. T., Novak, E., Brandt, E., and Skudlarek, M., 1978, Genetics of lysosomal functions, in: *Protein Turnover and Lysosomal Function* (D. Doyle and H. Segal, eds.), pp. 251–271, Academic Press, New York.

Templeton, A., 1979, The unit of selection in *Drosophila mercatorum,* II. Genetic revolution and the origin of coadapted genomes in parthenogenetic strains, *Genetics* **92**:1265–1282.

Thompson, J. N., and Kaiser, T. N. 1977, Selection acting upon slow migrating Adh alleles differing in enzyme activity, *Heredity* **38**:191–195.

Tsai, S., Tsai, M.-J., and O'Malley, B., 1981, Specific 5' flanking sequences are required for faithful initiation of *in vitro* transcription of the ovalbumin gene, *Proc. Natl. Acad. Sci. USA* **78**:879–883.

Turner, J., Johnson, M., and Eanes, W., 1979, Contrasted modes of evolution in the same genome: Allozymes and adaptive change in *Heliconius, Proc. Natl. Acad. Sci. USA* **76**:1924–1928.

Van Delden, W., and Kamping, A., 1980, The alcohol dehydrogenase polymorphism in populations of *Drosophila melanogaster*, IV. Survival at high temperature, *Genetica* **51:**179–185.

Vigue, C. L., and Johnson, F., 1973, Isozyme variability in species of the genus *Drosophila*. VI. Frequency-property-environment relationships of allelic alcohol dehydrogenases in *D. melanogaster, Biochem. Genet.* **9:**213–227.

Walker, P. R., 1977, The regulation of enzyme synthesis in animal cells, *Essays Biochem.* **13:**39–70.

Wallace, B., 1963, Genetic diversity, genetic uniformity and heterosis, *Canad. J. Genet. Cytol.* **5:**239–253.

Weintraub, H., Larson, A., and Groudine, M., 1981, α-Globin-gene switching during the development of chicken embryos: Expression and chromosome structure, *Cell* **24:**333–344.

Whitt, G. S., Shaklee, J. B., and Markert, C. L., 1975, Evolution of the lactate dehydrogenase isozymes of fishes, in: *Isozymes*, Volume IV, *Genetics and Evolution* (C. L. Markert, ed.), pp. 381–400, Academic Press, New York.

Whitt, G. S., Phillip, D. P., and Childers, W. F., 1977, Allelic expression at enzyme loci in an intertidal sunfish hybrid, *J. Hered.* **64:**55–61.

Wilson, A., 1975, Gene regulation in evolution, in: *Molecular Evolution* (F. J. Ayala, ed.), Sinauer, Sunderland, Massachussets.

Wilson, A. C., Carlson, S., and White, T., 1977, Biochemical evolution, *Annu. Rev. Biochem.* **46:**573–639.

Wilson, P. G., and McDonald, J., 1981, A comparative study of enzyme activity variation between α-glycerophosphate and alcohol dehydrogenases in *Drosophila melanogaster, Genetica* **55:**75–79.

Wold, F., 1981, *In vivo* chemical modification of proteins, *Annu. Rev. Biochem.* **50:**783–814.

Wu, C., Wong, Y.-C., and Elgin, S., 1979, The chromatin structure of specific genes: II. Disruption of chromatin structure during gene activity, *Cell* **16:**807–814.

Young, J. P. W., Koehn, R., and Arnheim, N., 1979, Biochemical characterization of "LAP", a polymorphic aminopeptidase from the blue mussel, *Mytilus edulis, Biochem. Genet.* **17:**305–323.

Young, R., Hagenbuchle, O., and Schribler, U., 1981, A single mouse α-amylase gene specifies two different tissue specific mRNA's, *Cell* **23:**451–458.

7

Evolution of Dermal Skeleton and Dentition in Vertebrates
The Odontode Regulation Theory

WOLF-ERNST REIF

Department of Geology and Paleontology
University of Tübingen
Tübingen, West Germany

INTRODUCTION

The starting point of comparative evolutionary studies of the dermal skeleton of vertebrates is Hertwig's series of papers (1874, 1876/1879/1882), which directly stimulated many dozens of papers, most of them in German and some of them long forgotten. The literature on comparative histology and histogenesis of the dermal skeleton and on regulatory and morphogenetic processes of the vertebrate integument is so voluminous that it can hardly be summarized. Ever since Hertwig, attempts have been made not only to contribute descriptive and experimental data, but also to arrive at a synthesis. Most of the synthetic papers, however, address only a small section of the theoretical problems, and it seems that some important questions have never been asked.

In this paper I propose a model of the evolution of the dermal skeleton based on phylogenetic, morphogenetic, histologic, and functional data and considerations. It is impossible to discuss in detail here the merits of the hypotheses which have appeared in the last 100 years. However, in two review sections relevant observations from the literature are brought together. There are fundamentally two alternative possibilities to account for evolutionary changes in the dermal skeleton. One hypothesis, called the "differentiation hypothesis," explains changes by modifications of the differentiation processes taking place in the unmineralized tissue. Each germ of a tooth or a scale is the result of an inductive stimulus. All germs are equivalent to each other; they simply differ with respect to their differentiation program. The other hypothesis explains morphologic changes (e.g., an increase in complexity) by the addition of elements (germs, cell clusters) to the unmineralized tissue, which are then completely integrated and lose their individuality ("concrescence hypothesis"). Within the scope of the differentiation hypothesis it is feasible to assume that two mineralizing tooth germs can be fused or welded together and still retain their individuality. Hence the fused teeth would be homologous to two teeth. The concrescence hypothesis goes one step further and assumes that two tooth germs can be fused in such a way that the components of the resulting tooth can no longer be distinguished (except by the number of cusps of the crown). The new tooth would hence be homologous to the original two teeth.

Stensiö's (1961) Lepidomorial Theory is a concrescence theory. The Lepidomorial Theory is widely accepted and has entered numerous textbooks, although its empirical basis has never been published. The "Odontode Regulation Theory," which is introduced here, is developed from Ørvig's (1967a) odontode concept. The word "regulation" indicates that regulatory processes (which we reconstruct in fossil animals and which we experiment with in living animals) are the sole basis of our understanding of the evolution of the dermal skeleton.

The odontode concept as it is understood here leads to a differentiation hypothesis. Hence I do not agree with Ørvig's (1977) statement that the odontode concept can be worked with irrespective of whether or not one accepts the Lepidomorial Theory. My view is that the Odontode-Regulation Theory is an alternative to the Lepidomorial Theory and that it has a better empirical basis. The Odontode Regulation Theory leads to conclusions different from the theories of Hertwig (1874, 1876/1879/1882), Goodrich (1907), Ørvig (1951), Gross (1966), and others. Some of these theories are based on the differentiation hypothesis and some on the concrescence hypothesis. An important basis for the model developed here is provided by the finding of Schaeffer (1977) that the dermal skeleton of vertebrates develops from a single, integrated, modifiable morphogenetic system that is initiated by an interaction between the epithelium and the adjacent mesenchyme.

PHYLOGENY OF THE VERTEBRATES

A discussion of the evolution of the dermal skeleton has to take place with a cladogram of the vertebrates as a background. As yet, no diagram of the interrelationships of the vertebrates has been unanimously accepted. The diagram used here (Fig. 1) is based on new suggestions for the Agnatha by Janvier (1979) and Janvier and Blieck (1979) and for the Gnathostomata by Denison (1979), Schaeffer (1975), Schaeffer and Williams (1977), and Rosen *et al.* (1981). It is important to note that, according to this diagram, the Agnatha are a paraphyletic group. The status and the interrelationships of the ''Crossopterygii'' are quite controversial (Miles, 1977; Schultze, 1977a,b, 1981; Rosen *et al.*, 1977). Most authors agree that they are most likely not a monophyletic group (even if we exclude the problem of the tetrapod ancestry). It should be stressed, however, that the diagram used here is quite eclectic. An account of the evolution of the dermal skeleton would not become contradictory if minor changes in the cladogram were made, e.g., Osteolepidida (''Crossopterygii'') as sister group of the Tetrapoda instead of the Dipnoi; exclusion of the

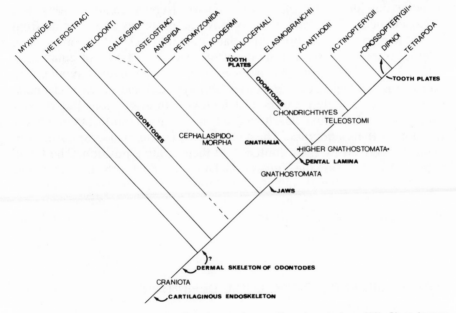

FIG. 1. Cladogram of the Craniota based on various authors (see text, p. 289). Since known stratigraphic ranges are of no importance for the present discussion, they have not been included. While the diagram does not imply that, as commonly held by numerous systematists, it is impossible to find ancestors in the fossil record, it should be stressed that in the present context no craniote group as currently defined can be regarded as ancestral to any other group. The problem of whether Myxinoidea are primitively or secondarily naked is not discussed here. ''Crossopterygii'' seem to be a poorly defined, polyphyletic group; hence they are put in quotation marks. Thelodonti and Chondrichthyes have retained the primitive odontode squamation. Tooth plates have evolved convergently at least three times: Dipnoi, Bradyodonti (Holocephali), and Chimaeroids (Holocephali). Gnathalia are not tooth plates but biting dermal bones.

Brachiopterygii (*Polypterus*, etc.) from the Actinopterygii. (The question, however, remains as to where the "lophosteiform" genera belong.) Other changes in the cladogram would be much more essential and would indeed lead to contradictions, e.g., grouping elasmobranchs and placoderms together as Elasmobranchiomorphi or the exclusion of the acanthodians from the teleostomes and relegating them to a position between Placodermi and Chondrichthyes. If the Ptyctodontida (Placodermi) and the chimaeroids (Chondrichthyes) would be shown on anatomical grounds other than the biting apparatus to be closely related, this could result in a reinterpretation of the biting apparatus of the Ptyctodontida. However, it would not imply that one has to interpret the placoid squamation of the chimaeroids as the product of the reduction of a placoderm macromeric plate skeleton. Rather, one can still assume that the chimaeroid placoid squamation exhibits a primitive condition.

While I fully appreciate the virtues of a cladogram in that it helps us to point out ideas in a clear way and to discover contradictions in a phylogenetic model, I am aware of possible problems which come from an all too direct "translation" of a cladogram into a classification. In such a direct translation the Tetrapoda would become a taxon of equal rank with the Dipnoi and would be included in the Osteichthyes. I avoid giving ranks to individual taxa (e.g., order, family, etc.) and I treat Tetrapoda in sections separately from the Osteichthyes. In discussions like the following no hard and fast classification is necessary. It suffices completely to refer to the cladogram. The taxa shown in the cladogram are automatically defined by their content and their relative position. The taxon names which I use are the conventional ones, except for the taxon "higher Gnathostomata," which includes the Chondrichthyes and the Teleostomi. This taxon is of interest in our discussion because a common ancestor of this group (not shared with the Placodermi) must have evolved a dental lamina.

HISTOLOGIC AND MORPHOLOGIC DEFINITIONS

The occurrence of toothlike structures is a very important characteristic of the dermal skeleton in lower vertebrates and it provides a basis for the model proposed here. Ørvig (1967a, 1977) calls these structures *odontodes*. This term is adopted here, but Ørvig's definition (1977) is modified: An odontode is an isolated superficial structure of the dermal skeleton which consists of a dentinous tissue (mesodentine, semidentine, or metadentine; Fig. 2). A hypermineralized cap of enamel or enameloid

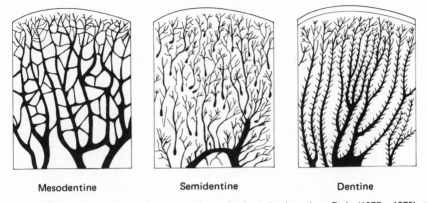

Mesodentine   Semidentine   Dentine

FIG. 2. Schematic drawings of vertical sections of odontodes based on Ørvig (1977a, 1978b,c). Cap on the dentine: enameloid.

is either present or can be lacking. Vascular supply takes place through basal canals and/or neck canals. The base of the odontode consists of (acellular or cellular) bone, which functions as an attachment tissue. Anchoring of the odontode occurs either by anchoring fibers which originate in the bony base, or the odontode can be ankylosed to an underlying bone. Formation of an odontode takes place in a single, undivided dental papilla of mesenchyme which is bounded at its outer surface by an epithelial dental organ.

Two kinds of odontodes are met with: (I) Isolated *dermal denticles*, usually cone or crest-shaped individual elements or dentinous increments which are added to preexisting odontodes during growth. These dentinous increments are often simply cone-shaped; in other cases they are ring-shaped and surround the whole preexisting crown of the scale. (II) Teeth.

To distinguish "dermal denticles" from teeth, a new definition is required, because both "teeth" and "dentition" have had a very vague meaning. All dentitions of Recent vertebrates share this common feature: their elements are formed in a deep epidermal invagination called the *dental lamina*. A major characteristic of the dental lamina is that it "prefabricates" replacement teeth before the old teeth are shed (Fig. 3). Hence, only those organ systems will be called *dentitions* that are formed by a dental lamina. The elements of the dentition are called *teeth*. (There may be special cases in which this terminology cannot be applied, but these are beyond the scope of this paper.) In contrast to teeth, "dermal denticles" are always formed superficially, directly at the epithelium/ mesenchyme interface, without a deep invagination.

Most dentitions of gnathostome groups are polyphyodont; their teeth are replaced throughout life. The dentitions are usually organized in *tooth*

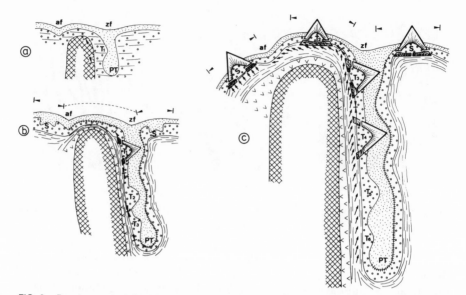

FIG. 3.   Development of the dental lamina. The diagram is largely based on embryologic studies of elasmobranchs, but can be regarded as the primitive condition of all "higher Gnathostomata" (Fig. 1). (a) The dentition is demarcated at the surface by an outer furrow ("äussere Grenzfurche," af) and a tooth furrow ("Zahnfurche," zf). (b) As the invagination (dental lamina) grows deeper during embryogenesis, new tooth germs are induced at the anterior interface between ectoderm and mesenchyme. The material of the tooth germs is derived by proliferation from the primordial tissue (PT) at the base of the invagination. (c) At the end of embryogenesis the invagination practically comes to a halt and the tooth movement mechanism starts. For further explanations see Fig. 4 and p. 324.

families. These consist of one functional tooth and all its successors. Mammals have a diphyodont dentition; the teeth of the milk dentition have one replacement tooth each. The newcomers in the permanent dentition have no replacement tooth at all.

The basal part of the dental lamina from which the tooth-forming cells proliferate is called *primordial tissue* (Reif, 1976). The dental lamina is developed differently in different groups. A *continuous dental lamina* is one infolding which stretches all along one jaw or another tooth-bearing bone. In other words, the tooth families are connected by epidermal tissue. The basal end of the continuous dental lamina is organized into cell clusters that are competent to proliferate tooth-forming cells [these clusters are called *protogerms* (Reif, 1976)] and intermediate cell clusters that are not competent to contribute to the tooth formation. Each protogerm is the site for the production of a whole tooth family. Continuous dental laminae occur in elasmobranchs and tetrapods. In *discontinuous dental laminae* each tooth family is formed by its own epidermal invagination. There is no epidermal tissue between neighboring tooth families. Discontinuous dental laminae develop (at least in some cases) from embryonic continuous dental laminae. They occur in actinopterygians and in the frilled-shark *Chlamydoselache* (p. 338 and Friedmann, 1897).

Another way of grouping the dental laminae is to distinguish between *permanent dental laminae* and *nonpermanent dental laminae*. A permanent dental lamina forms at least one, but in most cases many, replacement teeth (all continuous and many discontinuous dental laminae belong into this category). A nonpermanent dental lamina forms only one tooth. The replacement tooth will then be formed by a new invagination, etc.

Growth takes place in the dermal skeleton in two ways. Growth is either *continuous*, i.e., occurs by a more or less continuous deposition of very small growth layers (e.g. round scales of teleosts) or growth can be *discontinuous*. In this case a complete newly formed odontode is welded to another odontode (Reif, 1978a).

Elements of the dermal skeleton that do not grow, i.e., consist of a single odontode, will be called *microsquamose* elements. *Mesosquamose* elements grow (by addition of new odontodes), but are shed once they reach a certain size, or are inserted in later stages of ontogeny [see *Ctenacanthus costellatus*-type of dermal skeleton in sharks (Reif, 1978a, 1980a)]. *Macrosquamose* elements are never shed; they are formed at early ontogenetic stages, they grow throughout ontogeny, and their number remains constant.

The terms *micromeric*, *mesomeric*, and *macromeric* dermal elements were loosely defined by Ørvig (1968); they coincide more or less with Gross' (1966) distinction between "Kleinschuppen," "Grosschuppen," and dermal plates. Wherever necessary the sharply defined terms microsquamose, etc., which have a clear morphogenetic basis, will be used. In other cases the other set of terms is used, which relate simply to small, medium-sized, and large elements.

Stensiö (1958, p. 303; 1964, p. 220) suggests that the morphogenesis of a dermal plate which is ornamented with odontodes can be inferred by an analysis of the surface pattern. This, however, is by no means the case. The geometry of the surface ornamentation can be absolutely misleading. Morphogenesis of a dermal element can only be reconstructed on the basis of thin sections (Reif, 1978a, 1979a).

Following Shellis (1982), three different types of tooth and denticle attachment are distinguished: (I) Ankylosis: A completely mineralized union between tooth and bone. (II) Fibrous attachment by partly unmineralized collagen fibers. (III) Socketed (= thecodont) attachment. For a review of the comparative anatomy of tooth attachment the reader is referred to Shellis (1982) and Fink (1981).

In addition to the morphologic terms explained above, the histologic terms cellular and acellular bone will be distinguished; whenever possible more special terms will be avoided. These include isopedine, aspidine, cellular cementum, acellular cementum, and crown cementum (Ørvig, 1967a; Poole, 1967, 1971). The three types of dentinous tissue, dentine (= metadentine), mesodentine, and semidentine are distinguished following Ørvig's (1967a) definitions (Fig. 2). Terms for subtypes are avoided whenever possible (e.g., orthodentine = a subtype of dentine). Pleromin (= pleromic hard tissue) is a dentinous tissue which occurs convergently in several groups. It is hypermineralized compared to normal dentinous tissue. It differs strongly from enameloid with respect to its structure and its site of formation (Ørvig, 1976a, 1980b; Reif, 1973a).

Enamel is an ectodermal product which is formed after dentine formation has started. Enameloid [as defined by Halstead Tarlo and Halstead Tarlo (1965) and Poole (1967)] is a product of ectoderm and mesenchyme [see also Herold *et al.* (1980)]. It is formed *before* the dentine formation has started. Enameloid formation has changed into enamel formation probably in several lineages of the gnathostomes. Nothing is known about a possible reversed trend from enamel to enameloid. As histogenetic processes cannot directly be inferred from fossils, structural criteria have to be developed to distinguish fossil enamel from enameloid. Light-microscopic as well as SEM data are required as a basis. The following criteria are used by M. M. Smith and W. E. Reif (in preparation):

A. *Enamel:* Crystals more or less perpendicular to the tooth surface or organized in enamel prisms. Growth lines occur, which indicate that enamel is not mineralized in a single event, but is formed gradually over several days or weeks by more or less continuous deposition. The unmineralized matrix does not contain significant amounts of collagen. In SEM pictures the boundary between enamel and dentine is usually very sharp.

B. *Enameloid:* Crystals are arranged randomly, or parallel, or perpendicular to the tooth surface. Very often they form fiber strands. Growth lines are missing because mineralization of the soft matrix takes place in one event, which lasts no more than several hours. The unmineralized matrix contains large amounts of collagen, which are degraded during mineralization, but very fine filaments persist (Shellis and Miles, 1976). The structure of the mature enameloid is often clearly influenced by the directions of the preexisting collagen fibers. The enameloid–dentine boundary is usually not sharp in SEM pictures.

Using these criteria we find enamel in:

A1. *Tetrapoda:* Teeth of most groups, except for the larval teeth of urodeles, which have an enameloid cap (Kerr, 1960; M. M. Smith and Miles, 1969, 1971).

A2. *Crossopterygii:* Teeth (M. M. Smith, 1978, 1979c; Shellis and Poole, 1978); surface layer of the cosmine in dermal bones and scales (no recent data are available).

A3. *Dipnoi:* Tooth plates (M. M. Smith, 1979c); surface layer of cosmine in dermal bones and teeth [SEM pictures in M. M. Smith (1977); M. M. Smith (1979c) assumes that this surface layer is homologous to ganoin; see below].

A4. *Actinopterygii:* Collar enamel of the teeth. The thick collar enamel of chondrosteans and holosteans has crystals which are perpendicular to the surface. The tissue clearly grows away from the underlying surface. This is shown by the site of formation of the collar enamel and by a very dense spacing of growth lines (Reif, 1979b, and unpublished). In *Lepidotus* (Holostei, Jurassic) the enameloid cap is formed slightly prior to the dentine while the tooth germ is still in a bone cavity of the jaw. The collar tissue, however, is deposited after the tooth has erupted and has become ankylosed to the jaw (H.-P. Böss, Töbingen, personal communication). In Recent teleosts the collar tissue is a very thin layer of a few micrometers. Histochemical and autoradiographic studies have shown that in some genera the collar tissue is enameloid. In *Anguilla*, on the other hand, the tissue is only 2 μm thick and is double-layered. The inner layer is enameloid and the outer layer is enamel (Shellis, 1975; Shellis and Miles, 1974, 1976). These findings are quite compatible with the assumption that the thick collar tissue of earlier actinopterygians is enamel.

The collar tissue was first carefully studied by Peyer (1919, 1937), and then by Guttormsen (1937) and Thomasset (1930). This group of authors [see also Peyer (1968)] maintained that the collar tissue was enamel, whereas numerous other authors claimed that it was enameloid (Schmidt 1958; Schmidt and Keil, 1971; Ørvig, 1978a; Shellis, 1978; Reif, 1979b; Kerebel and Le Cabellec, 1980).

Ganoin, the hypermineralized surface layer of ganoid scales, also has crystals which are perpendicular to the scale surface. It consists of thick lamellae deposited on top of each other. Within each lamella numerous growth lines have been found [Ørvig (1978a), Fig. 34, "secondary lamellation"]. It is unknown whether the scales of the pre-*Palaeoniscus*-type scales (lophosteiform genera, see below) have enamel or enameloid.

Enameloid occurs in:

B1. *Amphibia, Urodela:* Teeth of larvae (Kerr, 1960; Smith and Miles, 1971).

B2. *Acanthodii:* Scales (Denison, 1979).

B3. *Crossopterygii, Actinistia:* M. M. Smith *et al.* (1972) assume that the hypermineralized cap on the odontodes of *Latimeria* is enameloid. Further studies are desirable.

B4. *Actinopterygii:* Cap enameloid of the teeth (Shellis, 1978; Ørvig, 1978a); ?hypermineralized caps on the toothlike structures of the scales of *Polypterus, Lepisosteus,* and the armored catfishes (Loricariidae and Callichthyidae).

B5. *Chondrichthyes:* Teeth and scales (Reif, 1973a, 1980b; Shellis, 1978).

B6. *Osteostraci:* Dermal plates (Obruchev, 1967).

B7. *Heterostraci:* Dermal plates (Reif, 1979b).

"Ganoin" and "cosmine" are not histologic terms in the strict sense. "Ganoin" denotes the surface cap of a ganoid scale, which consists of numerous lamellae of enamel [enameloid in the opinion of other authors, e.g., Ørvig (1967a, 1978a)]. With each new growth stage of the scale a new layer is deposited. "Cosmine" is used for a combination of (a) a surface layer of enamel, (b) a middle layer of dentine which houses a pore canal system, and (c) a bottom layer of bone. Cosmine occurs in dipnoans and "crossopterygians." An analogous combination occurs in osteostracans. If the misleading term "cosmine" cannot be abolished altogether, it should be reserved for the three-layered complex in the "crossopterygians" and dipnoans and should never be used for the osteostracans nor as a synonym for dentine.

EVOLUTION OF THE DERMAL SKELETON: A REVIEW

Heterostraci

The dermal skeleton is meso- to macromeric. It consists of large plates covering the head, and of (probably growing) scales covering the body. Isolated scales are assigned to the heterostracans called *Polymerolepis* by Karatajute-Talimaa (1968, 1977; Obruchev and Karatajute-Talimaa, 1967). Thin sections in Karatajute-Talimaa (1968, 1977) show that *Polymerolepis* scales did not grow!

The mesomeric and macromeric elements are three-layered with a basal laminated bone layer, a middle layer of spongy bone, and an outer layer of odontodes, consisting of orthodentine and enameloid. The bone tissues are acellular (Denison, 1963; Gross, 1930, 1936, 1961; Marss, 1977; Moy-Thomas, 1971; Ørvig, 1967a; Reif, 1979b).

In primitive heterostracans the head is covered by a mosaic of small polygonal plates (Bryant, 1936; Halstead Tarlo, 1964a, 1967; Lehtola, 1973; Ørvig, 1958; Repetski, 1978). In these groups the dermal skeleton probably grew by marginal growth of the polygonal plates, but no data are available.

As far as is known, the growth patterns of the other families vary considerably (Obruchev, 1967). Some families had a nongrowing head capsule which was acquired at the end of the ontogenetic growth of the body. In other families the head was covered by a set of growing plates which later fused and formed a solid capsule. In Drepanaspidae the large, probably growing head plates are separated by a mosaic of small plates ("tesserae"). It is unknown whether the tesserae were formed at the same time in ontogeny as the large plates or whether they were inserted during growth to compensate partly for the increasing body surface (Fahlbusch, 1957; Gross, 1963a; Lehmann, 1967; Moy-Thomas, 1971; White, 1958).

White (1946, 1958, 1973) studied growth stages of the body armor of different genera and came to the conclusion that the plates were formed fairly late in ontogeny, in some genera only after the animals had reached maturity and did not continue to grow. He assumed that the young growth stages did not have a dermal skeleton. This, however, is by no means necessarily the case. Rather, one can assume that the young growth stages had a microsquamose dermal skeleton of scales like *Polymerolepis*, which compensated for body growth by the insertion of new nongrowing scales (i.e., odontodes).

The outer layer of plates was formed first, then the middle layer, and then the basal layer (Denison, 1973). In contrast to White (1958, 1973), Denison came to the conclusion that the plates of Pteraspidae were formed early in ontogeny and continued to grow throughout life. Influenced by the Lepidomorial Theory (Stensiö, 1961), Denison assumed that the plate primordium, which has numerous surface ridges and tubercles, was formed "synchronomorially," i.e., in a single morphogenetic step. It is very doubtful, though, that this is a correct interpretation. The final formation of the plate primordium, namely by fusion of the odontodes with their bases, probably was a single morphogenetic step. Each odontode, however, was induced and formed separately and we do not know whether they were all induced and formed at the same time. Halstead Tarlo (1964) showed that resorption took place in the middle (= spongy bone layer) of the plates of some genera.

Abrasion of the surface layers of plates is quite common in several heterostracan families. The reason for abrasion is not known in most cases [mechanical abrasion in the natural habitat; scraping of the animals to remove parasites (Denison, 1973; Gross, 1930, 1936; Halstead, 1974; Halstead Tarlo, 1964b; Tarlo and Tarlo, 1961)]. Healing of the loss of the surface layers occurs in two ways: (a) formation of "blisters" of new generations of odontodes on top of the scales and (b) filling of the spongy bone layer with "secondary" hypermineralized dentine (pleromic dentine = pleromin) (Denison, 1973; Gross, 1930; Halstead Tarlo, 1964b; Ørvig, 1961, 1967a; Tarlo and Tarlo, 1961; Halstead Tarlo and Halstead Tarlo,

1965). The formation of a dentinous tissue not directly below the basement membrane, but in a deep layer of the corium (= inner layer of the skin), is a process which has never been observed in any Recent vertebrate. In addition to the heterostracans, "deep" dentine has also been found in the bony base of large nongrowing scales of Tertiary rays (Reif, 1979a). The histogenetic processes that are the basis of this tissue formation are completely obscure. They require a migration of odontoblasts in a basal direction.

By extrapolation from the available data, Ørvig (1976a) assumes that the formation of pleromin began before any abrasion of the surface odontodes. He also assumes that the formation of "blisters" of secondary and tertiary odontode generations also took place as part of the normal growth program without prior irritation of the dermal skeleton. Ørvig (1976a) does not explain what caused the abrasion. Denison (1973), on the other hand, assumes that the "healing" did not take place without prior irritation of the surface layer of the dermal skeleton.

Gross (1961) found that the formation of secondary generations of odontodes in *Traquairaspis* was preceded by resorption rather than by abrasion. The cracks in the plates (which had probably resulted from bites of predators) were healed by filling of the gap with new odontodes and acellular bone (Halstead Tarlo, 1964b). More information, however, is very much needed.

Thelodonti

All *thelodonts* (Thelodontida and Katoporida) had a dermal skeleton of nongrowing scales; the scales consist of orthodentine or mesodentine and a bony base. A surface layer of enameloid is probably lacking. Vascular supply of the pulp cavity occurred through a basal pore, which may be closed at the end of morphogenesis (Gross, 1967b, 1968b, 1971c; Karatajute-Talimaa, 1978; Turner, 1976). The reason for the closure of the basal pore is unknown. During morphogenesis of the individual scales the outer layer of the crown is formed first and then mineralization proceeds in a basal direction (Gross, 1967b).

No information about the morphogenesis of the thelodont dermal skeleton is available from the literature. Different specimens of the same species all seem to have more or less the same body size (Kiaer and Heintz, 1932, Figs. 1–4; Marss, 1979, Fig. 1). In other words, no ontogenetic series is known in any species. This could mean either that the dermal skeleton was formed late in ontogeny and the juveniles were scaleless or simply that young growth stages have not yet been discovered. (Articulated thelodonts are very rare!) Unfortunately, no information is available as to whether all scales of a single fish are at the same stage of their morphogenetic development.

Despite the lack of positive evidence, one has to assume that the squamation of thelodonts had the same morphogenetic characteristics as the squamation of Recent sharks. In this group, the squamation is formed early in ontogeny; scales are replaced, and growth of the body surface is compensated for by an increasing number and/or an increasing size of the scales (Reif, 1978a, 1980a). Photographs by Richie (1968) clearly show that the scales of thelodonts are not arranged in diagonal rows but are distributed rather randomly. [The drawings by Gross (1968a) and Moy-Thomas (1971), regrettably, show diagonal rows!] The implications of this observation are not yet clear. In young sharks the placoid scales are randomly distributed; in adult sharks they are either arranged randomly or in diagonal rows. The occurrence of diagonal rows results simply from continuous replacement and eventual close-packing of the scales (Reif, unpublished). The fact that scales of the thelodont *Phlebolepis elegans* have pores which belong to the lateral line system (Gross, 1968a; Marss, 1979) does not contradict the assumption that these scales were replaced. In Recent sharks two specialized scales cover each of the numerous pit organs (free neuromasts); these scales are replaced in the same way as normal scales despite the fact that they serve an important protective function for the sense organs (Reif, unpublished).

Osteostraci

Osteostraci have a mesomeric to macromeric dermal skeleton which consists of three layers. The uppermost layer is made up of odontodes, which consist of mesodentine [with an enameloid cap (Obruchev, 1967)]; in the middle there is vascular bone and the basal layer consists of dense lamellated bone (Gross, 1968b; Moy-Thomas, 1971; Ørvig, 1967a, 1975). The bone is a cellular bone, except for some of the youngest (Upper Devonian) genera, where the bone is acellular (Ørvig, 1965). Denison (1947, 1951, 1952) showed that the outer layer was formed first and then calcification proceded inward.

It is well known that most Osteostraci had a nongrowing, sutureless head capsule, which could have formed only after the animals had reached the mature stage (Moy-Thomas, 1971). Only a few of the latest Osteostraci had a dermal skeleton of the head consisting of growing polygonal plates (Ørvig, 1968). The information concerning these genera is, however, very limited.

Body and tail are covered with large scales, which also have a three-layered histology (Stensiö, 1932). Their morphogenesis is unknown. Long, narrow scales have often disintegrated into small, rectangular pieces. According to Stensiö, this was due to sediment compaction. The scales broke at weak zones, namely where they are pierced by canals. Obruchev (1967) interpreted the same finding by the assumption that the long, narrow scales are the product of fusion of rectangular units *intra vitam*.

Some authors assumed that the young growth stages of the Osteostraci had no dermal skeleton at all [see Moy-Thomas (1971) for references]. However, it is much more likely that the young growth stages had a flexible microsquamose dermal skeleton. This dermal skeleton would have been capable of growth through the insertion of new odontodes. The question as to whether these odontodes underwent replacement cannot be answered. As soon as the adult stage was reached the nongrowing scales fused with their bases in the head region and formed a solid capsule. As long as nothing is known about the morphogenesis of the large scales which covered the body and the tail, it remains open whether these scales were formed early in ontogeny and grew, or were formed late in ontogeny (by fusion of isolated odontodes) and never grew.

Remarkably little is known about the question of whether the dermal skeleton of the bottom-dwelling adult osteostracans were abraded and how they healed these abrasions. Gross (1968b, Fig. 6A) shows a scale of a probably nongrowing osteostracan (from the Silurian) with two generations of odontodes. Ørvig's (1968) figures of the head plates of the growing genus *Alaspis* seem to suggest that small odontodes were overgrown by large odontodes, but no histologic details are available. In *Tremataspis* the odontodes form a single generation, show no overgrowth, and are very densely spaced (Denison, 1951). In *Cephalaspis* (Ørvig, 1951, Fig. 11; Denison, 1952) and *Zenaspis* (Gross, 1961, Figs. 13–17) and on the dorsal side of *Dartmuthia gemmifera* (Gross, 1961, Figs. 21–26) there is a first generation of widely spaced odontodes. A second generation of odontodes fills (at least partly) the spaces between the odontodes of the first generation. The function of this type of morphogenesis is completely unknown. It does not necessarily have something to do with the abrasion of denticles. Before overgrowth the odontodes of the preceding generation are partly resorbed (Ørvig, 1951).

Some Osteostraci [especially the tremataspids; see Denison (1947), Gross (1956), Stensiö (1927, 1932)] have a dentinous surface layer in which odontodes are very closely integrated and cannot be separated morphologically. The dentinuous layer (and the spongy bone underneath) of the head shield and the scales contain a pore canal system, the function of which is discussed by Bölau (1951), Denison (1947), Moy-Thomas

(1971), Thomson (1977), Wängsjö (1944), and others. It is most likely that the pore canal system housed a sense organ. Ørvig (1969a) and Thomson (1977) call this tissue complex "cosmine," though it is very doubtful whether it is in any way homologous to the cosmine of "Crossopterygii" and Dipnoi (see below). Denison (1952) found indications of resorption in the basal layer of the exoskeleton of *Tremataspis mammillata*, which consists of laminated bone (= isopedine). This observation is confirmed by Gross (1956).

Galeaspida

The histology of the dermal skeleton of this group is unknown (Liu, 1975; Janvier, 1979; Janvier and Blieck, 1979; Halstead, 1979; Halstead *et al.*, 1979). Like the osteostracans, the galeaspids had a solid head capsule which most probably did not grow.

Anaspida

Most anaspids had a mesomeric squamation which covered the body and part of the head. The scales grew and their number remained constant [at least in the later stages of ontogeny (Moy-Thomas, (1971)]. Parrington (1958) assumed that the earliest growth stages had a microsquamose dermal skeleton of nongrowing scales. This type of skeleton was retained in the later ontogenetic stages of *Lasanius*, but in the other anaspids these nongrowing scales must have fused at their bases to form mesomeric scales, which then continued to grow.

Nothing is known, however, about the morphogenesis of anaspid scales. Surface figures and sections provided by Gross (1958a) do not contradict the view that the mesomeric scales were first formed by coalescence of nongrowing odontodes and then continued to grow by adding odontodes at the side. It is remarkable that only one layer of odontodes exists in anaspid scales; this means that overgrowth of odontodes does not occur. According to Gross (1958a), the scales, including the odontodes at the surface, consisted of acellular bone. This is questioned by Denison (1963), who assumed that the odontodes consisted of dentine.

Placodermi

Placoderms have a macromeric to mesomeric dermal skeleton consisting of growing scales, growing "tesserae" [small platelets (Gross,

1959)], and large bony plates. The earliest known arthrodires did not yet have plates, but had a mesomeric skeleton of tesserae and medium-sized plates (Gross, 1958b). Gross (1963a) and Denison (1978) assume that growing tesserae are primitive (= plesiomorphic) and large plates are derived (= apomorphic). No histologic evidence exists to prove that placoderms had nongrowing scales (= odontodes), though the morphology of the very small scales of Stensioellidae (Gross, 1962, Fig. 6) and Gemuendiniformes [*Tyriolepis* (Karatajute-Talimaa, 1968, Plate 5)] indicates that these scales did not grow. In other words, these genera probably had a dermal skeleton on the body which was very similar to that of thelodonts and sharks. On the cranial roof up to three dermal bones were identified.

The dermal plates of placoderms consist of three layers: a basal laminated bone layer, a middle vascular bone layer, and a superficial layer of odontodes, which consist of semidentine (Gross, 1930, 1936, 1959; Ørvig, 1967a, 1969b, 1975). In the arthrodires the dentinous tissues cover the whole range from semidentine to mesodentine (Ørvig, 1967a). The bone is cellular bone with the possible exception of some Ptyctodontida (Denison, 1979). An enameloid cap has not yet been described. The spongy layer seems to be lacking in small scales and tesserae (Gross, 1959). In later genera the superficial meso- or semidentine layer of the plates is lost (Denison, 1975, 1979).

The evolution of the dermal skeleton is unknown. Denison (1975) gives the following account: In *Radotina* there are tesserae and plates on the head. In the Rhenanida the tesserae can be shown to be homologous to the scales which cover the body (this is no surprising finding). The tesserae "may be considered remnants of the dermal scales that were the only exoskeleton of ancestral placoderms, and as such are comparable in general to chondrichthyan scales" (p. 11). (As the tesserae clearly grew, they can only be homologous to growing chondrichthyan scales.) "When bones [i.e., bony plates] first appeared in placoderms, they apparently arose deeper in the dermis quite independently of the tesserae and also of the lateral line system" (p. 11). It is not obvious to me on which observations this assumption is based. As far as I know, tesserae have never been found to overlie dermal plates. Also, if the dermal plates have a surface ornamentation of a dentinous tissue, as they primitively do, they cannot have formed in a deep layer of the corium, unless they were secondarily covered with odontodes. It seems much more likely that the small, growing, ornamented tesserae were gradually replaced in phylogeny by large ornamented plates through a process of fusion. The plates gradually lost their surface ornamentation and remained restricted to the deeper layer of the corium.

Overgrowth by a second generation of odontodes can be seen in sections by Gross (1973, Fig. 1) and Ørvig (1969b, Fig. 10). Before overgrowth, the old odontodes are partly resorbed. The adaptive significance of this resorption is unknown. The placoderm dermal skeleton grew by marginal growth of the plates, "tesserae," and scales. It is unknown whether the number of the elements remained constant, or whether they increased in number during ontogeny to compensate for body growth. No growth series of whole skeletons or histogenesis of any larger plate has been described. Hence it is impossible to ascertain whether the macrosquamose dermal skeleton was formed early or late in ontogeny. It is also unknown whether the macrosquamose dermal skeleton was preceded in ontogeny by a microsquamose skeleton in which the odontodes eventually fused at their bases to form "tesserae," scales, and plates.

Chondrichthyes

Recent sharks, skates, rays, and holocephalians have nongrowing scales which cover the body and in many, probably most, species also the integument of the mouth cavity (Schauinsland, 1903; Patterson, 1965; Peyer, 1968; Reif, 1973b, 1974, 1978a, 1979a, 1980a). The scales consist of an enameloid cap, a dentine crown, and a basal plate of acellular bone. Zangerl (1966, 1968) found cellular bone in scales of *Ornithoprion hertwigi, Orodus*, and *?Holmesella*. As a rule there is one vascular canal in the base and four neck canals. Pre-middle Jurassic elasmobranchs are either scaleless or they have growing and/or nongrowing scales (Gross, 1938, 1973; Ørvig, 1966; Reif, 1978a, 1979a, 1980a; Reif and Goto, 1979; Wells, 1944; Woodward and White, 1938; Zangerl, 1968). Stensiö's (1961, 1962) statement, which forms a major basis of his Lepidomorial Theory, namely that pre-Upper Permian elasmobranchs had only growing scales, is incorrect. Nongrowing scales are well known from the Carboniferous and also the Upper Devonian (Gunnell, 1931, 1933; Harlton, 1933; Twai, 1979; Wells, 1944; Zangerl and Case, 1976). The oldest growing scales of elasmobranchs have been reported by Gross (1973) from the Lower Middle Devonian. Still older (Upper Silurian) nongrowing scales, which are assigned to the elasmobranchs, were described by Karatajute-Talimaa (1973) as *Elegestolepis grossi*. The taxon *Ellesmereia schultzei* was introduced by Vieth (1980) for nongrowing scales from the Gedinnian (Lower Devonian).

Growth of the scales takes place by addition of complete odontodes (Gross, 1938; Ørvig, 1966; Reif, 1978a, 1980a; Woodward and White, 1938; Zangerl, 1966, 1968). In some genera the basal plate of the scale remains fairly thin; in others the base is thickened secondarily.

Zangerl (1966) found rods and complete sheaths of cellular dermal bone on the snout and the mandibular rostrum of *Ornithoprion hertwigi*. All of these elements seem to be covered with odontodes, though this is not clearly stated. Zangerl explains the formation of these elements by lateral extension and secondary thickening of the bases of the scales.

The number of neck canals range from one to four in different types of scales. *Elegestolepis* and *Ellesmereia* have one neck canal. Individual odontodes of growing scales of *Ctenacanthus costellatus* have three or more neck canals. Nongrowing scales (= placoid scales) of modern sharks (Jurassic to Recent) usually have four neck canals. In the large placoid scales of the late ontogenetic stages of the sharks and batoids the number can, however, be considerably greater. Oral cavity scales of *Heterodontus* (Recent) have only three neck canals (Reif, 1973b, 1974). In the Permian edestids (Holocephali) nongrowing scales and individual elements (odontodes) of growing scales have four or more neck canals (Reif, unpublished). Consequently, the first part of the Lepidomorial Theory (Ørvig, 1951; Stensiö, 1961), namely, that growing scales of elasmobranchs are primitively composed of "lepidomoria," cannot be agreed with. According to Stensiö, lepidomoria are morphologically very simple elements with crown and base, and one neck canal and one basal canal. Thus far, lepidomoria remain completely hypothetical elements; their real existence has never been demonstrated. In this context it is important to mention that *Elegestolepis* and *Ellesmereia* (with one neck canal) have a complicated morphology, especially in the crown. Zangerl (1966) found very simple cone-shaped scales in *?Agassizodus*. Unfortunately, the exact morphology and the number of neck canals are unknown. Additionally, it is very unlikely that these element had no bony bases at all, as was stated by Zangerl.

Stensiö's second hypothesis is that the complicated morphogeny of a growing scale may be abbreviated to one single morphogenetic step; this would imply a transition from a "cyclomorial" (growing) scale to a "synchronomorial" (nongrowing) scale. The "synchronomorial" scale, which consists of a single element, would be homologous to a growing scale which consists of numerous elements. This transition by abbreviation of the ontogeny, however, has never been shown in any lineage of elasmobranchs and it remains hypothetical. Stensiö based his hypothesis primarily on the scales of the Permian edestids from East Greenland. They were restudied by me (Reif, unpublished; the material was kindly provided by Prof. T. Ørvig, Stockholm, and Dr. S. E. Bendix-Almgren, Copenhagen). The three genera have either nongrowing plus growing scales or nongrowing scales only. There is no character of the scales which justifies the assumption of an evolutionary transition from a "cyclomorial" to a "synchronomorial" scale! Another transition, though, has been shown in a shark lineage: Lower Jurassic hybodontids had growing scales; Middle and Upper Jurassic hybodontids had nongrowing scales (Reif, 1978a). In this case, however, *one* odontode of a growing scale has to be regarded as homologous to *one* nongrowing scale. This transition is brought about by a change in a regulation process. In a skeleton of growing scales new elements are welded *to* existing scales, and in a skeleton of nongrowing scales new elements are inserted *between* older scales as separate scales. It is easily conceivable that this change in regulation could also be reversed, though this reversal has never taken place in any elasmobranch since the Middle Jurassic.

From this discussion it follows that the major premises of Stensiö's Lepidomorial Theory are incorrect and its major conclusions are unnecessarily complicated. It is much more parsimonous to assume that each odontode is an independent organ, which originated by an induction process. Complicated shapes of scales result from differentiation of the scale germ rather than from a "synchronomorial" morphogeny.

Resorption has never been observed in the dermal skeleton of any Recent elasmobranch. Scales and teeth are shed by a resorption of the anchoring collagenous fibers. However, as far as is known, this does not affect the hard tissues. Budker (1971, Figs. 22 and 23) pictured Recent shark teeth, which have an open pulp cavity and in which the bone and the dentine of the root is partly missing. He ascribes this situation to the resorbent activities of "osteophagous monocaryocytes." There is, however, no doubt that these teeth are newly formed teeth, which are not yet completely mineralized, rather than partly resorbed teeth.

Many fossil and Recent sharks and holocephalians have growing dorsal finspines. Fossil holocephalians also have other growing spines (Moy-Thomas, 1971; Patterson, 1965). As far as is known, all spines consist of enameloid and dentine. In Recent shark embryos the finspines develop within a single, enlarged papilla (Markert, 1876; Maisey, 1979). Hence Maisey's conclusion that each finspine has to be regarded as a single unit (which is in opposition to the concrescense hypothesis of numerous authors) is quite acceptable. Many finspines have tubercle-shaped, dentinous surface ornamentations. The suggestion that tubercles represent secondarily fused scales is refuted by the observation in *Squalus* of tubercles arising by scleroblast disorganization and clustering, not by new dermal papillae (Maisey, 1979).

Finspines are obviously highly derived organs. Their homology with other elements of the dermal skeleton is difficult to determine: "It is stressed that there is no merit in comparing finspines and simple scales, nor in trying to derive one from the other. Instead, finspines, teeth and scales are regarded as differentiation products in their own right. Each is the ultimate product of development within a dermal papilla, and each is therefore unique. Attempts to trace back the origin of finspines to placoid, or any other type of full-grown scale, are futile'' (Maisey, 1979, p. 181). Maisey is right in stating that at the present stage of our knowledge it may be futile to ask for an answer. However, the question itself has to be put in a slightly different way. We assume that the very distant, agnathous ancestors of sharks already had a more or less undifferentiated dermal skeleton of small nongrowing scales. Now, the real question is, did the gnathostome ancestors evolve teeth and finspines *de novo* or by a local modification of the morphogenetic program for scale formation? It will be shown in the second half of this paper that it is very likely that teeth did not evolve *de novo*, but rather by a modification of the morphogenetic program for scales. It is very likely that the same is true for finspines, though there is as yet no model available to define a continuous series of transitions during the evolution of finspines. Hence the exact homology between scales and finspines remains unknown.

Poisonous spines on the tail of several batoid families (Dasyatidae, Gymnuridae, Urolophidae, Myliobatidae, Rhinopteridae) differ considerably from the dorsal finspines. They are nongrowing spines, which are replaced. Thus they are clearly homologous to ordinary nongrowing placoid scales. They consist of vascular dentine with an enameloid cap. Their histogenesis has never been studied (Rauther, 1940; Bigelow and Schroeder, 1953).

Rostral teeth of sawfishes and sawsharks are also derived from normal placoid scales, but intermediate stages between placoid scales and rostral teeth are unknown. The histogenesis of these organs is also poorly known. In the Recent sawfishes (genus *Pristis*; batoids) the number of rostral teeth remains constant throughout ontogeny. As far as is known, the rostral teeth are persistently growing by addition of new dentine at the bases. In contrast, Recent sawsharks (*Pristiophorus* and *Pliotrema*) have nongrowing rostral teeth which are replaced by others of increasing size and number during ontogeny. The condition in the sawfishes is clearly the advanced condition compared to that in the sawsharks (growth and no replacement versus no growth and replacement). This is nicely illustrated by the fact that in some Cretaceous sawfishes (*Sclerorhynchus* and *Onchopristis*) indications of replacement have been found (Engel, 1910; Rauther, 1940; Bigelow and Schroeder, 1953; Schaeffer, 1963; Slaughter and Springer, 1968; Shellis and Berkovitz, 1980).

A very specialized type of placoid scale is the gillraker of *Cetorhinus*. These form a fine-meshed sieve for gathering plankton. Since a continuous replacement of the scales would impede the function of the sieving apparatus, the scales are probably replaced as a whole set during a feeding pause in winter (Parker and Boeseman, 1954).

Acanthodii

Acanthodians have dermal bones and "tesserae" (small bony plates) on the head which are ornamented, but whose histology is unknown. The scales covering the body and the tail have no pulp cavity and are very small ["Kleinschuppen" (Gross, 1971b, 1973)] compared with the scales of Osteichthyes. So far, only growing scales have been found. The crown of the scales consists of mesodentine or dentine depending on the family, with an enameloid cap (Denison, 1979; Gross, 1966, 1971b, 1973; Moy-Thomas, 1971; Ørvig, 1967b; Vieth, 1980). The base is thickened and consists of cellular or acellular bone. Growth takes place by adding new odontodes. Individual odontodes can have very different shapes. In many genera each increment (apparently consisting of a single odontode) covers the scale on all sides. The tesserae are ornamented with odontodes [Denison (1979), Thorsteinsson (1973)]. Schultze (personal communication) indicates that Gross regarded Thorsteinsson's *Pilolepis* as identical with the acanthodian genus *Nostolepis*. It is obvious that some of them grew by lateral addition of new odontodes and a lateral extension of the bony base. No histologic study of these tesserae is known to me. Denison (1979) assumes that the normal body squamation extended with little changes to the head in primitive acanthodians. There followed a trend toward the formation of larger tesserae and plate-shaped dermal bones which, as far as is known, are restricted to the lateral sides of the head. The later acanthodians show an opposite tendency toward reduction or loss of the scales of the head, except perhaps along the lateral lines.

In some families of Acanthodii there are two types of large dermal bones, which cannot be derived from ornamented tesserae directly: (1) a mandibular dermal bone attached to the ventral edge of the meckelian cartilage (as far as is known, it is not ornamented by odontodes); and (2) dermal jaw bones which bear ankylosed teeth or which serve as an attachment surface for the fibrous attachment of teeth or tooth spirals (see below) (Denison, 1979).

It is very likely that acanthodian scales were never shed but continued to grow until they reached a certain size. According to Watson (1937), the formation of scales in some body regions was delayed until late in ontogeny, body growth was very advanced. The first scales to appear on the body of *Mesacanthus* are those bordering the main lateral-line canal in the rear part of the body, new scales being added both above and below the lateral line and its anterior end. The addition of new scales is described by Watson (1937) in a growth series of complete specimens. His results are nevertheless very puzzling. He states that "the size of the scales is on the whole similar over the whole flank of the fish, but they decrease a little toward the dorsal and ventral surfaces. . . . This mode of development of the squamation gives the explanation of the remarkable fact that the scales of large specimens of *Acanthodes* may be no larger than those of small individuals" (Watson, 1937, p. 112). If the scales appear earlier in the posterior part of the body than in the anterior part, they should be larger in the posterior part, unless their growth is delayed and unless increase in body size is compensated by insertion of additional scales in the posterior body part. If the scales of small and large specimens of *Acanthodes* are more or less of the same size, we must conclude that their number increased during ontogeny, thus compensating for the increase in body size. If this is the case, as Watson himself assumes (p. 117), one should find newly inserted scales surrounded by scales that have reached their final size. These have never been found.

The scales of a few acanthodian genera contain a pore-canal system. It seems to resemble the pore-canal system in heterostracans, rhipidistians, and dipnoans (Gross, 1956; Denison, 1979). Its function is discussed by Thomson (1977); it is most probably a sense organ.

Acanthodians had numerous spines, several of which supported the fins (Moy-Thomas, 1971; Harper, 1979). Primitively they consist of mesodentine and bone; both tissues were replaced early in phylogeny by true dentine (Denison, 1979). They grew throughout their ontogeny. The evolutionary origin of the spines is as mysterious as in the elasmobranchs (see above).

Osteichthyes

In Osteichthyes there is a fairly sharp distinction between growing dermal bones on the head and growing dermal scales on the trunk. Nongrowing tesserae or nongrowing scales resembling placoid scales have never been found. Much more is known about the histology of scales than of dermal bones in the Osteichthyes. As far as can be known, evolutionary changes in the histology of the dermal bones of Actinopterygii differ from those of the scales.

Schultze (1977a; see also Schultze, 1966) showed that the whole spectrum of scales of Ostheichthyes can be derived from the scales of *Lophosteus* (Gross, 1966, 1968c, 1969, 1971a; Schultze, 1968). This genus must be closely related to the common ancestor of all Ostheichthyes (or even Teleostomi!). The *Lophosteus* scale consists of three layers: lamellar (cellular) bone, vascular (cellular) bone, and, on top, dentine tubercles. Schultze (1977a) assumed that there was no cap of enamel or enameloid. However, this has never been confirmed by scanning electron microscope studies. In scales of an old morphogenetic stage the odontodes are overgrown by a second and a third generation of dentine tubercles with underlying vascular bone. The function of this overgrowth is to provide larger tubercles in order to meet different functional needs from those in the young ontogenetic stages. In Schultze's (1977a) schematic drawings the second generation of odontodes has the same size as that of the first generation, which is a misrepresentation of Gross's (1969) results.

The evolution of the scales in Actinopterygii is very much different from that of the other Osteichthyes.

Scales of Actinopterygii

In addition to *Lophosteus*, the Upper Silurian and Devonian genera *Andreolepis, Orvikuina, Dialipina*, and *Ligualepis* (here collectively called "lophosteiforms") have individual odontodes, each of which is covered by a layer of "ganoin" (= enamel). Overgrowth of odontodes was observed in *Andreolepis* (Gross, 1968c, Fig. 8E) and *Orvikuina* (Gross, 1953, Fig. 9; Schultze, 1968, Fig. 19). In *Andreolepis* the odontode generations are separated by vascular bone in the same way as in *Lophosteus*.

In the more advanced genera *Orvikuina* and *Dialipina* the odontode generations overlie each other directly. In some of the primitive Devonian species the cellular bone, which seems to be a plesiomorphic character for all Osteichthyes, is replaced by acellular bone [e.g., *Dialipina salgueiroensis* and *Orvikuina vardianensis* (Gross, 1953; Schultze, 1977a)]. In the most advanced of the "lophosteiform" genera, namely in *Dialipina* and in *Ligualepis*, overgrowth was not observed; new odontodes are added only laterally to the scales as increments (Schultze, 1968, Figs. 5, 6, 8–10; Schultze, 1977a, Fig. 4).

The five "lophosteiform" genera, *Lophosteus, Andreolepis, Orvikuina, Dialipina*, and *Ligualepis*, are almost exclusively known from scales (their scales are united here as pre-*Palaeoniscus* type). Available descriptions of the few larger dermal bones of head and shoulder girdle and of the finspines (Gross, 1968c, 1969; Janvier, 1978) indicate that these elements were organized in the same way as the scales. They consist of lamellar bone, vascular bone, and odontodes, as ornamentation at the surface.

The typical *Palaeoniscus*-type scale (Goodrich, 1907) can be derived from *Dialipina* and *Ligualepis* (Schultze, 1977a). The most primitive of a *Palaeoniscus*-type scale are the scales of *Cheirolepis* (Middle to Upper Devonian; Chondrostei, Palaeonisciformes). Gross (1966) classifies them as "Kleinschuppen" in contrast to all other ganoid scales, which are regarded as "Grosschuppen." Morphogenesis of the *Cheirolepis* scales was fully documented by Goodrich (1907), Aldinger (1937), and Gross (1953, 1966). Growth is concentric; each increment consists of (1) a bone lamella on the base, (2) a more or less ring-shaped odontode which is added at the periphery of the scale, and (3) a ganoin (= enamel) layer, which covers not only the new odontode but the whole crown. In very early morphogenetic stages the new odontode forms the periphery of the scale but also covers the ganoin layer of the preceding increment. From a morphogenetic point of view, *Cheirolepis* scales are organized similarly to acanthodian scales; there are, however, histologic differences (Gross, 1966).

Palaeoniscus-type scales (in the strict sense), which are Grosschuppen in Gross' (1966) classification, differ from *Cheirolepis* scales only by their size and by certain morphologic characters, related to the articulation of neighboring scales. Unfortunately, morphogenesis has been documented only in a few *Palaeoniscus*-type scales as fully as in *Cheirolepis* (Williamson, 1849, Fig. 8; Aldinger, 1937, Fig. 8). From the available evidence, however, it seems to be certain that during each growth step a more or less ring-shaped odontode is added together with a new bone layer (covering the whole base) and a new ganoin (= enamel) layer which covers the whole crown. In other words, the dentine layer of this scale type consists of discrete elements which do not grow, but which are formed individually at consecutive growth steps.

For the sake of the present discussion the fish classification of Nelson (1976) is used. It may not be regarded as up to date in certain respects by paleoichthyologists. I know, however, of no classification on which the majority of paleoichthyologists would agree. The *Palaeoniscus*-type scale occurs in numerous Chondrostei but also in the Brachiopterygii (= Cladista), which include the Recent genera *Calamoichthys* and *Polypterus* (Hertwig, 1876/1879/1882; Goodrich, 1907; Sewertzoff, 1932; Ermin *et al.*, 1971). [In contrast to Nelson (1976) and in agreement with Rosen *et al.* (1981), I regard the Cladista, or Cladistia, as a side branch within the Actinopterygii rather than a group outside the Actinopterygii of unclear systematic relationships.] The scales on the shoulder girdle and the fins of *Polypterus* have a special feature, namely cone-shaped thorns on the outer surface. These cones consist of dentine and enamel and can be traced down into the dentine layer. They are either secondary modifications of the dentine organs of the whole growing scale or have evolved *de novo*. They form independently from the scale and later become fused to it (Sewertzoff, 1932).

Resorption was observed neither in the pre-*Palaeoniscus* type, in many primitive genera of the *Palaeoniscus*-type, nor in *Polypterus* (Kerr, 1952). In the more advanced genera, however, the outer rim of the scales is partly resorbed before the new increment is laid down (Aldinger, 1937, Figs. 20, 21, 59, 79).

Gross (1966) distinguished between a *Palaeoniscus* type in the strict sense (the primitive type, without canals in the bony base) and a *Lepidotus* type (the advanced type, with Williamson's canals in the bony base). While this distinction is valid, it is of no relevance for the present discussion. The *Palaeoniscus*-type scale (*sensu lato*) is modified in numerous Chondrostei [e.g., *Elonichthys, Acropholis, Plegmolepis, Acrolepis, Boreosomus, Stegotrachelus* (Aldinger, 1937)].

Overgrowth of odontodes occurs in the scales of *Scanilepis* (Aldinger, 1937, Fig. 62). As early as the Lower Carboniferous the Platysomoidea (Chondrostei) lose the dentine and the ganoin layer and retain only a bony scale. Other chondrosteans lose the squamation altogether.

The main trend in the modification of the *Palaeoniscus*-type scale leads to a reduction of the dentine layer. The *Lepisosteus*-type scale (Goodrich, 1907) consists only of bone lamellae in the base and a sequence of enamel lamellae on the crown. Each enamel + bone lamella covers the whole scale and contributes to the increases in size. *Lepisosteus*-type scales occur in some Chondrostei (*Gyrolepis*) and numerous Holostei. Details of the evolution of the *Palaeoniscus*-type scale and the *Lepisosteus*-type scale are unknown. *Lepisosteus* itself, however, does not lose the dentine tissue completely. Hertwig (1876/1879/1882) found growing scales ornamented with tooth-shaped dentine spikes on the trunk, fins, head, and shoulder girdle of this genus. On the ventral side of the head he found isolated dentine spikes which clearly are individual odontodes; each of the odontodes is anchored separately in the corium by a bony base. Bony bases with several odontodes occur also, but it is unknown whether they arose by fusion of neighboring bony bases or by real growth. Hertwig regards these odontodes as a very primitive character which was inherited from shark-like ancestors and thus is directly homologous to placoid scales. In the light of the model presented here, it is, however, much more likely that either (1) these tooth-shaped odontodes are modifications of odontodes which occurred in the *Paleoniscus*-type scales of the ancestors, or (2) that the direct ancestors of *Lepisosteus* had no dentine in the scales at all and that the odontodes are newly developed as a sort of "atavism." It is very important to note in this context that the bony base develops long before one can see the epidermal–mesodermal papillae that form the tooth-shaped odontodes. The odontodes are later fused with the scale (Klaatsch, 1890, Nickerson, 1893; Kerr, 1952).

The transition of the scales of Holostei to the bony scales of Teleostei was documented by Schultze (1966). Within the Teleostei the thin bony scale (cycloid scale), which allows a high degree of body flexibility, is the primitive condition.

In advanced groups numerous modifications of the cycloid scale can be found:

1. Ctenoid scales with small bony spikes (Hase, 1911b; Rosén, 1914).

2. A mosaic of growing plates which form a solid armor in the Agonidae, Ostraciontidae, Gasterosteidae, and Syngnathidae (Williamson, 1851; Hertwig, 1876/1879/1882; Rauther, 1940; Bertin, 1958; Harder, 1975);

3. Thick, enlarged scales, some ornamented with spines in Doradidae, Callichthyidae, Loricariidae, Balistidae, Triacanthidae, and Molidae (Williamson, 1851; Hertwig, 1876/1879/1882; Gegenbaur, 1898/1901; Rosén, 1913; Goetsch, 1921; Bhatti, 1938; Rauther, 1940).

4. In Diodontidae, Tetraodontidae, Dactylopteridae, Cyclopteridae, Ogcocephalidae, Antennariidae, and on the rostrum of Xiphiidae the scales are modified into growing spines (Hertwig, 1876/1879/1882; Rosén, 1913; Goetsch, 1921; Rauther, 1940). *Cyclopterus* has composite spines consisting of many cusps. They are formed by concrescence of individual cusps and by marginal growth. According to Hase (1911a) and Goetsch (1921), the cusps consist of dentine. This is questioned by Tretjakoff and Chinkus (1927) and Rauther (1940).

5. *Centriscus* (Centriscidae) has, simply as a morphologic convergence, scales which resemble closely the placoid scales of modern sharks. They consist, however, solely of bone and grow by deposition of bone lamellae (Hertwig 1876/1879/1882).

Because the phylogeny of the Teleostei is still poorly known, it is impossible to outline the evolutionary changes of the scales; one can only list the different types of scales. It is certain that all the organs listed under 1–5 are derived characters rather than primitive characters with respect to the Teleostei. However, it is not quite certain that all the organs are really derived from cycloid scales. It could also be that some of them are newly evolved calcifications occurring in species with scaleless ancestors.

All the dermal elements listed above consist of bone. With the possible exception of *Cyclopterus* and *Xiphias gladius*, dentine and enamel (or enameloid) occur only in the toothlike elements (odontodes) on the large scales of Callichthyidae and Loricariidae (Williamson, (1851; Hertwig, 1876/1879/1882; Goetsch, 1921; Bhatti, 1938; Rauther, 1940). Schmidt and Keil (1971) consider the hypermineralized cap to consist of enameloid. The skull bones of *Denticeps clupeoides* (Denticipitidae, Clupeomorpha) are covered with tooth-shaped structures, which are referred to as odontodes by Greenwood (1968). I know, however, of no histologic study which confirms that the spikes consist of dentine rather than bone. The odontodes of the Loricariidae and Callichthyidae are either firmly ankylosed or anchored by connective fibers, depending on the species and on the body region. It also depends on species and body region whether the odontodes undergo replacement or are never replaced (Hertwig, 1876/1879/1882; Bhatti, 1938). The size of the bony scales in the dermis varies from very small scales carrying a single tooth to very large scales with numerous teeth. All scales grow by accretion of bone layers; according to Goetsch (1921), the very small scales can also fuse to form larger scales.

There is no doubt that the odontodes in Loricariidae and Callichthyidae (armored catfishes) were absent in the ancestors of the two closely related families and have newly evolved. Hertwig's (1876/1879/1882) attempt to homologize the very small scales with the bases of placoid scales and the odontodes with the crown of placoid scales is wrong. Each odontode in the two families is a unit of its own and has its own anchoring mechanisms, including bone tissue (Bhatti, 1938). Regan (1924) regards the odontodes of the scales of the armored catfishes as an exception to Dollo's law of evolutionary irreversibility, because they are a redevelopment of placoid scales. This is certainly an overemphasis of the evolutionary significance of these odontodes. Though the ancestors of teleosts have lost dentine in their scales, they have retained the capacity to form dentine in their oral teeth. Hence, what has redeveloped are not placoid scales *per se*, but simply the dentine formation outside the mouth. Histologic sections show clearly that the odontodes develop independently of the underlying growing bony scale and are only later anchored to the scale (Goetsch, 1921).

It is quite likely that the hypermineralized cap of these odontodes is true enamel (in other words, an ectodermal product), because Bhatti (1938) reports that dentine is formed first and then this hypermineralized tissue.

Teleost scales are formed in the corium. They are hence a mesenchymal product. Fach's (1936) assumption that the epidermis is directly involved in the scale formation must be doubted (Neave, 1936; Rauther, 1940; Kassner, 1965; Harder, 1975). It is, however, very important to note that the scale primordium is formed by a papilla directly below the basal membrane which separates epidermis and corium (Nardi, 1936). Hence it is very likely that the first stimulus for the scale formation comes from an inductive interaction between ectoderm and mesenchyme.

Fin rays (dermotrichia), which occur in different groups of fishes, are part of the dermal skeleton and according to some authors are probably derived from some type of scale. Several types of finrays are largely uncalcified, and hence need not be mentioned specifically here [ceratotrichia of the Elasmobranchii, camptotrichia of the Dipnoi, actinotrichia of the Actinopterygii; see discussion in Patterson (1977)]. Lepidotrichia are segmented and consist of bone; they occur in Actinopterygii and Crossopterygii. Hard rays of Actinopterygii are unsegmented. They can be derived from round scales (Rauther, 1940; Harder, 1975; Starck, 1979).

Skull of Actinopterygii

Like all other Osteichthyes but remarkably unlike Acanthodii, the endocranium and the viscerocranium of the Actinopterygii is covered by a mosaic of dermal bones which form a fairly stable pattern. The dermatocranium evolves largely independently of the squamation. This is true from the point of view of morphology and osteology, but also with respect to histologic and histogenetic features.

Except for a few examples, the histology of the dermal skull bones of fossil Chondrostei and Holostei is largely unknown. Gross (1936) described odontodes on jaw bones of *Hypsocormus* (Holostei). The histology of the dermal bones of *Andreolepis* ("lophosteiforms") agrees well with that of the scales (Gross, 1968c). Ørvig (1978a,b,c) studied the histology of skull bones in seven genera of Permian and Triassic Chondrostei. *Plegmolepis, Boreosomus, Scanilepis, Nephrotus*, and *Colobadus* have body scales of a modified *Palaeoniscus* type. The *Gyrolepis* scale is intermediate between the *Palaeoniscus* type and the *Lepisosteus* type. *Birgeria* has no scales at all. All seven genera have dermal bones with well-developed odontodes. Usually each odontode has its own enamel cap. The advanced condition, which is exhibited in the ganoin scales, namely that the enamel lamellae of successive increments are stacked on top of each other, is rare in the dermal bones. On the other hand, overgrowth of successive generations is common; up to nine odontode generations have been found in a single bone, with the preceding generation of odontodes being partly resorbed before overgrowth. This overgrowth contributes considerably to the secondary thickening of the dermal plates, all of which have a vascular layer in the middle and a dense layer below. Whereas resorption is common in the odontode layer, remodeling of the bone is very rare.

Ørvig (1978b) found secondary osteons in *Plegmolepis*. Morphogenetic processes in the upper layer of the dermal plates can only be related to secondary growth in thickness, rather than to areal growth of the plates. Ørvig does not describe the morphogenesis of whole dermal plates, and hence it remains open as to how histogenesis and morphogenesis of individual plates are interrelated with the ontogenetic growth of the whole skull.

Schaeffer (1977), Ørvig (1978a), and Meincke (1980) showed that the histology of the dermal skull bones of *Polypterus* does not differ from that of the scales, which are three-layered (see above).

In the advanced Actinopterygii, dentine and ganoin have also been lost in the skull. However, there is no detailed information about this process.

"Crossopterygii" and Dipnoi

Many modern authors regard the Crossopterygii as a di- or polyphyletic taxon (Rosen *et al.* 1981; Schultze, 1977a, 1981). Five groups of Crossopterygii are generally accepted: Actinistia (= Coelacanthiformes), Holoptychiida (= Porolepiformes), Osteolepidida (= Osteolepiformes), Rhizodontida (= Rhizodontiformes), and Onychodontida (= Struniiformes).

According to Schultze (1977a), one has to assume that the common ancestor of "Crossopterygii" and Dipnoi had scales which were covered with several generations of odontodes. These scales were probably similar to the scales of *Lophosteus*. It is quite likely, according to Schultze (1977a) and Miles (1977), that this primitive condition is retained in the Actinistia. However, it is also true that on the basis of scale histology alone, one cannot tell whether this is still the primitive condition or whether it is a secondary simplification of the advanced cosmoid-type scale (Rosen *et al.*, 1981). Superposition of odontodes occurs, e.g., in the Jurassic actinistian *Undina* (Gross, 1936, 1966) and in the Recent actinistian *Latimeria* (Roux, 1942; M. M. Smith *et al.*, 1972; Ørvig, 1977 and M. M. Smith, 1979a). Ørvig found up to four generations of odontodes on top of each other. M. M. Smith (1979a) showed that the complete odontodes with enameloid, dentine, and a bony base develop independently from the main body of the scale, which consists of lamellar bone. After calcification the odontodes are welded to the scale. As the scale grows, new odontodes are welded to its growing margin. The fact that odontodes and bony scales develop independently (as in *Lepisosteus, Polypterus*, and the armored catfishes) and that the bony scale is formed in a deeper layer of the corium indicates that the bony base of the odontodes can justifiably be interpreted as a secondary condition. It should be noted in passing that the dermal skull bones of *Latimeria* are also ornamented with odontodes. No further details are known (Bernhauser, 1961).

"Cosmine" is a complex tissue consisting of a superficial layer of enamel, a layer of dentine, a layer of spongy bone, and a layer of dense bone. The dentine layer seems to be continuous, but detailed studies are lacking concerning whether or not it consists of discrete but densely fused odontodes. The dentine layer houses a pore canal system. Cosmine occurs in scales as well as in dermal skull bones. It is a tissue which can be derived from the primitive odontode-bearing dermal elements. It occurs in Holoptychiida, Osteolepidida, and Dipnoi. It has never been convincingly shown to occur in Onychodontida (Rosen *et al.*, 1981). No information is available on whether it occurs in Rhizodontida.

Several authors (Schultze, 1977a; Rosen *et al.*, 1981) have suggested that cosmine evolved only once, namely in a common ancestor of Holoptychiida, Osteolepidida, and Dipnoi. This assumption is based on the complexity of the cosmine structure, which is similar in Dipnoi and Crossopterygii. There are, however, to my knowledge, no independent phylogenetic criteria to either support or refute this hypothesis.

The function of cosmine is unknown. Cosmine, and also the porecanal system of Osteostraci and Acanthodii (see above), may have housed an electroreceptor organ (Thomson, 1977). Cosmine could not grow (Gross, 1956; Ørvig, 1969a; Thomson, 1975). Whole plaques of cosmine formed synchronously, often covering the sutures between individual bones. In order to increase the size of the plaque or let the underlying bone grow in area, the plaques were resorbed. In view of the observation (often mentioned in this paper) that the bone of attachment is an integral part of a dermal dentinous structure, Thomson's (1975) observations that the spongiosa takes part in the resorption and redeposition process and that it has a different structure underneath the cosmine than in a naked dermal bone are not very surprising.

Cosmine is clearly a derived tissue. As yet, however, no model has been proposed as to how cosmine evolved from a dermal element consisting of discrete odontodes which are fused by bone at their bases. It is very important in the present context that some scales of Rhipidistia and Dipnoi have isolated odontodes which were formed at early stages of morphology and were later covered by a uniform sheet of cosmine [Rhipidistia: *Laccognathus* (Gross, 1966, Fig. 6A) and *Porolepis* (Gross, 1966, Fig. 5A); Dipnoi: *Uranolophus* (Denison 1968, (Fig. 23)].

The cosmine layer is lost independently in several lines of the Rhipidistia and in the Dipnoi (Gross, 1966; R. S. Miles, 1977; Schultze, 1977; M. M. Smith, 1977; Rosen *et al.*, 1981). Advanced genera of these groups have scales and dermal bones consisting only of bone (Klaatsch, 1890; Rauther, 1940; Schultze, 1980). An intermediate stage which is known from two genera of the Lower Middle Devonian of Australia (M. M. Smith, 1977) is a reversion back to a surface ornamentation of scales and dermal bones consisting of normal isolated odontodes. According to M. M. Smith (1977), this type of surface ornamentation cannot be considered a disintegration product of cosmine but rather the retention of an earlier ontogenetic condition. In other words, the common ancestor of Rhipidistia and Dipnoi must have had superimposed denticles beneath a cosmine covering. Secondarily, the cosmine covering was lost during phylogeny, so that denticles were the only type of surface structure. M. M. Smith (1977) showed that odontodes were not only covered by superpositional growth (she found up to five generations of odontodes), but were also shed by the resorption of the basal bone tissue and replaced.

In close analogy to Stensiö's (1961) Lepidomorial Theory, M. M. Smith (1977) assumes that denticles with complex morphology have to be considered as composite structures in which every pulp canal with the circumpulpal dentine is a unit of its own, i.e., a lepidomorium. This is not very convincing. I would rather assume that a whole tubercle or denticle has to be regarded as a single unit of its own, i.e., an odontode. Comparative morphologic and histologic analogs of shark scales show that within one species a simple shape can grade into a complicated shape, and a simple histology (with one pulp canal) can grade into a complicated histology (Reif, 1973b, 1974, 1978a).

Another question, however, is what kind of homology exists between individual denticles (odontodes) and large plaques of dentine, which obviously do not consist of denticles and form synchronously. Two different types of plaque are known to occur in dipnoans, one with a pore canal system, i.e., true cosmine plaques, and one without a pore canal system (M. M. Smith, 1977).

Tetrapoda

In our current state of knowledge the question of whether the tetrapods are a sister group of the Dipnoi (Rosen *et al.*, 1981) or are most closely related to the Osteolepidida (Janvier, 1980; Thomson, 1980; Westoll, 1980; Schultze, 1981) has no direct bearing on the discussion of the evolution of the dermal skeleton. It is certain that the Tetrapoda inherited from the fish ancestors a mosaic of dermal bones in the skull and the shoulder girdle. In the common ancestor of the Tetrapoda, the dentine and enamel (or enameloid) layers were already lost. It is very likely that the amphibians also inherited a body squamation of bony scales. A post-cranial dermal skeleton is common in fossil amphibian groups, i.e., Ichthyostegalia, Rhachitomi, Plagiosauria, Anthracosauria, and Micro-sauria, but in none of these groups is it certain whether this is a primitive character or a secondary acquisition. In the Recent Amphibia only the Caecilia (= Gymnophiona) have a squamation of bony scales (Klaatsch, 1980; Gegenbaur, 1898/1901; Markus, 1934; Lawson, 1963; Casey and Lawson, 1978).

These scales in Gymnophiona consist of a layer of fibrous bone, which undergoes areal growth and is ornamented with lens-shaped elements. These also consist of bone, are formed independently, and are anchored in depressions of the bony plate. Most authors who discuss the evolutionary significance of the scales agree that they are most likely derived from fish ancestors rather than newly evolved organs (Klaatsch, 1890; Gegenbaur, 1898/1901; Markus, 1934; Casey and Lawson, 1978). In the attempt of the Hertwig's school to derive all scales more or less directly from placoid scales of sharks, Marcus (1934) goes much too far and tries to show that the ectoderm plays the leading role in the formation of the gymnophionid scale. Marcus' observations have never been repeated. Provided that the scale and its ornamentation really consist of bone, it is unlikely that the ectoderm plays any role in secreting the scale at all. Rather, its role is confined to a possible inductive interaction with the mesenchyme to form the mesenchymal papilla that secretes the scale primordium.

Postcranial dermal skeletons also occur in numerous fossil and Recent Reptilia; e.g., Chelonia, Squamata (Anguidae, Gerrhosauridae, Cordylidae), Crocodilia, Placodontia, Phytosauria, Araeoscelidia, Ankylosauria, Stegosauria, and Pseudosuchia (Peyer, 1931). In practically none of these groups is it clear whether the dermal skeleton was inherited from the fish grade or whether it evolved *de novo*. Both Gegenbaur (1896/1901) and Romer (1976) assume that the gastralia of the Tetrapoda are a fish heritage. The only group where the evolution of the dermal skeleton can possibly be traced from naked ancestors are the Placodontia (Westphal, 1975, 1976). The phylogeny of this group is still largely enigmatic. The present evidence of the fossil record, however, suffices to make the argument convincing that advanced armored groups evolved from primitive naked groups.

Only one mammal group, the Loricata, has a well-developed postcranial dermal skeleton. Bony plates also occur in the skin of whales (Romer, 1976). Both types of dermal element have evolved *de novo* (Peyer, 1931).

As far as is known, the dermal skull bones, the dermal pectoral bones, and the postcranial dermal skeleton consist only of bone tissue. In the tetrapods, dentine and enamel or enameloid have never reappeared in the dermal skeleton outside the mouth cavity.

EVOLUTION OF THE DERMAL SKELETON

Discussion

The validity of one of the most basic premises of this discussion was reaffirmed by Patterson's (1977) very important review article: The dermal skeleton and the endoskeleton of vertebrates are two independent organ systems. They have been independent of each other throughout the phylogeny of the vertebrates, i.e., throughout the ontogenesis of each individual vertebrate. Nowhere in the history of the vertebrates has an integration or an exchange taken place. Bones of the dermal skeleton were never incorporated in the endoskeleton or vice versa. The endoskeleton consists of cartilage, cartilage bones, and membrane bones; the dermal skeleton consists of dermal bones (primitively covered with dentine and enameloid) and, only in the birds and the mammals, of adventitious cartilage. Ceratinous and collagenous dermal-epidermal products need not concern us here.

A cartilaginous endoskeleton (including a branchial basket, a brain caspule, and a notochord) most probably preceded all other skeletons in the vertebrate ancestors.

The complete or almost complete lack of an ossified endoskeleton in heterostracans, galeaspids, thelodonts, anaspids, and chondrichthyans suggests that the mineralized dermal skeleton evolved before the mineralized endoskeleton.

There has long been a discussion over whether bone evolved *before* dentine or vice versa, or whether bone evolved *from* dentine or vice versa. The history of this discussion will not be traced here. It seems to me that one can extrapolate the following points from comparative histologic and histogenetic studies:

1. Mesenchyme derived from neural crest cells plays a dominant role in the development of the dermal skeleton (Grant, 1978; Le Douarin, 1975; Johnston, 1975). There may be phylogenetic information hidden in the fact that the neural crest cells not only contribute to the nervous system, but also develop into pigment cells and contribute to visceral cartilages, other endoskeletal cartilages, and the dermal skeleton. The meaning of this information is, however, completely obscure. The neural

crest is a unique vertebrate feature (Moss, 1968a,b,c; 1969; Maderson, 1972, 1978; Graver, 1974; Hall, 1975).

2. Bone, dentine, enamel, and enameloid are separate tissues. While it is likely that enameloid evolved into enamel (see below), all available evidence points to the fact that dentinous tissues (dentine, mesodentine, semidentine) and bone are independent of each other and that one did not evolve from the other. [For a contrasting view see Ørvig (1967a).] Except for cases which are clearly derived [e.g., pleromin formation in heterostracans, ptyctodonts, etc.; dentine in the bony base of batoids (Reif, 1979a)], dentine always forms directly below the basement membrane that separates epidermis (ectoderm) from dermis (mesenchyme). If an enameloid cap is present, dentine is formed in continuity below the enameloid. If dentine and enameloid are present, bone develops below the dentine. If dentine and enameloid are lacking, dermal bones primitively start from a papilla which differentiates directly below the basement membrane. It is a derived condition if bones develop in deeper layers of the dermis. [These statements should not be taken as support of the Holmgren–Jarvik delamination hypothesis, which are discussed by Jollie (1968) and criticized by Patterson (1977) and Schaeffer (1977).]

3. Enameloid, dentinous tissue, and bone are the products of a continuous histogenetic and morphogenetic process. For this reason and for comparative histologic reasons they are supposed to have evolved together. Enameloid, however, is not as essential as the other two tissues. It is lacking in several agnathan and fish groups. This may mean that enameloid is not of the same age as bone and dentinous tissues. During evolution it may also have been lost and reevolved several times.

4. Whether cellular bone or acellular bone evolved first is an important question, but it has no direct bearing on the model presented here. In teleosts it is well known that acellular bone evolved from cellular bone. Ørvig (1965, 1967a) assumed the same trend for the heterostracan ancestors. [In Heterostraci only acellular bone, "aspidine," is known; see also Denison (1963), Halstead Tarlo (1963, 1964b), Halstead (1969b).] However, other authors (e.g., Schaeffer, 1977) suggest that the problem cannot be solved at this time.

5. Enameloid is assumed to be the primitive and enamel the advanced tissue. This assumption is based on the respective distribution of the tissues among vertebrate taxa and on histogenetic criteria. (Enamel formation seems to be more specialized.) Enamel evolved from enameloid by a heterochronous shift in the activities of ectoderm and mesenchyme [Schaeffer (1977), Reif (1979); for a somewhat different model see Poole (1971)].

6. In the dentinous tissue Ørvig (1967a) assumes an evolutionary

sequence mesodentine–semidentine–metadentine (dentine in the strict sense). Again, this assumption is not readily testable with the data available (Schaeffer, 1977). The distribution of tissues among the vertebrates does not directly support it: Mesodentine in the Osteostraci, Thelodonti, Placodermi, Acanthodii; semidentine in the Placodermi; metadentine in the Heterostraci, Thelodonti, Placodermi, Chondrichthyes, Acanthodii, Osteichthyes (Ørvig, 1967a). In the present context the question is of no great relevance. Another part of Ørvig's (1967a) model, namely to regard odontoblasts (dentine-forming cells) as specialized osteoblasts (bone-forming cells), is not accepted here. Pleromic (= hypermineralized) dentine is clearly a secondary development. It must have evolved convergently in the various groups where it occurs (Heterostraci: dermal plates; Ptyctodontida: gnathalia; Holocephali: tooth plates; Dipnoi: tooth plates; Bradyodonti: tooth plates; Selachii: teeth).

The comparative survey of dermal skeletons of agnathans and fishes leads to the following conclusions: Microsquamose dermal skeletons occur in Thelodonti and Chondrichthyes, but probably also in a few Heterostraci and one genus of Anaspida. In Heterostraci, Galeaspida, Osteostraci, Anaspida, and Placodermi only middle-aged to adult specimens are known. These have macrosquamose dermal skeletons, which in some cases grew by marginal growth, and in other cases were formed only at the end of ontogeny. In several groups (especially Heterostraci, Osteostraci, and some primitive Placodermi) there are strong indications that the adult skeleton consisting of plates and scales was preceded in ontogeny by a microsquamose dermal skeleton consisting of individual odontodes. From the scales of Acanthodii and the "lophosteiform" genera it is obvious that in the Teleostomi the odontode is the most primitive element of the dermal skeleton. [Very little can be said about the laterosensory component of the dermal skeleton, except that there is some sort of causal relationship involving neuromast organs and the initiation of dermal bone ossification (Schaeffer, 1977; see also Aumonier, 1941; Ørvig, 1972).] Hence we conclude that odontodes (consisting of dentine and bone, and probably also enameloid) are the most primitive element of the dermal skeleton in vertebrates. Ørvig (1968) and Nelson (1970) seem to arrive at a similar conclusion, but unfortunately they use Ørvig's (1960) vague terminology ("micro-, meso-, macromeric") and do not distinguish between growing and nongrowing elements.

Before the first dermal skeleton evolved in early vertebrates two regulatory processes must have evolved: (1) migration of neural crest cells (see above); and (2) inductive interaction between ectoderm and mesenchyme. Probably all dermal–epidermal products (scales, dermal bones, teeth, hairs, feathers, fingernails, etc.) are the result of these mutual interactive inductive stimuli between mesenchyme and ectoderm. Hence, odontodes can only form at the ectoderm/mesenchyme interface. The inductive interaction was probably inherited from prevertebrate ancestors. Kollar and Baird (1969), Kollar (1972), Kollar and Fisher (1980), and others have shown that the dermal papilla plays the leading role in the inductive interaction.

If one tries to arrive at a model of the evolution of the first mineralized elements of the dermal skeleton, two aspects are important: (1) Secretion of enameloid and dentine matrices takes place into the space between epidermis and mesenchyme; this is the site of the basement membrane complex. This basement membrane complex is secreted by both tissue layers (Bernfield *et al.*, 1973). Thus, secretion of the matrices could just be a temporal extension of the secretory activities of ectoderm and mesenchyme. (2) Conditions that led to the evolution of connective tissue were preadaptive for the evolution of mineralized tissue (Richey, 1977).

The highly specialized mode of formation makes it very likely that odontodes were the earliest constituents of the dermal skeleton of Craniota and makes it unlikely that odontode formation evolved convergently in various groups of Craniota.

A Model for the Evolution of the Dermal Skeleton

Step 1

If odontodes produced by a papillary organ are the oldest calcified elements of the dermal skeleton, it is likely that they were preceded in evolution by uncalcified organs also produced by a papillary organ. The primary function of the odontodes may have been protection of the skin against abrasion of predators or storage of phosphate [see discussion in Halstead (1974)]. (The assumption that calcified elements function as a phosphate storage has been widely discussed, but it should be kept in mind that resorption has rarely been found in the dermal skeleton of agnathans.) In light of the assumption that odontodes protected the skin against abrasion and/or parasites, it is very important to note that dentine is a sensitive tissue (Halstead Tarlo and Halstead Tarlo, 1965; Halstead, 1969a). The structural basis for this sensitivity are intradentinal nerve endings of peripheral nerve fibers. These nerve endings lose their Schwann cell covering, their basement membrane, and their myelin sheath. The odontoblasts seem to replace the Schwann cell in their supporting function (Frank *et al.*, 1972). It is remarkable that this relationship is very similar to that observed in the epidermis and dermis, where nerve endings approaching the epidermis surface lose their Schwann cells and become invested by other cell types (Frank *et al.*, 1972). The conclusion which can be drawn from this observation is that the nerve supply of the dentine reflects the primitive condition is an unmineralized stage. Structure and function of these hypothetical unmineralized dermal elements remain completely obscure. Recent cyclostomes have no structured dermal products at all.

The dermal skeleton of the early Craniota was probably an undifferentiated skeleton of nongrowing odontodes (i.e., placoid-like scales) of uniform size and shape covering the whole integument (body surface and mouth cavity). It was probably similar to the dermal skeleton of thelodonts and Recent chondrichthyans.

I do not assume that the dermal skeleton was formed only at the end of ontogeny, but rather that scale formation started early in ontogeny. Growth of the body was compensated by insertion of new scales, each consisting of a single odontode. A basic prerequisite for this insertion is a regulatory process which has been inferred in vertebrate dentitions and shark squamations (Osborn, 1970, 1971, 1973a,b, 1974, 1977, 1978; Reif, 1974, 1976). It is assumed that each element of the dermal skeleton (or dentition, for that matter) forms an inhibitory field around itself, thus preventing the induction of more elements within the field. Outside the field new dermal elements are automatically induced. Consequently, as the animal grew, its scales were moved apart and the gaps between the scales were filled with new scales (Fig. 4).

If the primary function of the microsquamose elements was to protect the skin, it is not unlikely that replacement of the scales evolved very early, simply by resorption of the anchoring fibers and by insertion of replacement scales according to the Inhibitory Field Model. It is important to note that replacement scales could only be induced once the original scale was shed (Fig. 5). A prefabrication of scales is not possible in this type of skeleton in contrast to the vertebrate dentitions.

The hypotheses developed here and the cladogram (Fig. 1) lead to the assumption that microsquamose dermal skeletons occurred in the ancestors of Heterostraci, Thelodonti, Cephalaspidomorpha, Placodermi, and "higher Gnathostomata." Growing and nongrowing dermal plates and growing scales consequently must have evolved independently in the various groups.

Step 2

If the inhibitory field was short-lived in comparison with the functional life of an odontode, a new odontode was induced directly adjacent to the old one and welded to it (Fig. 6). This leads to growing scales. At this stage of evolutionary development, which is exemplified by certain fossil elasmobranchs, growth of the body can be compensated by (1) insertion of new scales, (2) size increase of newly introduced odontodes during ontogeny, and (3) growth of the scales, while (4) replacement of scales may or may not take place. The combination of these four different processes leads to a wide spectrum of different types of squamation, some of which have been found in fossil elasmobranchs [for details see Reif (1978a, 1980a)]. In acanthodians a combination of processes 1 and 3 must have occurred.

Step 3

Step 2 grades into the next step, in which the number of dermal elements remains constant during ontogeny and growth of the body is compensated solely by scale growth (Figs. 7 and 8). Each scale primor-

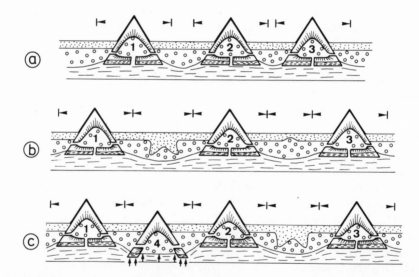

FIG. 4. Vertical sections of the skin of a primitive craniote. Skin layers from top to bottom: epidermis (ectoderm); vascular dermis layer (mesenchyme); fibrous dermis layer (mesenchyme). Tissues in the scales from top to bottom: enamel (or enameloid; shown in black); dentine (dentine tubules are indicated); bone (anchoring collagen fibers are indicated). The model expounded in the text assumes that simple nongrowing dentine organs (odontodes) (1, 2, 3) formed in a single morphogenetic step were the first elements in the dermal skeleton of Craniota. They consisted of enameloid, dentine, and bone (which functioned as an anchoring tissue). Each odontode was surrounded by an inhibitory field (arrows), which prevented the induction of new odontode germs. When the skin grew, new germs (4) were automatically induced in the gaps between the inhibitory fields by a mutual inductive stimulus between ectoderm and mesenchyme. This type of squamation persists today in the Chondrichthyes.

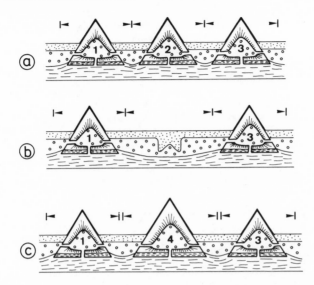

FIG. 5. If nongrowing scales (= odontodes) are shed, the gaps are automatically filled with new scales. The diagram illustrates a morphogenetic mechanism in which a small scale is replaced by a large scale. Thus, the scales of the young growth stages are gradually replaced by bigger ones. This regulatory mechanism occurs in most elasmobranchs. For further explanations see Fig. 4.

dium consists of a single odontode which is formed very early in ontogeny. The scales continue to grow as long as the body grows. Abrasion of the scales may be healed by overgrowth (Fig. 9) or by pleromin (Fig. 10).

Step 4

Large plates are formed by the fusion of a large number of isolated odontodes at their bases (Figs. 9 and 10). Once these plates are formed they are either capable of marginal growth (as in Heterostraci, Placodermi, Actinopterygii) or they do not grow (as in most Osteostraci). The formation of solid plates can take place at any stage of ontogeny. During plate formation calcification proceeds downward. Repair of the plate occurs by overgrowth or by pleromin secretion. Secondary thickening of the plate occurs by addition of basal bone layers or by overgrowth. Remodeling of the plate to allow changes in plate shape has not been found.

Step 5

No special mechanisms are required to explain the evolution of the ganoid scale. The ganoid scale is practically included in step 3. An important problem, however, is posed by the evolution of cosmine, which

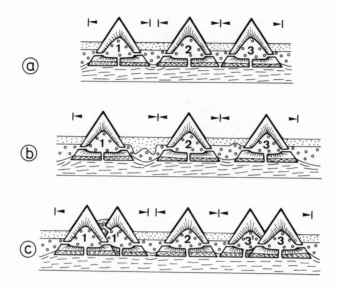

FIG. 6. Growth of micromeric scales. In some groups the odontodes grow by welding of new odontodes to the preexisting ones. A gap in the inhibitory field or degeneration of the field (a) leads to the induction of a new papilla (b). Odontode 1^1 is welded (c) to odontode 1. This mechanism is the basis for the morphogenesis of numerous types of growing scales, "tesserae" (small platelets), and dermal plates of anaspids, heterostracans, osteostracans, placoderms, acanthodians, elasmobranchs, actinopterygians, "crossopterygians," and dipnoans; it is compatible with the differentiation hypothesis. The contrasting concrescence hypothesis goes one step further and assumes that two developing papillae can completely fuse and lose their identity. For futher explanations see Fig. 4.

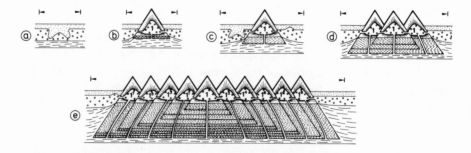

FIG. 7. Growth of macromeric scales. The growth mechanism explained in Fig. 6 can lead to large plates. For further explanations see Fig. 4.

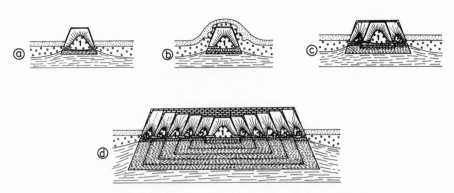

FIG. 8.   Growth of a ganoid scale. In ganoid scales of the *Palaeoniscus* type a ring-shaped odontode is added at each growth step. At the same time the whole crown is covered with a new lamella
of enamel; this ultimately leads to a thick layer of "ganoin." The functional significance is probably
to guarantee a continuous control of the surface of the scale crown. The ganoid scales are either
smooth or finely ridged. Both types of surfaces probably have a direct relationship to drag control
during swimming. In the advanced ganoid scale type (*Lepisosteus* type) the dentine layer is lost.
Acanthodian scales have a morphogenesis similar to that of the *Palaeoniscus* type scales. For
further explanations see Fig. 4.

exhibits a large continuous layer of dentine that is not obviously organized
into odontodes.

A gradual fusion of small units (odontodes) to form large units during
phylogeny is very difficult to prove, though it must have taken place many
times in phylogeny. As yet no satisfactory model has been proposed of
the morphogenetic processes involved. Generally there are two different
ways to proceed from a large number of small elements to a small number
of large elements. (1) Either certain germs grow at the expense of others
and suppress them, or (2) individual germs actually fuse completely and
form one large, uniform germ. As long as a dermal element is fairly small
(even if it has a complicated internal structure), one cannot be sure
whether it resulted from fusion of odontodes or whether it represents one
odontode.

Another situation arises with the large dentine plaques of Dipnoi,
which either have no pore-canal system or have a pore-canal system and
then are called cosmine. It is very likely that they evolved by process 2,
gradual fusion of odontodes. This would then be a case of concrescence.
However, much more paleohistologic work is required to provide a complete sequential model of the origin of cosmine.

Step 6

A next step is exemplified by the scales of *Latimeria*, in which a
scale of laminated bone is formed during embryogenesis and is then cov-

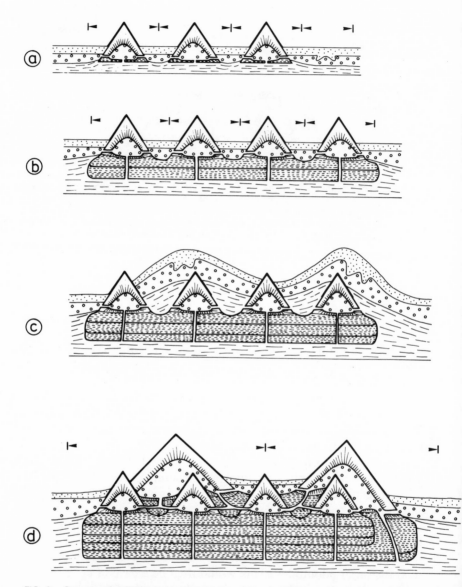

FIG. 9.   Overgrowth of odontodes is a very common mechanism. It is found in heterostracans, osteostracans, placoderms, early actinopterygians, "crossopterygians," and dipnoans, and also, in rare cases, in elasmobranchs. There are probably several different functional reasons for overgrowth; among them are secondary increases of the plates, ontogenetic changes of the surface sculpture, and repair of a worn scale surface. Note that originally the plate or scale in this case was formed by simultaneous fusion of the odontodes, rather than by growth. For further explanations see Fig. 4.

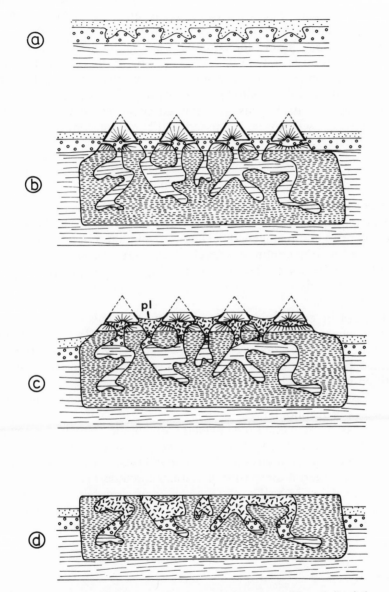

FIG. 10. Worn scales and plates are secondarily strengthened by a filling of all soft tissue space with pleromic hard (pl) tissue. This tissue is most probably a hypermineralized dentine, i.e., a tissue which is formed by odontoblasts, but secondarily loses its collagen fibers during mineralization. Pleromic hard tissue occurs in dermal plates of heterostracans, gnathalia of ptyctodontids (Placodermi), and teeth and tooth plates of holocephalians, some Mesozoic sharks, and dipnoans. For further information see Fig. 4.

ered with odontodes (each having a small bony base of their own) in its nonoverlapped part. The scale continues to grow throughout ontogeny and is continuously covered with odontodes. The scale has most probably evolved *de novo*, whereas the odontodes are an old heritage.

It is important to note in this step that the activity of bone formation has migrated deeper into the corium than its original site. The formation of the bony bases of the odontodes seems to be independent of the formation of the scale itself. This is reminiscent of the delamination principle, which, however will not be discussed in the present context (Patterson, 1977; Schaeffer, 1977).

Step 7

Steps 1–4 involved only changes in the morphogenetic program, some of which may be as simple as heterochronic shifts. Steps 5 and 6 involved morphogenetic and histogenetic changes (evolution of a new bone layer; evolution of a large continuous layer of dentine). The next major events are reduction processes. These can probably not be explained as simple heterochronic shifts.

They include:

1. Loss of enameloid. These are numerous cases reported in the literature where the odontodes lack an enameloid cap. In many cases this can be ascribed to secondary loss. However, many cases deserve to be restudied by SEM because a thin layer of enameloid often cannot be identified with the light microscope.

2. Loss of dentine. The most important example is the *Lepisosteus*-type scales of Actinopterygii.

3. Loss of enameloid (or enamel) and dentine. This is the common process which (together with step 6, *de novo* evolution of dermal bones) leads to bony scales in advanced "Crossopterygii," Dipnoi, Actinopterygii, and Placodermi and to the nonornamented dermal bones of fishes and Tetrapoda. The scales of Gymnophiona (Amphibia) are probably analogous to the scales of *Latimeria*, with ornamentations fused to a secondarily evolved dermal bone scale. However, gymnophionid ornamentations are simple bone elements and not complete odontodes.

Step 8

Reevolution of odontodes. Dermal denticles consisting of dentine and enameloid (or enamel) have reappeared in the scales of Teleosts (armored catfishes) other reported cases deserve to be restudied (to find out whether they really consist of dentine and enameloid), and probably also in Holostei (*Lepisosteus*) and even in the Cladistia (*Polypterus*). This reevolution of odontodes required only minor histogenetic changes, as is shown by the experiments of Kollar and Fisher (1980). They grafted chick epithelium with mouse molar mesenchyme and produced a variety of dental structures, including perfectly formed crowns with differentiated ameloblasts depositing enamel matrix.

Conclusions

The spectrum of dermal elements in vertebrates can be derived by a small number of morphogenetic and regulatory processes from a simple microsquamose skeleton consisting of isolated odontodes. The question of what the uncalcified evolutionary precursors of odontodes were like is far from being solved. A close connection between dermal sense organs (pore-canal system, lateral line system) and the dermal skeleton has often been observed (Denison, 1966). Neave (1936), for example, demonstrated that the spacing of the scales in teleosts is determined by the spacing of the neuromasts in the lateral line canal. Graham-Smith (1978a,b) suggested that the pattern of the dermal plates on the skull of placoderms, crossopterygians, and dipnoans is at least partly controlled by the lateral line system. Ørvig (1972) concluded that there are two components in the dermal skeleton, namely a "laterosensory" and a "membranous" component, which have independent evolutionary origins. In the light of the, admittedly speculative, conclusions drawn above, it is much more likely to assume that the dermal skeleton consists of one component only and had only one evolutionary origin. There is hardly any doubt that this origin had something to do with sensory structures in the skin and with protective functions.

Significance of Scales for Aquatic Locomotion

The evolution of large, articulating rhomboid scales of Heterostraci, Osteostraci, Anaspida, Placodermi, primitive Ostheichthyes, etc. also had consequences for the locomotor apparatus. They stabilize the shape of the body, stiffen the body cross section, prevent torsion of the body, and induce sigmoidal movements in the horizontal plane, in the absence of an ossified axial skeleton (Pearson, 1981). While this is true, one should not forget that there were anaspids and chondrosteans which had neither a squamation of rhomboid scales nor an ossified axial skeleton (*Lasanius*,

Anaspida; *Birgeria, Chondrosteus*, Chondrostei, etc.). Yakovlev (1966) and Gutmann (1967, 1975, 1977) use the undisputed importance of the dermal skeleton for swimming (in certain groups) as an argument for their hypothesis that locomotion rather than protection was the primary function of the dermal skeleton in vertebrates. Gutmann assumes that rhomboid, articulating scales were the primary elements of the dermal skeleton. According to his model, they must have evolved rather rapidly and fully endowed with peg and socket for articulation (Gutmann, 1975, Figs. 3A, 4K; 1977, Fig. 3). The dentine and enamel (or enameloid) cover evolved later. While offering a biomechanical explanation for the almost instantaneous evolution of rhomboid scales. Gutmann does not explain the dentine and enamel (or enameloid) coating. Gutmann's model seems to be in conflict with the methodology of "functional gradualism" (Anderson, 1967), because it claims that the complex rhomboid scales evolved in a single step. It does not explain the outer layers of the rhomboid scales. It also neglects the histogenetic finding that enamel (or enameloid), dentine, and bone form an integrated complex that must have evolved in the common ancestors of all Craniota (with the possible exclusion of the Myxinoidea).

As an adaptation to swimming (including control of body cross section) in the first swimming craniotes the collagen-fiber network of the dermis was used in the same way as in present-day sharks (Motta, 1977; Wainwright *et al.*, 1978). Nongrowing small scales (= placoid-like scales) of thelodonts, chondrichthyans, and some anaspids and heterostracans and growing small scales of chondrichthyans, acanthodians, some placoderms, and *Cheirolepis* (Actinopterygii) did not interfere negatively with the body undulation during swimming. Articulating (or at least overlapping) large scales which are part of the locomotor apparatus evolved only later in a gradual fashion (see above). Gutmann (1975) is correct in pointing out that they occur in Anaspida, Osteostraci, Heterostraci, Placodermi, and Osteichthyes (but not in Acanthodii!) and that they permitted the evolution of a wide variety of body cross sections. The alternative model introduced here is not in conflict with functional gradualism. The thick collagen layer in the dermis prevented an inhibition of body flexibility and muscle contraction by the growing scales. Thus the really essential aspect of the fully evolved large-scale squamation, namely that segmentation of the musculature and segmentation of the squamation are in accordance (Ryder, 1893), could have evolved gradually. As a final conclusion, it should be emphasized that Gutmann and Bonik's (1981) statement that mineralization can only take place in nonmobile areas is not contradicted by the model outlined here.

EVOLUTION OF THE VERTEBRATE DENTITION: A REVIEW

As yet there is no agreement about the time at which the dental lamina evolved and whether it evolved only once or convergently in several lineages.

Evolution of the Gnathostome Jaw

The most important prerequisite for the evolution of the vertebrate dentition is the evolution of a jaw apparatus in the common ancestors of Placodermi and "higher Gnathostomata." Most authors assume that feeding was the primary function when a pair of jaws evolved from a gill arch (Romer, 1966; Gutmann, 1967; Starck, 1979). This assumption is not satisfactory, because it is in conflict with Anderson's (1967) principle of "functional gradualism." Nobody has ever designed a functional gradualistic model starting from the gill apparatus of agnathans and leading directly to a tooth-bearing jaw capable of "pinching fairly large pieces of prey" (Gutmann, 1967), as was hypothesized for the early gnathostomes. An alternative model which is in accordance with the principle of functional gradualism was proposed by Wahlert (1966a,b, 1968, 1970). In his view the gill chamber of generalized agnathans is a hose pump, which operates peristaltically by compression. This original apparatus is probably exemplified in the Anaspida. In the adult Osteostraci and Heterostraci the solid capsule of the head prevented such a pumping mechanism. The two groups consequently must have had different adaptations for feeding and gill ventilation (Denison, 1961). In the evolution of gnathostomes this pump was supplemented by a suction and compression pump in front of the gill chamber. The membrane of this second pump was the floor of the mouth cavity. In this evolutionary process the mandibular arch played an important role because it stiffened the rim of the mouth. During the opening (suction) phase of the anterior pump it supported the rim of the mouth against deformation, while in the closure (compression) phase it allowed an effective closure. This is a very important mechanism for pumping the water in a posterior direction. The pressure for this pump is generated by the floor of the mouth. [Primitive living sharks, e.g., nurse sharks, recent dipnoans, and certain actinopterygians have been shown to be largely suction feeders (Wahlert, 1968; Lauder, 1980a,b; Reif, unpublished).]

This model leads to the conclusion that the first gnathostomes were suction feeders, more effective but not very different from agnathans

(Fig. 11). Tearing, cutting, crushing, and chewing of prey can have been evolved only later. This again leads to the question of the stage at which the dental lamina evolved.

The mode of replacement of nongrowing scales was no suitable preadaptation for elements situated at the rims of the jaws. Prefabrication of elements was not possible; consequently the shedding of an odontode led to a gap. The formation of a new element had to take place in this gap, which meant that the uncalcified papilla was very vulnerable (Fig. 12). In order to evolve a biting apparatus from the dermal skeleton two basic designs are possible: (1) Growing plates of dermal bone are anchored to the mandibular arch and work in a scissor-like manner. Growth is necessary to accommodate the size of the plates to the size of the growing elements of the mandibular arch and to compensate for the loss of material at the working edges (gnathalia of Placodermi and beaks of some advanced teleosts). (2) Nongrowing teeth are individually replaced. They have to

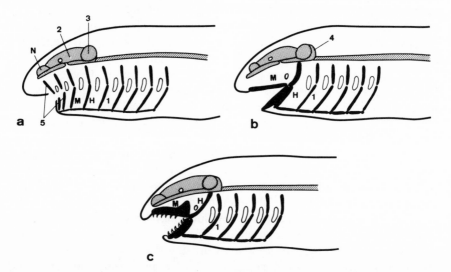

FIG. 11. Evolution of jaws and dentition of gnathostomes (modified and redrawn after Starck (1979): (a) Agnathous ancestor; (b) early gnathostome (without dentition); (c) gnathostome with dentition. Dotted: neurocranium; diagonal hatching: chorda dorsalis; black: viscerocranium. N, Nasal capsule. M, Mandibular arch. H, Hyoid arch. 1, First gill arch. 2, Brain capsule. 3, Labyrinth capsule. 4, Occipital region. 5, Two hypothetical premandibular arches. The classical hypothesis illustrated by Starck (1979) is here modified in several ways. (1) I assume that the agnathans were primitively suction feeders. In the evolution of the gnathostomes the mandibular arch was simply transformed into a toothless frame for the suctional mouth, with the hyoid arch as a mechanical support. (2) The common assumption that the hyoid arch was transformed at a much later stage is rejected, because the hypothetical "aphetohyoid" condition (with free hyoid arch) has never been found. (3) The dentition, however, evolved at a later stage in the common ancestor of Chondrichthyes and Teleostomi (Fig. 1).

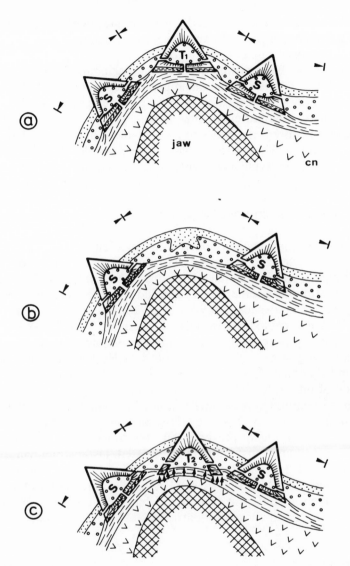

FIG. 12. The evolutionary steps that led to a dental lamina are unknown. It is certain, however, that the dentition evolved from a simple squamation on the jaws and in the mouth cavity. This required not only a morphologic change of the odontode into elements which could catch and chop prey (S evolved into T in the diagram). The mode of replacement also had to change. In (c), T still erupts vertically like a scale. (In this and the following figures the lingual side (oral cavity) is on the right; the labial side is on the left.) This has two drawbacks: The replacement time is long because teeth cannot be prefabricated in this case and the developing tooth germ is easily damaged by the antagonistic tooth (b). cn, Undifferentiated connective tissue; arrows indicate the direction of the forces which cause the teeth and scales to erupt; for further explanation see Fig. 4.

be prefabricated in order to speed up the replacement phase. Their fabrication has to take place in an area where they cannot easily be damaged (dentition of most "higher Gnathostomata").

Most authors have assumed that the dental lamina evolved synchronously with the evolution of the mandibular arch. In the light of Wahlert's theory, this is not very likely. Rather, it is much more likely that the dental lamina evolved after the mandibular arch.

The major difference between a tooth and a dermal denticle is their locus of formation. Theoretically it should be possible in fossil vertebrates to find out where individual elements of the dermal skeleton were formed.

"Biting Apparatus" in Fossil Agnatha

As this discussion deals only with calcified dermal products, the horny rasping tongues of Recent Agnatha will not be referred to. All three major groups of fossil Agnatha (Osteostraci, Anaspida, and Heterostraci) have at one time or another been regarded as large-particle feeders with a biting apparatus, especially by Kiaer (1928) [see discussion in Denison (1961)]. With respect to Osteostraci and Anaspida, Stensiö (1968) weakens the arguments of his earlier papers and now assumes that the two groups had a nonbiting mouth. For some groups of Heterostraci, mainly Pteraspida and Cyathaspida, however, Stensiö (1968) maintains that they had an arcade of biting dermal plates ("tooth plates") on the ventral side of the mouth opening ("lower jaw") which bit against hypothetical tooth plates on the "upper jaw." White (1968), in a discussion, emphasizes that despite fine conditions of preservation, the "upper dentition" has never been discovered and most likely never existed. Osteostraci had a protrusible mouth and the oral plates most likely functioned to select and pick up small food (Denison, 1961, 1969). In conclusion, the fossil agnathans had a spectrum of different feeding mechanisms, but none of them had a biting apparatus to chop food particles.

Placodermi

In Stensioellida the palatoquadrates and meckelian cartilages bear only small denticles, in Pseudopetalichthyida they carry small, thin, scale-like plates, and in Rhenanida they are covered by small "tesserae" with toothlike denticles (Gross, 1962; Denison, 1978). Morphogenesis, replacement, and regeneration of these elements are completely unknown. It is beyond doubt, however, that they neither are, nor bear, real teeth, but that they are all scalelike, tesseraelike platelets covered with odontodes and hence do not differ from the rest of the dermal skeleton. There is no indication at all that Stensioellida, Pseudopetalichthyida, and Rhenanida had a dental lamina. In the light of the cladogram of Fig. 1 and the fact the Denison (1978) considers these groups the most primitive Placodermi, it is very likely that they are very close to ancestral microphagous Gnathostomata!

In the Acanthothoraci, Petalichthyida, and Phyllolepida the biting apparatus is completely unknown. Ptyctodontida, Arthrodira, and Antiarcha have a biting apparatus consisting of large, plate-shaped elements, called "gnathalia." Ptyctodontida and Antiarchi have one pair of gnathalia. Arthrodira have two gnathalia in the upper jaw and one in the lower jaw (Gross, 1957, 1967b; Ørvig, 1973a; Denison, 1979; Dennis and Miles, 1980).

The gnathalia of Arthrodira and Antiarcha consist of bone (Gross, 1957, 1967b; Denison, 1979), sometimes with superficial layers of semidentine (Ørvig, 1980a). The gnathalia of the Ptyctodontida, on the other hand, consist of spongy bone, which was secondarily filled with pleromin (hypermineralized dentine) (Ørvig, 1973a, 1980b; see also Gross, 1957, 1967b). This mode of tissue formation is probably analogous to the pleromin formation in Heterostraci.

In most families of the three groups the gnathalia are ornamented with odontodes on the working edge and sometimes on their lingual side. These ornamentations are surface tubercles and were not formed in a dental lamina, though they might have functioned like teeth. Once they were abraded, the gnathalia themselves worked as tritural cutting or shearing elements. Abrasion of the gnathalia was compensated for by marginal addition at their basal side. Growth took place by marginal addition at the posterior side. In many groups this mode of growth had the effect that new odontodes were continuously added to the posterior end of the cutting edge (Ørvig, 1973a).

An ancestor–descendant relationship between Ptyctodontida and Holocephali was proposed by some authors [especially Ørvig (1960, 1962, 1973a, 1976a); Westoll (1962)]. An important part of Ørvig's argument is the similarity between the gnathalia of Ptyctodontida and the tooth plates of Holocephali. Most recent authors, however, agree that a close relationship between Ptyctodontida and Holocephali does not exist (Moy-Thomas, 1971; Lund, 1977; R. S. Miles and Young, 1977). Additionally, structural similarities between gnathalia of Ptyctodontida and Holocephali are clearly convergences. In contrast to gnathalia, the holocephalian toothplates are formed in a dental lamina (Bargmann, 1933; Brettnacher, 1939; see below).

To summarize, placoderms did not evolve a dental lamina. Their primitive members seem to have been microphagous suction feeders without any special elements on the jaws to cut, tear, or crush food. More advanced members developed large dermal plates, which functioned as tooth plates. In the current context it is of no importance whether these tooth plates evolved convergently in the ptyctodontids on the one hand and arthrodires and antiarchs on the other [see cladogram in R. S. Miles and Young (1977)] or whether they are homologous in all three groups (Denison, 1978). The biting apparatus of the advanced placoderms shows this group to be a side branch in the evolution of gnathostomes. Hence it is not surprising that the morphology of the palatoquadrate and the hyomandibula are very different from that of the other gnathostomes (Schaeffer, 1975). Schaeffer (1975), on the basis of osteologic comparisons, comes to the conclusion that "the distinctive placoderm feeding mechanism arose independently from that in the chondrichthyans or in the teleostomes."

Chondrichthyes

A dental lamina occurs in the Chondrichthyes and in the Teleostomi and there is no indication that it evolved independently in these two groups. Rather, one has to assume that it evolved in the common ancestors of both groups after the line leading to the Placodermi had branched off.

The ontogenetic development of the dental lamina in Recent shark embryos was described by Laaser (1900, 1903) and Reif (1980b). The dental lamina is an ectodermal fold which forms very early in embryogenesis immediately behind the jaw cartilage. As the fold becomes deeper, tooth germs are induced at the *anterior* interface between ectoderm and mesenchyme, but not at the posterior interface. The fully mineralized teeth are anchored by collagen fibers to a fibrous and vascular layer, which later becomes the tooth-transporting conveyor belt. This tooth transport starts only at the end of embryogenesis and it continues throughout ontogeny (Reif, 1978b; Reif *et al.*, 1978; Grady, 1970a).

All Recent sharks studied so far have a continuous dental lamina, i.e., an uninterrupted fold reaching from one end of the jaw ramus to the other. A specialization which has escaped notice so far (Reif, unpublished) is found in *Chlamydoselache anguineus* [see Compagno (1973, 1977), Gudger and Smith (1933), B. G. Smith (1937) for morphologic information]. In this species no continuous dental lamina is present in postnatal specimens. Rather, each tooth family is formed in a fold of its own. The tooth families are widely spaced in *Chlamydoselache* and the ectodermal folds are separated by undifferentiated connective tissue. At the surface, the gaps between the tooth families are filled with normal pigmented epidermis, which produces placoid scales in adult specimens. In other words, there is a regular alternation on the jaw between a set of teeth (belonging to one tooth family) and a set of placoid scales.

Two embryos of *Clamydoselache anguineus* were studied at the department of Ichthyology of the American Museum of Natural History (New York): No. 26273 (length 205 mm) and No. 38148 (length 333 mm). The specimens show that initially the dental lamina forms a continuous infolding as in any other shark and only during gradual deepening of the fold do the pockets for the individual tooth families become separated.

The dentition of all Elasmobranchii is organized into discrete tooth families, each of which consists of a functional tooth and all its successors. Growth of the whole dentition is accomplished either (1) by regular increase in the size of successive teeth within the tooth families, (2) by splitting of tooth families, or (3) by insertion of new tooth families (Reif, 1976). A fusion of neighboring tooth families has never been observed in any Recent shark.

A reduced tooth number occurs in two families of rays, Myliobatidae and Rhinopteridae (Bigelow and Schroeder, 1953). Myliobatidae have one broad symphyseal tooth and three lateral teeth on either side. This tooth formula remains constant from the first tooth generation throughout ontogeny (Garman, 1913, Plate 49, Figs. 4–6; Studnička, 1942; Bigelow and Schroeder, 1953). *Rhinoptera* also has a broad symphyseal tooth and two to four lateral teeth, depending on the species. In this genus the tooth formula also remains constant from the first tooth generation throughout ontogeny. [A few additional teeth seem to be formed during the first tooth generation, but these have no successor (Garman, 1913, Plate 48, Fig. 4).] This can also be observed in the dentitions of shark embryos (Reif, unpublished). *Aetobatis* (Myliobatidae) has only one tooth in the upper and one in the lower jaw. The first one or two tooth generations, however, consist of two teeth, one on either side of the symphysis (Garman, 1913, Plate 4a, Figs. 1 and 2; Bigelow and Schroeder, 1953, Figs. 106E and F). After the first or second generation both families of each jaw fuse. (This is a clear case of concrescence!)

No information is available concerning how the reduced tooth number of Myliobatidae and Rhinopteridae evolved. There are two different possibilities, either gradual fusion (concrescence) of protogerms [in the sense of Reif (1976)] or gradual suppression of protogerms. Either mode excludes any attempt to homologize the large teeth of these families with normal-sized teeth of other elasmobranchs. Splitting of tooth families (in contrast to fusion of tooth families) can be found in many specimens of

Rhinoptera as an anomalous mode of development [Garman (1913); Reif (1976, Fig. 7c): in the specimen shown, one side of the jaw has eight instead of four lateral teeth].

Elasmobranch teeth consist largely of dentine and thus can be regarded as advanced odontodes. They are anchored by collagen fibers which originate in the bony base of the teeth. The hypermineralized cap is enameloid. This enameloid is single-layered in all primitive sharks and is called single-crystallite enameloid; it is double-layered (tangle-fibered enameloid plus parallel-fibered enameloid) in all modern sharks (Neoselachii). The skates and rays (batoids) are probably derived from early Neoselachii. Their teeth have a secondarily reduced enameloid layer (Poole, 1956; Preuschoft *et al.*, 1974; Reif, 1973a, 1977, 1978c, 1979b, 1980c). In Neoselachii the enameloid is covered with a "shiny layer" of a few micrometers; this tissue is probably ectodermal enamel.

In all elasmobranchs, teeth are restricted to the upper and the lower jaw. In the rest of the oral cavity, scales occur in numerous genera (Reif, unpublished), but never teeth.

Chimaeroids (Holocephali) have tooth plates which grow throughout ontogeny and are never replaced. These plates (three or fewer pairs in the upper jaw and a single pair in the lower jaw) are formed in a dental lamina (Bargmann, 1933; Brettnacher, 1939; Patterson, 1965) and have nothing to do with the gnathalia of any placoderm. The histologic descriptions of the plates in the literature (Bargmann, 1933; Brettnacher, 1939; Peyer, 1968; Ørvig, 1976b) are inadequate. A light microscopy and SEM study of a mandibular plate of *Pachymylus* sp. (Oxfordian, Fletton, U.K.) showed it to consist of hypermineralized "osteodentine" (= trabecular dentine) throughout. Concentric structures are clearly seen. The vascular canals are very narrow in most areas of the plates. Normal mineralized dentine was found only in two columns in the mesial and in the distal part of the plates (Reif, unpublished).

Nothing is known about the evolution of these plates, their histogenesis during the embryogenetic stage, or their growth regulation (Schauinsland, 1903; Peyer, 1968). There is hardly any doubt that the plates evolved from normal polyphyodont dentitions (Schnakenbeck, 1962). No model, however, can be proposed to show how the tooth plates evolved. In any case the evolution must have started with a dentition with reduced tooth numbers (analogous to the batoids mentioned above). Two different possibilities can be suggested: (1) The tooth plates formed by a gradual fusion (concrescence) of a functional tooth with its replacement teeth: in this case one tooth plate would be homologous to a whole tooth family. (2) The tooth germ of the first tooth generations gradually suppressed the induction of germs of later generations and continued to grow. No decision between these possibilities can be made, although the second possibility is more likely. In other words, it is likely that the chimaeroid tooth plates can be described as very large, rootless, growing teeth. They are completely abraded at their apical end and grow at their basal end.

Tooth plates of a completely different type occur in some Paleozoic holocephalians ("Bradyodonti"), the phyletic position of which is unclear (Patterson, 1965, 1968; Bendix-Almgren, 1968). Because of this constructional difference between chimaeroid and bradyodont tooth plates I do not regard the bradyodonts as ancestors of the chimaeroids, as was suggested by Patterson (1965) and Lund (1977). In this context it is important to note that the chimaeroids can now be traced as far back as Lower Carboniferous (Lund, 1977). The cross section of bradyodont tooth plates is a spiral (Jaekel, 1924/1926; Patterson, 1968, Fig. 12). In chimaeroids basal addition to the plate takes place in the same measure as the apical end is completely abraded. In the spiral tooth plates of bradyodonts, on the other hand, only the crowns of the plates are subject to wear. As new material is added to the lingual end of the plate, the labial end of the tooth plate is automatically shifted outward and below under the functional part of the crown. Hence, the whole morphogeny can be traced in the tooth plate of an adult specimen. There is hardly any doubt that these tooth plates also evolved from polyphyodont dentitions. It is commonly assumed that they evolved by a gradual fusion of replacement teeth.

Tooth plate transport in chimaeroids seems to be "vertical" (= basal–apical) with respect to the axis of the dental lamina if one interprets Schauinsland's (1903) drawings and Peyer's (1968) plate 24A correctly; tooth plate transport in bradyodonts, on the other hand, is "lateral" (= lingual–labial).

Because of these highly significant constructional differences, it is unlikely that the chimaeroid dentition evolved from the bradyodont dentition. Rather, one has to assume that both types of dentition have independent origins in groups with normal polyphyodont dentitions (Reif, 1975).

Acanthodii

Acanthodii have three different types of teeth: (1) single teeth anchored by fibers, (2) single teeth ankylosed to jaw bones, and (3) tooth spirals (Denison, 1979; Gross, 1957, 1967a; Ørvig, 1967b, 1973a). It is very likely that both types of single teeth were regularly replaced (Gross, 1957), although direct evidence is missing. There is also no information as to whether the single teeth were formed in a dental lamina. Among the tooth spirals there are two types: (1) short spirals: bow-shaped elements consisting of four teeth in a curved row fused at their bases; and (2) complete spirals with 1½ whorls consisting of up to 16 teeth fused at their bases. The size of the teeth increases regularly from the inner to the outer end of the spiral. There is no doubt that these spirals were formed in a dental lamina by a gradual addition of replacement teeth at the basal end of the spiral. The *short* spirals were shed once they consisted of four teeth and were replaced by a larger spiral which was gradually formed. It could very well be that the *complete* spirals were never shed, and that only 16 replacement teeth were formed in one tooth family during ontogeny.

It is not very likely that the spirals are a primitive character within the Acanthodii. Rather, one has to assume that they evolved from normal tooth families. The functional significance of the spiral probably was that the functional teeth are mechanically supported by a large base and that two or three teeth of one tooth family could function at the same time. The evolutionary importance of the tooth spirals is that one can show that a dental lamina was present in the Acanthodii and must have evolved in an ancestor of the teleostomes (see below).

All teeth of Acanthodii consist of dentine (Denison, 1979) and have a bony base. Either the bony base is attached directly to the tooth-bearing jaw bone, or it provided the collagen fibers that anchored the individual teeth or the tooth spirals. Enameloid was not discovered in thin sections. No SEM study to demonstrate the presence or absence of enameloid has been undertaken. Teeth are restricted to the dentigerous bones of the upper and the lower jaw. It is doubtful whether the ancathodid *Homalacanthus* had gill rakers consisting of dentine (Ørvig, 1973a).

Ostheichthyes

Actinopterygii

Tooth formation in actinopterygians has been fairly well studied (Lühmann, 1954; Kerr, 1960; Peyer, 1968; Poole, 1967; Berkovitz, 1975, 1977, 1978a,b; Berkovitz and Shellis, 1978). Actinopterygians primitively have a polyphyodont dentition which is organized in tooth families. Friedmann (1897) found that the dental lamina in the jaws of *Esox lucius* and in the pharynx of *Cyprinus carpio* is permanent but discontinuous. In other parts of the dentition of *Esox* and in other genera the dental lamina seems to be nonpermanent, i.e., a new infolding occurs for every new replacement tooth (Kerr, 1960). Sufficient data, however, are clearly lacking. Although some teeth are formed rather superficially, especially on the palate, replacement teeth are—in marked contrast to the shark squamation—always formed before the functional teeth are shed. In the Gadidae (Gadiformes) Holmbakken and Fosse (1973) found a rather irregular mode of tooth replacement. They could not identify tooth families. The teeth were formed superficially and not in a dental lamina. The situation in Gadidae seems to be a derived one in comparison with the rest of the Actinopterygii.

A permanent dental lamina developed in Tetraodontidae (Pflugfelder, 1930) and probably also in Diodontidae and Scaridae [see histologic data given by Boas (1879)]. The differences between permanent and nonpermanent dental lamina should not be overestimated. It is almost impossible to find out whether the permanent or the nonpermanent dental lamina is the more primitive one for the actinopterygians.

The hypermineralized outer layer of actinopterygian teeth consists of two different tissues:

1. Enameloid, which forms a cap. This is present in all actinopterygian teeth studied so far. Ørvig (1973b) called it "acrodin." Generally it has a tangle-fibered structure (the "fibers" consist of strands of crystallites); in two very advanced families, Characidae and Sphyraenidae, parallel-fibered enameloid has evolved convergently and also in convergence to the tissue found in modern sharks (Shellis and Berkovitz, 1976; Reif, 1978d, 1979b).

2. Collar enamel. This occurs as a thick layer in numerous chondrostean and holostean genera. It has been found as a very thin layer in teleosts, but it is not known in which families it occurs.

The bulk of the teeth consists of various kinds of dentine [see discussions in Ørvig (1967a) and Harder (1975)]. The modes of attachment of the teeth with their bony bases, is either with fibers or by ankylosis and is discussed by Fink (1981) and Shellis (1982).

Teeth can occur on all dermal bones in teleosts (Schnakenbeck, 1962; Harder, 1975), but they are usually confined to a small number of bones. The wide spectrum of different types of dentitions was discussed by Rauther (1940).

To accommodate for strong wear, the teeth of some advanced families (Scaridae, Diodontidae, Triodontidae, Tetraodontidae, and Molidae) have evolved "beaks" (a similar functional necessity was met in certain mammal groups by the evolution of rootless growing teeth). As far as is known, the beaks are formed by a permanent dental lamina; they are attached to jaw bones and consist at any given growth stage of a large number of tooth generations stacked upon each other and "welded" together by mineralized connective tissue. It must be highly emphasized that the

beaks consist of numerous individual teeth. Whereas the beaks are growing in size, the individual teeth are formed in a single event and, like any
other actinopterygian tooth, are not growing. Each quadrant of a beak
contains numerous tooth families in Scaridae Triodontidae, and Diodontidae. In Tetraodontidae each quadrant contains only one very long and
slender tooth and all its successors. In the four families tooth replacement
is automatic and is accomplished by abrasion of the previous tooth. The
beaks of the Molidae differ markedly from the constructions mentioned
above; the body of the beaks consist of pure bone (without any imbedded
teeth). Sharpening of the bony beaks is accomplished by a self-sharpening
mechanism. Several families of dentine teeth are ankylosed at the lingual
side of the beak. The individual teeth are moved to a functional position
in the same measure as the functional edge of the bony beak is abraded.
There is no shedding mechanism, but the teeth are completely abraded
(Boas, 1879; Pflugfelder, 1930; Reif, unpublished). In summary, all fish
beaks consist in one way or another of dentine teeth with a thick enameloid
cap. The dentine teeth are formed in a single morphogenetic step, they
do not grow, and they are organized in tooth families, and thus do not
differ from any other actinopterygian teeth. Because descriptions of these
beaks in the literature (e.g., Nelson, 1976) are often misleading, they have
been discussed rather extensively in this context.

"Crossopterygii"

The Recent *Latimeria* (Actinistia) has a well-developed dentition
which is restricted to the jaw bones. The teeth consist of an enamel cap,
dentine, and basal bone and are ankylosed to dermal bones (Shellis and
Poole, 1978; Smith, 1978). They are formed in a fairly superficial region,
although deeper than placoid scales. A permanent dental lamina is probably not developed. Nothing is known about the general organization and
the replacement patterns of the dentition (Miller and Hobdell, 1968;
Miller, 1969; Bernhauser, 1961; Isokawa *et al.*, 1968; Grady, 1970b; Shellis and Poole, 1978).

A well-developed dentition is also present in the other crossopterygians, Holoptychiida, Osteolepidida, Rhizodontida, and Onychodontida
(= Struniiformes) (Schultze, 1969). The teeth occur on several different
dermal bones, to which they are ankylosed (Moy-Thomas, 1971). Because
all four groups are extinct, there is no direct evidence of a dental lamina.
However, two groups, Holoptychiida (Jarvik, 1972) and Struniiformes
(Jessen, 1966), have tooth spirals. As was shown in the section on acanthodian dentitions, tooth spirals can only be formed by a dental lamina.

Morphogenesis of the tooth spirals has never been studied. Reconstructions by Jarvik (1972) and Jessen (1966) seem to indicate that the bases of the spirals were resorbed at their labial ends and the teeth were thus shed. It is unlikely that the spirals were shed as a whole.

Dipnoi

Recent genera have large tooth plates. Early (Devonian) Dipnoi, on the other hand, had three different dental structures: "buccal denticles," tooth ridges, and tooth plates (Denison, 1968, 1974; Ørvig, 1976b; M. M. Smith, 1977). "Buccal denticles" were replaced (Denison, 1974; M. M. Smith, 1977). Replacement denticles were formed only after functional denticles had been shed. Hence, it is very likely that they were not formed in a dental lamina.

The histology of tooth ridges was studied by Denison (1968), Ørvig (1976b), and M. M. Smith (1977). However, morphogenesis of these ridges is by no means clear. "There is convincing evidence that the 'tooth ridges' were periodically shed or resorbed and replaced as the fish grew" (Denison, 1974, p. 34). The information available does not suffice, however, to tell whether they are homologous to teeth or whether they have to be considered as products of fusion (concrescence) of denticle germs.

"Buccal denticles" and tooth ridges seem to be primitive features within the dipnoans, and the tooth ridges are unique for the group compared with the rest of the vertebrates (Miles, 1977; M. M. Smith, 1977, 1979b). Tooth plates are an advanced feature which evolved only once within the Dipnoi (Miles, 1977; M. M. Smith, 1977). Morphogenesis and histogenesis of Recent tooth plates were studied by Röse (1892), Semon (1899), Lison (1941), Kemp (1977, 1979), and others. The studies showed that the adult tooth plates develop during ontogeny from originally discrete and separate unicusped teeth which are fused at their bases. The plates can be compared to a polyphyodont dentition. Each plate consists of radiating ridges. Each ridge results from the fusion of numerous cusps, and is homologous to one tooth family. At the side of the plate to which the ridges diverge new material is added throughout morphogenesis. At the side where the ridges converge material is resorbed. This process is equivalent to the shedding of individual teeth. Kemp also observed addition and bifurcation of ridges; this is equivalent to addition and bifurcation of tooth families in polyphyodont dentitions (Reif, 1976). The angle between the ridges is an expression of the ontogenetic growth of the dental lamina. It is thus an automatic result of the fusion of the ridges. The fact that "material transport" in dipnoan tooth plates takes place in a posterior–mesial direction instead of the generally anterior–lateral direction in sharks is of little relevance in the comparison of dipnoan tooth plates with polyphyodont dentitions.

To summarize, the morphogenetic analysis of dipnoan tooth plates leads to the conclusion that the tooth plates of dipnoans evolved from a well-organized polyphyodont dentition, rather than from the fusion of scattered denticles which were enlarged and arranged in rows, as was assumed by Denison (1974). I agree with Denison (1974), however, that tooth plates in which the individual plate elements are not completely fused are more primitive than plates that have continuous ridges.

Tetrapoda

Tetrapoda have a permanent continuous dental lamina which has been well studied in Amphibia, Reptilia, and Mammalia (Peyer, 1968; A. E. W. Miles and Poole, 1967; Gaunt and Miles, 1967; Clemen, 1978a,b; Graver, 1973, 1974, 1978; Chibon, 1977). It is not necessary to discuss details here. The morphogenetic organization and the control of replacement patterns are discussed by Edmund (1960a, 1960b, 1969), De Mar (1972, 1974), Bolt and De Mar (1975), and Osborn (1970, 1973a, 1973b, 1974, 1978).

The teeth are usually formed in a single morphogenetic event. The only exception are the rootless, growing teeth of rodents, elephants, and several other groups of mammals where enamel, dentine, and cement are continuously laid down. These teeth grow throughout the creature's life.

Tooth plates do not occur. Tooth number can vary considerably. It is doubtful, however, whether reduction of tooth numer in evolution was ever brought about by real fusion of neighboring tooth germs or simply by suppression of tooth germs. At least there are no empirical data which would justify the assumption that large teeth in a dentition with a small tooth number are the product of fusion (concrescence) of odontodes. At present the only justifiable model is the assumption that each individual tooth represents one odontode.

In amphibians and reptiles there are numerous tooth-bearing bones in the oral cavity. In the mammals the dentition is restricted to the maxillary, intermaxillary, and dental (Stadtmüller, 1936a,b; Versluys, 1936).

Dermal denticles, which were not formed in a dental lamina but which are analogous to the denticles on the plates of "fishes," occur on the roof of the oral cavity of stegocephalians. In the Carboniferous and Permian groups they occur on all bones; in the Triassic genera, however, they are restricted to parasphenoid and pterygoid. Bystrow (1938) studied the denticles of the Triassic *Benthosuchus* and found that four generations of denticles occur on top of each other. They were separated by cellular bone, to which they were firmly ankylosed. Before overgrowth the dermal denticles were partly resorbed.

Other aspects of the organization of the tetrapod dentition are well known and can be found in textbooks (A. E. W. Miles, 1967; Peyer, 1968; Schmidt and Keil, 1971).

EVOLUTION OF THE VERTEBRATE DENTITION: DISCUSSION AND A MODEL

A major prerequisite for the evolution of a dentition was the evolution of a jaw apparatus. However, as was shown on p. 333, the jaws at first did not evolve to provide a biting mechanism but to increase the efficiency of the suction-feeding mechanism. [In a recent paper Gutmann and Bonik (1981) elaborate on this model without referring to the original papers of Wahlert (1968, 1970).] In the present discussion the exact homology between the mandibular arch and a gill arch in the agnathans is of no importance. In other words, it does not matter whether there were one or two premandibular arches in the agnathous ancestors (see Fig. 11). According to Wahlert's theory, the first gnathostomes were suction feeders, which probably had no dentition at all. It is very likely that the most primitive known placoderms were such primary suction feeders. The placoderms never evolved a dentition but instead evolved large dermal bone plates (gnathalia) which bit in a scissor-like manner.

The evolution of the jaw apparatus involved not only the mandibular arch but also the hyoid arch (Gegenbaur, 1872). The hyoid was transformed into a support of the jaws and the gill slit between mandibular arch and hyoid arch remained as a small spiracle. The condition in which jaws are already present but the hyoid arch is free was called "aphetohyoid" by Watson (1937). Recent studies have shown that no aphetohyoid gnathostomes have been found and that the aphetohyoid condition remains completely hypothetical [for placoderms see Denison (1978); acanthodians are not aphetohyoid—cf. R. S. Miles (1968), in contrast to Watson (1937); the reconstruction of a Carboniferous aphetohyoid shark by Zangerl and Williams (1975) and Zangerl and Case (1976) is doubted by Maisey (1980)]. Hence it is most likely that the evolutionary changes of the mandibular arch and the hyoid arch were synchronous.

The comparative survey of the dentitions of the "higher Gnathostomata" has shown that a polyphyodont dentition consisting of numerous tooth families is the primitive condition and is met with in most groups.

Except for the rootless growing teeth of some mammalian groups, teeth are always formed in a single morphogenetic step. The growing beaks of some advanced teleost families are complexes consisting of dermal bone (= ossified connective tissue, which undergoes growth by the addition of incremental layers) and of nongrowing dentine + enameloid teeth. The tooth plates of dipnoans and bradyodonts can be easily derived from polyphydont dentitions. For the chimaeroid tooth plates direct evidence of a derivation from a polyphyodont dentition is missing, except for the well-developed dental lamina itself, which has been clearly documented.

Changes in morphology and element number in the squamation and in the dentition can in all cases be explained by two theories, either the concrescence theory or the differentiation theory. Stensiö's (1961) Lepidomorial Theory (which is rejected above) is a concrescence theory of the squamation. It has been shown [see above and Reif (1973b, 1974, 1979a, 1980a)] that the differentiation theory is much more successful and easier to deal with, because any concrescence theory is notoriously difficult to test. A complicated scale or tooth crown is explained by the concrescence theory as the fusion of numerous germs, and by the differentiation theory as the differentiation of a single germ.

In mammals the discussion between the concrescence theory and the differentiation theory is classic (Peyer, 1968). It is not surprising that the differentiation theory has proved to be the more successful one.

The differentiation theory can only explain changes in shape of scales and teeth but not changes in number. An increase in number can either be ascribed to a real fission of germs or to *de novo* induction of germs. A decrease in number can be ascribed to a real fusion of germs or to the suppression of germs. If direct evidence is lacking, no testable conclusions can be drawn and the alternatives are left open by the differentiation theory. Hence the most parsimonous solution in the differentiation theory is to regard each odontode as the product of *one* papillary organ which resulted from *one* interactive inductive stimulus between ectoderm and mesenchyme. This is why I do not agree with Ørvig (1977) that one can accept the odontode concept without taking sides for or against the Lepidomorial Theory.

The dental lamina is a unique organ which produces replacement teeth while the functional tooth is still in place. In the morphogenetic and histogenetic analysis of the dermal skeleton of agnathans no regulatory process was found which could have been preadaptive for the evolution of the dental lamina. Figure 12 shows that an *in situ* replacement of teeth has two drawbacks: replacement takes a long time and the unmineralized tooth germ is easily damaged by hard food particles or by the antagonistic tooth.

We need much more information about regulatory processes in the vertebrate integument before a functional-gradualistic model of dental lamina evolution can be constructed. Hypothetical intermediate stages are depicted in Figs. 13 and 14. Both are based on the inhibitory field concept. In Fig. 13 the scales (S) and the functioning tooth (T1) are surrounded by inhibitory fields which prevent the induction of new scales. A specialized area is found on the lingual side of T1, where a new tooth is induced which then undergoes lateral transport onto the crest of the jaw to replace T1. In Fig. 14 the coverage of the integument with inhibitory fields is complete at the surface. An invagination of epithelial material on the lingual side of T1 leads to a field-free area in the invagination. Here a replacement tooth is automatically induced. Neither of the models

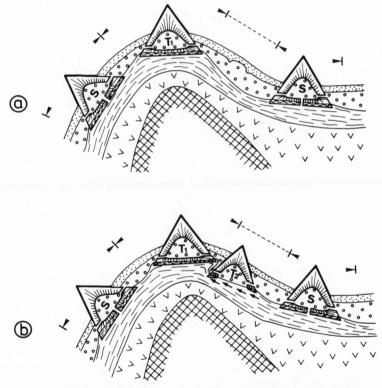

FIG. 13. Hypothetical stage during the evolution of the dental lamina. A replacement tooth (T2) is prefabricated in a gap between T1 and S, from which it can move laterally into a functional position. For further explanation see Figs. 4 and 12.

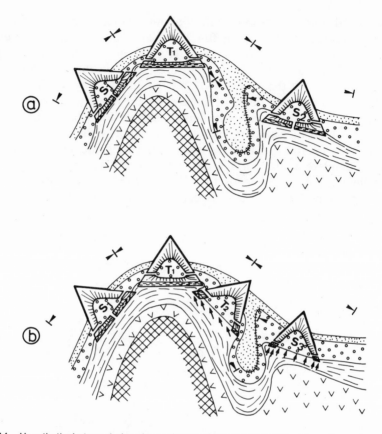

FIG. 14. Hypothetical stage during the evolution of the dental lamina. The outer surface of the integument is completely covered with inhibitory fields, but an ectodermal invagination on the lingual side of the jaw is beyond their influence. Here, a replacement tooth for T1 is formed. It is unexplained why only the anterior side but not the posterior side of the invagination is competent to form teeth. For further information see Figs. 4 and 12.

is quite satisfactory. Specific regulatory aspects of the fully evolved dental lamina remain to be explained (Fig. 3). Among these aspects is the fact that teeth are only formed on one side of the dental lamina (in the case of the jaw dentition, the labial side). The proliferation of ectoderm and mesenchyme in the primordial tissue at the base of the dental lamina is only poorly understood. In the scales the direction of eruption is vertical, while in the dental lamina tooth transport has a lateral component (at least in the primitive cases). The forces that cause scales and teeth to erupt can be either contraction forces or fluid pressures. Histologic observations of tooth and scale eruption in sharks suggest that the forces come from fluid pressures in vacuoles [Reif (1980b); see also reviews in A. E. W. Miles and Poole (1967) and Poole and Shellis (1976)].

SUMMARY

1. Dentine organs, consisting of the three tissues enameloid (or enamel), dentinous tissue, and bone, are regarded as the primary constituents of a dermal skeleton in the first craniotes. They are called "odontodes," and probably served primarily a protective function. Dentine is a sensitive tissue innervated similarly to the superficial layers of the skin, but it remains obscure how odontodes evolved from hypothetical unmineralized skin organs.

2. Except for the Myxinoidea and Petromyzonida, all known craniote groups have a mineralized dermal skeleton consisting of scales and plates. It has been shown that all scale and plate types can be derived from simple odontodes through a small number of morphogenetic regulatory processes.

3. Jaws evolved from a gill arch in the common ancestor of the Gnathostomata. Their primary function, however, was not to bite, but rather to support the suction feeding apparatus, which was inherited from the agnathans. Gnathostomes consist of three groups: (a) Placodermi, (b) Chondrichthyes, and (c) Teleostomi; the latter two are sister groups; together they form the sister group of the first. Primitive placoderms, being the most primitive group of all gnathostomes, were suction feeders and had no biting apparatus at all. Advanced placoderms had a biting apparatus, but it consisted of dermal bone plates instead of real teeth.

4. Dentitions in the proper sense are formed by a dental lamina, i.e., an ectodermal infolding which prefabricates successive teeth and thus reduces their replacement time. The dental lamina of the vertebrates evolved only in the common ancestors of Chondrichthyes and Teleostomi.

5. The dental lamina, being a soft tissue, cannot directly be demonstrated in fossils. Important indirect evidence, however, are tooth spirals, because they can be formed only by a dental lamina.

6. Regulatory processes in the integument of vertebrates are not sufficiently known to allow the formulation of a functional-gradualistic model of dental lamina evolution.

7. The Odontode-Regulation Theory is a differentiation theory. It regards all dentine organs (odontodes) as homologous whether they occur as denticles in the dermal skeleton or as teeth on the jaws or in the mouth cavity. Each odontode is produced by one papillary organ (mesenchyme capped by ectoderm), which in itself is the product of an inductive in-

teractive stimulus between mesenchyme and ectoderm. Differences in shape of individual odontodes are accounted for by differentiation of the papillary organ.

8. Since the Lepidomorial Theory of Stensiö (1961) is a concrescence theory rather than a differentiation theory, it is irreconcilable with the odontode concept. It is shown that the Lepidomorial Theory has a poor empirical basis. Major premises of this theory have been shown to be wrong; lepidomoria themselves are hypothetical and have never been found. The transition from a "cyclomorial" to a "synchronomorial" condition has never been shown directly. It invokes hypothetical intermediate stages and it is notoriously difficult to test.

ACKNOWLEDGMENTS

This is the final report of a project on the dermal skeleton of vertebrates which has been supported by the Sonderforschungsbereich 53 "Paleoecology" since 1970. Thanks are due to the Sonderforschungsbereich and to the Deutsche Forschungsgemeinschaft. This is Konstruktionsmorphologie No. 143. I thank several colleagues for their valuable suggestions during various stages of the preparation of the model outlined here: S. E. Bendix-Almgren (Copenhagen), H.-P. Böss (Tübingen), M. K. and B. M. Hecht (New York), J. Maisey (New York), T. Ørvig (Stockholm), B. Schaeffer (New York), A. Seilacher (Tübingen), P. Shellis (Bristol), M. M. Smith (London), S. Tarsitano (Tübingen/New York), R. D. K. Thomas (Lancaster, Pennsylvania), and F. Westphal (Tübingen).

Note Added in Proof. Janvier (1981), in a new revision of the phylogeny of the Craniota (= Craniata), comes to the conclusion that the Thelodonti (*sensu lato*) are the stem group of the Vertebrata. They are grouped into Thelodontida (with typical dentine) and Katoporida (with mesodentine). Heterostraci and Galeaspida are derived from the Thelodontida. Petromyzonida, Anaspida, Oseostraci, and Gnathostomata are derived from the Katoporida. This result agrees very well with the model developed here that the Thelodonti (*sensu lato*) show the primitive condition of the dermal skeleton of vertebrates.

DEDICATION

The concepts which form the basis of my model were taught to me by Prof. W. Gross (1903–1974). Unfortunately he died before I could discuss my own independent views with him. However, I am fully aware that I could not have written this chapter without the inspiration of his broad interests, knowledge, and warm-hearted personality. Hence, whatever value this chapter may have as a contribution to the literature on the evolution of the dermal skeleton in vertebrates, I dedicate it to the memory of Walter Gross.

REFERENCES

Aldinger, H., 1937, *Permische Ganoidfische aus Ostgrönland*, Medd. Grønland, Volume 102, C. A. Reitzels Forlag, Copenhagen.

Anderson, D. T., 1967, Morphological integration and animal evolution, *Scientia* **102**:83–88.

Aumonier, F. J., 1941, Development of the dermal bones in the skull of *Lepidosteus osseus*, *Q. J. Microsc. Sci.* **83**:1–33.

Bargmann, W., 1933, Die Zahnplatten von *Chimaera monstrosa*, *Z. Zellforsch. Mikrosk. Anat.* **19**:537–561.

Bendrix-Almgren, S. E., 1968, The bradyodont elasmobranchs and their affinities; a discussion, in: *Current Problems of Lower Vertebrate Phylogeny* Nobel Symposium 4 (T. Ørvig, ed.), pp. 153–170, Almqvist & Wisksell, Stockholm.

Berkovitz, B. K. B., 1975, Observations on tooth replacement in Piranhas (Characidae), *Arch. Oral Biol.* **20**:53–56.

Berkovitz, B. K. B., 1977, Chronology of tooth development in the rainbow trout (*Salmo gairdneri*), *J. Exp. Zool.* **200**:65–70.

Berkovitz, B. K. B., 1978a, Tooth ontogeny in the upper jaw and tongue of the rainbow trout (*Salmo gairdneri*), *J. Biol. Buccale* **6**:205–215.

Berkovitz, B. K. B., 1978b, The order of tooth development and eruption in the rainbow trout (*Salmo gairdneri*), *J. Exp. Zool.* **201**:221–226.

Berkovitz, B. K. B., and Shellis, R. P., 1978, A longitudinal study of tooth succession in piranhas (Pisces, Characidae), with an analysis of the tooth replacement cycle, *J. Zool. Lond.* **184**:545–561.

Bernfield, M. R., Cohn, R. H., and Banerjee, S. D., 1973, Glycosaminoglycans and epithelial organ formation, *Am. Zool.* **13**:1067–1083.

Bernhauser, A., 1961, Zur Knochen und Zahnhistologie von *Latimeria chalumnae* Smith und einiger Fossilformen, *Sitzungsber. Math.-Naturwiss. Kl. Abt. I* **170**:119–137.

Bertin, L., 1958, Écailles et sclérification dermiques, in: *Traité de Zoologie, Tome XIII (Agnathes et Poissons)* (P. P. Grassé, ed.), Volume 1, pp. 482–504, Masson, Paris.

Bhatti, H. K., 1938, The integument and dermal skeleton of Siluroidea, *Trans. Zool. Soc. Lond.* **24**(1):1–79.

Bigelow, H. B., and Schroeder, W. C., 1953, Sawfishes, guitarfishes, skates and rays, *Mem. Sears Found. Mar. Res.* **1**(2):1–514.

Boas, J. E. V., 1879, Die Zähne der Scaroiden, *Z. Wiss. Zool.* **22**:189–210.

Bölau, E., 1951, Das Sinnesliniensystem der Tremataspiden und dessen Beziehungen zu anderen Gefäßsystemen des Exoskelettes, *Acta Zool.* **32**:31–40.

Bolt, J. R., and De Mar, R., 1975, An explanatory model of the evolution of multiple rows of teeth in *Captorhinus aguti*, *J. Paleontol.* **49**:814–832.

Brettnacher, H., 1939, Aufbau und Struktur der Holocephalenzähne, *Z. Mikrosk.-Anat. Forsch.* **46**:584–616.

Bryant, W. L., 1936, A study of the oldest known vertebrates, *Astraspis* and *Eriptychius*, *Proc. Am. Phil. Soc.* **76**:409–427.

Budker, P., 1971, *The Life of Sharks*, Weidenfeld and Nicholson, London.

Bystrow, A. P., 1938, Zahnstruktur der Labyrinthodonten, *Acta Zool.* **19**:387–426.

Bystrow, A. P., 1959, The microstructure of skeleton elements in some vertebrates from lower Devonian deposits of the U.S.S.R., *Acta Zool.* **40**(1):59–83.

Casey, J., and Lawson, R., 1978, Amphibians with scales: The structure of the scale in the caecilian *Hypogeophis rostratus*, *Br. J. Herpetol.* **5**:831–833.

Chibon, P., 1977, Vitesse de croissance et renouvellement des dents cher les Amphibiens, *J. Embryol. Exp. Morphol.* **42**:43–63.

Clemen, G., 1978a, Aufbau und Veränderungen der Gaumenzahnleisten beim larvalen und metamorphosierenden *Salamandra salamandra* (L.) (Salamandridae: Amphibia), *Zoomorphologie* **90**:135–150.

Clemen, G., 1978b, Beziehungen zwischen Gaumenknochen und ihren Zahnleisten bei *Salamandra salamandra* (L.) während der Metamorphose, *Wilhelm Roux' Arch.* **185**:19–36.

Compagno, L. J. V., 1973, Interrelationships of living elasmobranchs, in: *Interrelationships of Fishes* (P. H. Greenwood, R. S. Miles, and C. Patterson, eds.), *Zool. J. Linn. Soc.* **53**(Suppl. 1):15–61.

Compagno, L. J. V., 1977, Phyletic relationships of living sharks and rays, *Am. Zool.* **17**:303–322.

De Mar, R. E., 1972, Evolutionary implications of Zahnreihen, *Evolution* **26**:435–450.

De Mar, R. E., 1974, On the reality of Zahnreihen and the nature of reality in morphological studies, *Evolution* **38**:328–330.

Denison, R. H., 1947, The exoskeleton of *Tremataspis*, *Am. J. Sci.* **245**:337–365.

Denison, R. H., 1951, The exoskeleton of early Osteostraci, *Fieldiana Geol.* **11**:199–218.

Denison, R. H., 1952, Early Devonian fishes of Utah, *Fieldiana Geol.* **11**(1):265–287.

Denison, R. H., 1961, Feeding mechanisms of Agnatha and early gnathostomes, *Am. Zool.* **1**:177–181.

Denison, R. H., 1963, The early history of the vertebrate calcified skeleton, *Clin. Ortop* **34**:141–152.

Denison, R. H., 1966, The origin of the lateral-line sensory system, *Am. Zool.* **6**:369–370.

Denison, R. H., 1968, Early Devonian lungfishes from Wyoming, Utah, and Idaho, *Fieldiana Geol.* **17**(4):353–413.

Denison, R. H., 1969, The origin of the vertebrates: A critical evaluation of current theories in: *Proceedings North American Paleont. Convention, September 1969*, Part H, pp. 1132–1146.

Denison, R. H., 1973, Growth and wear of the shield in Pteraspididae (Agnatha), *Palaeontogr. Abt. A* **143**:1–10.

Denison, R. H., 1974, The structure and evolution of teeth in lungfishes, *Fieldiana Geol.* **33**(3):31–58.

Denison, R. H., 1975, *Evolution and classification of Placoderm Fishes*, Breviora, Museum of Comparative Zoology, No. 432, Cambridge, Massachusetts.

Denison, R. H., 1978, Placodermi, in: *Handbook of Paleoichthyology* (H.-P. Schultze, ed.), Volume 2, Gustav Fischer Verlag, Stuttgart.

Denison, R. H., 1979, Acanthodii, in: *Handbook of Paleoichthyology* (H.-P. Schultze, ed.), Volume 5, Gustav Fischer Verlag, Stuttgart.

Dennis, K., and Miles, R. S., 1980, New durophagous arthrodires from Gogo, Western Australia, *Zool. J. Linn. Soc.* **69**:43–85.

Edmund, G., 1960a, Evolution of dental patterns in the lower vertebrates, in: *Evolution*, *its Science and Doctrine* (T. W. M. Cameron, ed.), pp. 45–62, University of Toronto Press.

Edmund, A. G., 1960b, *Tooth Replacement Phenomena in the Lower Vertebrates*, Royal Ontario Museum, Life Sciences Division, Contribution 52.

Edmund, A. G., 1969, Dentition, in: *Biology of the Reptiles*, Part 1: *Morphology A* (C. Gans, ed.), pp. 117–200, Academic Press, London.

Engel, H., 1910, Die Zähne am Rostrum der Pristiden, *Zool. Jahrb. Abt. A Anat.* **29**:51–100.

Ermin, R., Rau, R., and Reibedanz, H., 1971, Der submikroskopische Aufbau der Ganoidschuppen von *Polypterus* im Vergleich zu den Zahngeweben der Säugetiere, *Biomineralisation Forsch.* **3**:12–21.

Fach, M., 1936, Zur Entstehung der Fischschuppe, *Z. Anat. Entwicklungsgesch.* **105**:288–304.

Fahlbusch, K., 1957, *Pteraspis dunensis* Roemer. Eine Neubearbeitung der Pteraspidenfunde (Agnathen) von Overath (Ber. Köln), *Palaeontogr. Abt. A* **108**:1–56.

Fink, W. L., 1981, Ontogeny and phylogeny of tooth attachment modes in actinopterygian fishes, *J. Morphol.* **167**:167–184.

Frank, R. M., Sauvage, C., and Frank, P., 1972, Morphological basis of dental sensitivity, *Int. Dent. J.* **22**:1–19.

Friedmann, E., 1897, Beiträge zur Zahnentstehung der Knochenfische, in: *Morphologische Arbeiten* (G. Schwalbe, ed.), Volume 7, pp. 545–582 G. Fischer-Verlag, Jena.

Gardiner, B. G., 1973, Interrelationships of teleostomes, in: *Interrelationships of Fishes* (P. H. Greenwood, R. S. Miles, and C. Patterson, eds.), *Zool. J. Linn. Soc.* **53**(Suppl. 1):105–135.

Garman, S., 1913, *The Plagiostomia*, Mem. Mus. Comp. Zool. Harvard College, No. 34.

Gaunt, W. A., and Miles, A. E. W., 1967, Fundamental aspects of tooth morphogenesis, in: *Structural and Chemical Organization of Teeth* (A. E. W. Miles, ed.), Volume 1, pp. 151–198, Academic Press, New York.

Gegenbaur, C., 1872, *Untersuchungen zur vergleichenden Anatomie der Wirbeltiere. 3: Das Kopfskelett der Selachier*, Englemann, Leipzig.

Gegenbaur, C., 1898/1901, *Vergleichende Anatomie der Wirbeltiere mit Berücksichtigung der Wirbellosen*, Englemann, Leipzig.

Goetsch, W., 1921, Hautknochenbildungen bei Fischen, *Zool. Jahrb. Abt. 2* **42**:1–42, 435–528.

Goodrich, E. S., 1907, On the scales of fish, living and extinct, and their importance in classification, *Proc. Zool. Soc. Lond.* **77**:751–774.

Grady, J. E., 1970a, Tooth development in sharks, *Arch. Oral Biol.* **15**:613–619.

Grady, J. E., 1970b, Tooth development in *Latimeria chalumnae* (Smith), *J. Morphol.* **132**:377–388.

Graham-Smith, W., 1978a, On some variations in the latero-sensory lines of the placoderm fish *Bothriolepis*, *Phil. Trans. R. Soc. Lond. B* **282**:1–39.

Graham-Smith, W., 1978b, On the lateral lines and dermal bones in the parietal region of some crossopterygian and dipnoan fishes, *Phil. Trans. R. Soc. Lond. B* **282**:41–105.

Grant, P., 1978, *Biology of Developing Systems*, Holt, Rinehart and Winston, New York.

Graver, H. T., 1973, The polarity of the dental lamina in the regenerating salamander jaw, *J. Embryol. Exp. Morphol.* **30**:635–646.

Graver, H. T., 1974, Origin of the dental lamina in the regenerating salamander jaw, *J. Exp. Zool.* **189**:73–84.

Graver, H. T., 1978, Re-regeneration of lower jaws and the dental lamina in adult urodeles, *J. Morphol.* **157**:269–280.

Greenwood, P. H., 1968, The osteology and relationships of the Denticipitidae, a family of clupeomorph fishes, *Bull. Br. Mus. (Nat. Hist.)* **16**(6):213–273.

Gross, W., 1930, Die Fische des mittleren Old Red Süd-Livlands, *Geolog. Pälaontolog. Abhandl. Neue Folge* (J. F. Pompeckj and F. F. V. Huene, eds.) Fischer, Jena **18**:123–156.

Gross, W., 1936, Histologische Studien am Außenskelett fossiler Agnathen und Fische, *Palaeontogr. Abt. A* **83**:1–60.

Gross, W., 1938, Das Kopfskelett von *Cladodus wildungenesis* Jaekel. 2. Teil: Der Kieferbogen. Anhang: *Protacrodus vetustus* Jaekel, *Senckenbergiana* **20**:123–145.

Gross, W., 1953, Devonische Palaeonisciden-Reste in Mittel- und Osteuropa, *Paläontol. Z.* **27**:85–112.

Gross, W., 1956, Über Crossopterygier und Dipnoi aus dem baltischen Oberdevon im Zusammenhang einer vergleichenden Untersuchung des Porenkanalsystems paläozoischer Agnathen und Fische, *K. Svenska Vet. Akad. Handl.* (4) **5**(6):5–140.

Gross, W., 1957, Mundzähne und Hautzähne der Acanthodier und Arthrodiren, *Palaeontogr. Abt. A* **109**:1–40.

Gross, W., 1958a, Anaspiden-Schuppen aus dem Ludlow des Ostseegebiets, *Paläontol. Z.* **32**:24–37.

Gross, W., 1958b, Über die älteste Arthrodiren-Gattung, *Notizbl. Hess. Landesamtes Bodenforsch.* **86**:7–30.

Gross, W., 1959, Arthrodiren aus dem Obersilur der Prager Mulde, *Palaeontogr. Abt. A* **113**:1–35.

Gross, W., 1961, Aufbau des Panzers obersilurischer Heterostraci und Osteostraci Norddeutschlands (Geschiebe) und Oesels, *Acta Zool.* **42**:73–150.

Gross, W., 1962, Neuuntersuchung der Stensiöellida (Arthrodira, Unterdevon), *Notizbl. Hess. Landesamtes Bodenforsch.* **90**:48–86.

Gross, W., 1963a, *Gemuendina stuertzi* Traquair, *Notizbl. Hess. Landesamtes Bodenforsch.* **91**:36–73.

Gross, W., 1963b, *Drepanaspis gemuendensis* Schlüter. Neuuntersuchung, *Palaeontogr. Abt. A* **121**:133–155.

Gross, W., 1966, Kleine Schuppenkunde, *Neues Jahrb. Geol. Palaeontol. Abh.* **125**:29–48.

Gross, W., 1967a, Über das Gebiss der Acanthodier und Placodermen, *J. Linn. Soc. London Zool.* **47**:121–130.

Gross, W., 1967b, Über Thelodontier-Schuppen, *Palaeontogr. Abt. A* **127**:1–67.

Gross, W., 1968a, Porenschuppen und Sinneslinien des Thelodontiers *Phlebolepis elegans* Pander, *Paläontol. Z.* **42**:131–146.

Gross, W., 1968b, Die Agnathenfauna der silurischen Halla-Schichten Gotlands, *Geol. Förening. Stockh. Förhandl.* **90**:369–400.

Gross, W., 1968c, Fragliche Actinopterygier-Schuppen aus dem Silur Gotlands, *Lethaia* **1**:184–218.

Gross, W., 1969, *Lophosteus superbus* Pander, ein Teleostome aus dem Silur Oesels, *Lethaia* **2**:15–47.

Gross, W., 1971a, *Lophosteus superbus* Pander: Zähne, Zahnknochen und besondere Schuppenformen, *Lethaia* **4**:131–152.

Gross, W., 1971b, Downtonische und dittonische Acanthodier-Reste des Ostseegebietes, *Palaeontogr. Abt. A* **136**:1–82.

Gross, W., 1971c, Unterdevonische Thelodontier- und Acanthodier-Schuppen aus West Australien, *Paläontol. Z.* **45**:97–106.

Gross, W., 1973, Kleinschuppen, Flossenstacheln und Zähne von Fischen aus europäischen und nordamerikanischen Bonebeds des Devons, *Palaeontogr. Abt. A* **142**:51–155.

Gudger, E. W., and Smith, B. G., 1933, The natural history of the frilled shark *Chlamydoselachus anguineus,* in: *Archaic Fishes* (E. W. Gudger, ed.), Volume I, pp. 245–319, American Museum of Natural History, New York.

Gunnell, F. H., 1931, Conodonts from the Fort Scott limestone of Missouri, *J. Paleontol.* **5:**244–252.

Gunnell, F. H., 1933, Conodonts and fish remains from the Cherokee, Kansas City, and Wabaunsee groups of Missouri and Kansas, *J. Paleontol.* **7:**261–197.

Gutmann, W. F., 1967, Das Dermalskelett der fossilen "Panzerfische" funktionell und phylogenetisch interpretiert, *Senckenbergiana Lethaea* **48:**277–283.

Gutmann, W. F., 1975, Das Schuppenhemd der niederen Wirbeltiere und seine mechanische Bedeutung, *Nat. Mus.* **105:**169–185.

Gutmann, W. F., 1977, Phylogenetic reconstruction: Theory, methodology, and application to chordate evolution, in: *Major Patterns in Vertebrate Evolution* (M. K. Hecht, P. C. Goody, and B. M. Hecht, eds.), pp. 645–669, Plenum Press, New York.

Gutmann, W. F., and Bonik, K., 1981, Hennigs Theorien und die Strategie des stammes-geschichtlichen Rekonstruierens. Die Agnathen-Gnathostomen-Beziehung als Beispiel, Paläontol. Z. **55:**51–70.

Guttormsen, S. E., 1937, Beiträge zur Kenntnis des Ganoidengebisses, insbesondere des Gebisses von *Colobodus, Schweiz. Palaeontol. Abh.* **60:**1–42.

Hall, B. K., 1975, Evolutionary consequences of skeletal differentiation, *Am. Zool.* **15:**329–350.

Halstead, L. B., 1969a, *The Pattern of Vertebrate Evolution,* Oliver and Boyd, Edinburgh.

Halstead, L. B., 1969b, Calcified tissues in the earliest vertebrates, *Calcif. Tissue Res.* **3:**107–124.

Halstead, L. B., 1974, *Vertebrate Hard Tissues,* Wykeham, London.

Halstead, L. B., 1979, Internal anatomy of the polybranchiaspids (Agnatha, Galeaspida), *Nature* **282:**833–836.

Halstead, L. B., Liu, Y.-H., and P'an, K., 1979, Agnathans from the Devonian of China, *Nature* **282:**831–833.

Halstead Tarlo, L. B., 1963, Aspidin: The precursor of bone, *Nature* **199:**46–48.

Halstead Tarlo, L. B., 1964a, The origin of bone, in: *Proceedings of the First European Bone and Tooth Symposium, Oxford, April 1963,* pp. 3–15, Pergamon Press, Oxford.

Halstead Tarlo, L. B., 1964b, Psammosteiformes (Agnatha). A review with descriptions of new material from the Lower Devonian of Poland I. General Part, *Palaeontol. Pol.* **13:**1–135.

Halstead Tarlo, L. B., 1967, The tessellated pattern of dermal armour in the Heterostraci, *J. Linn. Soc. London Zool.* **47:**45–54.

Halstead Tarlo, B. J., and Halstead Tarlo, L. B., 1965, The origin of teeth, *Discovery* **26:**20–26.

Harder, W., 1975, *Anatomy of Fishes,* Schweizerbart'sche, Stuttgart.

Harlton, B. H., 1933, Micropaleontology of the Pennsylvanian Johns Valley shale of the Ouachita mountains, Oklahoma, and its relationship to the Mississippian Caney shale, *J. Paleontol.* **7:**3–29.

Harper, D. A. T., 1979, Ordovician fish spines from Girvan, Scotland, *Nature* **278:**634–635.

Hase, A., 1911a, Studien über das Integument von *Cyclopterus lumpus.* Beiträge zur Kenntnis der Entwicklung der Haut und des Hautskelettes von Knochenfischen, *Jen. Z. Naturwiss.* **47:**217–342.

Hase, A., 1911b, Die morphologische Entwicklung der Ktenoidschuppe, *Anat. Anz.* **40:**337–356.

Herold, R. C., Graver, H. T., and Christner, P., 1980, Immunohistochemical localization of amelogenins in enameloid of lower vertebrate teeth, *Science* **207:**1357–1358.

Hertwig, O., 1874, Über Bau und Entwicklung der Placoidschuppen und der Zähne der Selachier, *Jen. Z. Naturwiss.* **8**(N.F. 1):331–404.

Hertwig, O., 1876/1879/1882, Über das Hautskelett der Fische, *Morphol. Jahrb.* **2:**328–395; **5:**1–21; **7:**1–42.

Holmbakken, N., and Fosse, G., 1973, Tooth replacement in *Gadus callarias*, *Z. Anat. Entwicklingsgesch.* **143:**65–79.

Isokawa, S., Toda, Y., and Kubota, K., 1968, A histological observation of a coelacanth (*Latimeria chalumnae*), *J. Nihon Univ. School Dentistry* **10:**102–114.

Jaekel, O., 1924/1926, Zur Morphogenie der Gebisse und Zähne, *Vierteljahresschr. Zahnheilkunde* **1924:**313–349; **1926:**217–242, 354–6.

Janvier, P., 1978, On the oldest known teleostome fish *Andreolepis hedei* Gross (Ludlow of Gotland), and the systematic position of the lophosteids, *Eesti NSV Tead. Akad. Toim. Köide Geol.* **27**(3):88–95.

Janvier, P., 1979, D'etranges Poissons chinois, *Recherche* **100:**584–586.

Janvier, P., 1980, Osteolepid remains from the Devonian of the Middle East, with particular reference to the endoskeletal shoulder girdle, in: *The Terrestrial Environment and the Origin of Land Vertebrates* (A. L. Panchen, ed.), pp. 223–254, Academic Press, London.

Janvier, P., 1981, The phylogeny of the Craniata, with particular reference to the significance of fossil "agnathans," *J. Vertebr. Paleontol.* **1:**121–159.

Janvier, P., and Blieck, A., 1979, New data on the internal anatomy of the Heterostraci (Agnatha), with general remarks on the phylogeny of the Craniota, *Zool. Scr.* **8:**287–296.

Jarvik, E., 1972, *Middle and Upper Devonian Porolepiformes from East Greenland with Special Reference to Glyptolepis groenlandica n.sp.*, Medd. Gronland, Volume 187, No. 2, C. A. Reitzels Forlag, Copenhagen.

Jessen, H., 1966, Die Crossopterygier des Oberen Plattenkalkes (Devon) der Bergisch-Gladbach-Paffrather Mulde (Rheinisches Schiefergebirge) unter Berücksichtigung von amerikanischem und europäischem *Onychodus*-Material, *Ark. Zool.* (2) **18:**305–389.

Johnston, M. C., 1975, The neural crest in abnormalities of face and brain, *Birth Defects Orig. Art. Ser.* **11:**1–18.

Jollie, M., 1968, Some implications of the acceptance of a delamination principle, in: *Current Problems of Lower Vertebrate Phylogeney*, Nobel Symposium 4, (T. Ørvig, ed.), pp. 89–102, Almqvist & Wiksell, Stockholm.

Karatajute-Talimaa, V., 1968, Novye telodonty, heterostraci i artrodiry iz Cortkorskogo Gorizonta Podolii, Öcerki po filogenii i sistematike iskopaemych ryb i besceljustuych, in: *Otdelenic Obscej Biologii, pp. 33–42, Akad. Nauk SSSR.*

Karatajute-Talimaa, V., 1973, *Elegestolepis grossi* gen. et sp. nov., ein neuer Typ der Placoidschuppe aus dem oberen Silur der Tuwa, *Palaeontogr. Abt. A* **143:**35–50.

Karatajute-Talimaa, V., 1977, Stroenie i sistematiceskoe polozenie cesuj, *Polymerolepis whitei* Karatajute Talimaa, *Ocerki Filogenii Akad. Nauk SSSR* **1977:**46–60.

Karatajute-Talimaa, V., 1978, *Silurian and Devonian Thelodonts of the U.S.S.R. and Spitzbergen*, Mokslas, Vilnius.

Kassner, J., 1965, Studies on the early stages of scale morphogenesis in the rainbow trout, *Salmo irideus* Gibb., *Zool. Pol.* **15:**79–94.

Kemp, A., 1977, The pattern of tooth plate formation in the Australian lungfish, *Neoceratodus forsteri* Krefft, *Zool. J. Linn. Soc.* **60:**223–258.

Kemp, A., 1979, The histology of tooth formation in the Australian lungfish, *Neoceratodus forsteri* Krefft, *Zool. J. Linn. Soc.* **66:**251–287.

Kerebel, L.-M., and Le Cabellec, M.-Th., 1980, Enameloid in the teleost fish *Lophius*, *Cell Tissue Res.* **206:**211–223.

Kerr, T., 1952, The scales of primitive living actinopterygians, *Proc. Zool. Soc. Lond.* **122:**55–78.

Kerr, T., 1960, Development and structure of some actinopterygian and urodele teeth, *Proc. Zool. Soc. Lond.* **133:**401–422.

Kiaer, J., 1928, The structure of the mouth of the oldest known vertebrates, pteraspids and cephalaspids, *Palaeobiologica* **1:**117–134.

Kiaer, J., and Heintz, A., 1932, New coelolepids from the Upper Silurian of Oesel (Esthonia), *Eesti Loodustead. Ark.* **10:**1–8.

Klaatsch, H., 1890, Zur Morphologie der Fischschuppen und der Geschichte der Hartsubstanzgewebe, *Morphol. Jahrb.* **16:**97–202, 209–258.

Kollar, E. J., 1972, The development of the integument: Spatial, temporal, and phylogenetic factors, *Am. Zool.* **12:**125–135.

Kollar, E. J., and Baird, G. R., 1969, The influence of the dental papilla on the development of tooth shape in embryonic mouse tooth germs, *J. Embryol. Exp. Morphol.* **21:**131–148.

Kollar, E. J., and Fisher, C., 1980, Tooth induction in chick epithelium: Expression of qiescent genes for enamel synthesis, *Science* **207:**933–995.

Laaser, P., 1900, Die Entwicklung der Zahnleiste bei den Selachiern, *Anat. Anz.* **17:**479–489.

Laaser, P., 1903, Die Zahnleiste und die ersten Zahnanlagen der Selachier, *Jen. Z. Naturwiss.* **37:**551–570.

Lauder, G., Jr., 1980a, Evolution of the feeding mechanism in primitive actinopterygian fishes. A functional anatomical analysis of *Polypterus, Lepisostus,* and *Amia, J. Morphol.* **163:**283–317.

Lauder, G. V., 1980b, Hydrodynamics of prey capture by teleost fishes, *Biofluid Mechanics* **2:**161–181.

Lauder, G. V., and Liem, K. F., 1980, The feeding mechanism and cephalic myology of *Salvelinus fontinalis*: Form, function and evolutionary significance, in: *Charrs. Salmonid Fishes of the genus Salvelinus* (E. K. Balon, ed.), pp. 365–390, W. Junk, The Hague, The Netherlands.

Lawson, R., 1963, The anatomy of *Hypogeophis rostratus* Cuvier. Part I: The skin and skeleton, *Proc. Univ. Durham Philos. Soc.* **13A**(25):254–273.

Le Douarin, N., 1975, The neural crest in the neck and other parts of the body, *Birth Defects Orig. Art. Ser.* **11:**19–50.

Lehmann, J. P., 1967, Quelques remarques concernant *Drepanaspis gemuendensis* Schlüter, *J. Linn. Soc. London Zool.* **47:**39–43.

Lehtola, K. A., 1973, Ordovician vertebrates from Ontario, *Contrib. Mus. Palaeontol. Univ. Mich.* **24:**23–30.

Lison, L., 1941, Recherches sur la structure et l'histogenese des dents des Poissons Dipneustes, *Arch. Biol. Paris* **52**(3):279–320.

Liu, Y., 1975, Lower Devonian agnathans of Yunnan and Sichuan, *Vertebr. Palasiatica* **13:**202–216 (in Chinese, with English summary).

Lühmann, M., 1954, Die histogenetischen Grundlagen des periodischen Zahnwechsels der Katfische und Wasserkatzen (Fam. Anarrhichidae, Teleostei), *Z. Zellforsch.* **40:**470–509.

Lund, R., 1977, *Echinochimaera meltoni,* new genus and species (Chimaeriformes), from the Mississippian of Montana, *Ann. Carnegie Mus. Nat. Hist.* **46**(13):195–221.

Maderson, P. F. A., 1972, When? Why? and How?: Some speculations on the evolution of the vertebrate integument, *Am. Zool.* **12:**159–171.

Maderson, P. F. A., 1978, On the early evolutionary history of the neural crest, *Am. Zool.* **18:**586.

Maisey, J. G., 1979, Finspine morphogenesis in squalid and heterodontid sharks, *Zool. J. Linn. Soc.* **66:**161–183.

Maisey, J. G., 1980, An evaluation of jaw suspension in sharks, *Am. Mus. Novit.* **2706:**1–17.

Markert, F., 1876, Die Flossenstacheln von *Acanthias,* ein Beitrag zur Kenntnis der Hartsubstanz: Gebilde der Elasmobranchier, *Zool. Jahrb. Anat.* **9:**665–730.

Markus, H., 1934, Beitrag zur Kenntnis der Gymnophionen. Nr. 21: Das Integument, *Z. Anat. Entwicklungsgesch.* **103:**189–234.

Marss, T., 1977, Structure of *Tolypelepis* from the Baltic Upper Silurian, *Eesti NSV Tead. Akad. Toim. Köide Keem. Geol.* **26:**57–69.

Marss, T., 1979, Lateral line sensory system of the Ludlovian thelodont *Phlebolepis elegans* Pander, *Eesti NSV Tead. Akad. Toim.* Köide *Keem. Geol.* **28:**108–111.

Meincke, D. K., 1980, Structure and development of the dermal skeleton of *Polypterus* and fossil osteichthyans and acanthodians, Ph.D. Thesis, Yale University, New Haven, Connecticut (University Microfilms, Ann Arbor, No. 8026428).

Miles, A. E. W., 1967, *Structural and Chemical Organization of Teeth,* Academic Press, New York.

Miles, A. E. W., and Poole, D. F. G., 1967, The history and general organization of dentitions, in: *Structural and Chemical Organization of Teeth* (A. E. W. Miles, ed.), Volume 1, pp. 3–44, Academic Press, New York.

Miles, R. S., 1968, Jaw articulation and suspension in *Acanthodes* and their significance, in: *Current Problems of Lower Vertebrate Phylogeny,* Nobel Symposium 4 (T. Ørvig, ed.), pp. 109–127 Almquist & Wiksell, Stockholm.

Miles, R. S., 1973, Relationships of acanthodians, in: *Interrelationships of Fishes* (P. H. Greenwood, R. S. Miles, and C. Patterson, eds.), *Zool. J. Linn. Soc.* **53**(Suppl 1):63–103.

Miles, R. S., 1975, The relationships of the Dipnoi, *Colloq. Int. CNRS* **218:**133–148.

Miles, R. S., 1977, Dipnoan (Lungfish) skulls and the relationships of the group. A study based on new specimens from the Devonian of Australia, *J. Linn. Soc. London Zool.* **61:**1–328.

Miles, R. S., and Young, G. C., 1977, Placoderm interrelationships reconsidered in the light of new ptyctodontids from Gogo, Western Australia, in: *Problems in Vertebrate Evolution,* (M. S. Andrews, R. S. Miles, and A. D. Walker, eds.), *Linnean Soc. Symp. Ser.* **4:**123–198.

Miller, W. A., 1969, Tooth enamel of *Latimeria chalumnae* (Smith), *Nature* **221:**1244.

Miller, W. A., and Hobdell, M. H., 1968, Preliminary report on the histology of the dental and paradental tissues of *Latimeria chalumnae* (Smith) with a note on tooth replacement, *Arch. Oral. Biol.* **13:**1289–1291.

Moss, M. L., 1968a, Bone, dentin, and enamel and the evolution of vertebrates, in: *Biology of the Mouth,* AAAS Symposium, pp. 37–65.

Moss, M. L., 1968b, The origin of vertebrate calcified tissues, in: *Current Problems of Lower Vertebrate Phylogeny,* Nobel Symposium 4 (T. Ørvig, ed.), pp. 359–371, Almquist & Wiksell, Stockholm.

Moss, M. L., 1968c, Comparative anatomy of vertebrate dermal bone and teeth. 1: The epidermal co-participation hypothesis, *Acta Anat.* **71:**178–208.

Moss, M. L., 1969, Phylogeny and comparative anatomy of oral ectodermal–ectomesenchymal inductive interactions, *J. Dent. Res.* **48:**732–737.

Moss, M. L., 1972, The vertebrate dermis and the integumental skeleton, *Am. Zool.* **12:**27–34.

Motta, P. J., 1977, Anatomy and functional morphology of dermal collagen fibres in sharks, *Copeia* **1977**(3):454–464.

Moy-Thomas, J. A., 1971, *Palaeozoic Fishes,* 2nd ed., revised by R. S. Miles, Chapman and Hall, London.

Nardi, F., 1936, Zur Genese der Fischschuppe, *Biol. Zentralbl.* **56:**630–639.

Neave, F., 1936, Origin of the teleost scale-pattern and the development of the teleost scale, *Nature* **137**:1034–1038.

Nelson, G. J., 1970, Pharyngeal denticles (placoid scales) of sharks, with notes on the dermal skeleton of vertebrates, *Am. Mus. Novit.* **2415**:1–26.

Nelson, J. S., 1976, *Fishes of the World,* Wiley, New York.

Nickerson, W. S., 1893, Development of the scales of *Lepisosteus, Bull. Mus. Comp. Zool. Harv.* **24**:115–138.

Obruchev, D. V., 1967, Branch Agnata, in: *Fundamentals of Paleontology (Osnovy paleontologii),* Volume 9, pp. 36–167, translated from the Russian, Israel Program of Scientific Translations, Jerusalem.

Obruchev, D., and Karatajute-Talimaa, V., 1967, Vertebrate faunas and correlation of the Ludlovian-Lower Devonian in eastern Europe, *J. Linn. Soc. London Zool.* **47**:5–14.

Ørvig, T., 1951, Histologic studies of placoderms and fossil elasmobranchs 1: The endoskeleton, with remarks on the hard tissues of lower vertebrates in general, *Ark. Zool.* **2**:321–454.

Ørvig, T., 1958, *Pycnaspis splendens,* new genus, new species, a new ostracoderm from the Upper Ordovician of North America, *Proc. U.S. Nat. Mus.* **108**(3391):1–23.

Ørvig, T., 1960, New finds of acanthodians, arthrodires, crossopterygians, ganoids and dipnoans in the Upper Middle Devonian calcareous flags (Oberer Plattenkalk) of the Bergisch Gladbach-Paffrath Trough. 1, *Paläontol. Z.* **34**:295–335.

Ørvig, T., 1961, Notes on some early representative of the Drepanaspida (Pteraspidomorphi, Heterostraci), *Ark. Zool. Ser. 2* **12**(33):515–535.

Ørvig, T., 1962. Y a-t-il une relation directe entre les Arthrodires Ptyctodontides et les Holocephales? in: Problems actuels de Paleontologie: Evolution des Vertebres, *Colloq. Int. CRNS* **104**:49–61.

Ørvig, T., 1965, Palaeohistological notes. 2: Certain comments on the phyletic significance of acellular bone tissue in early lower vertebrates, *Ark. Zool.* **16**:531–556.

Ørvig, T., 1966, Histologic studies of ostracoderms, placoderms and fossil elasmobranchs. 2: On the dermal skeleton of two late Palaeozoic elasmobranchs, *Ark. Zool.* **19**:1–39.

Ørvig, T., 1967a, Phylogeny of tooth tissues: Evolution of some calcified tissues in early vertebrates, in: *Structural and Chemical Organization of Teeth* (A. E. W. Miles, ed.), Volume 1, pp. 45–110, Academic Press, New York.

Ørvig, T., 1967b, Some new acanthodian material from the lower Devonian of Europe, *J. Linn. Soc. London Zool.* **47**:131–153.

Ørvig, T., 1968, The dermal skeleton, general considerations, in: *Current Problems of Vertebrate Phylogeny,* Nobel Symposium 4 (T. Ørvig, ed.), pp. 373–398, Almquist & Wiksell, Stockholm.

Ørvig, T., 1969a, Cosmine and cosmine growth, *Lethaia* **2**:241–260.

Ørvig, T., 1969b, Vertebrates from the Wood Bay Group and the position of the Emsian-Eifelian boundary in the Devonian of Vestspitzbergen, *Lethaia* **2**:273–328.

Ørvig, T., 1972, The latero-sensory component of the dermal skeleton in lower vertebrates and its phyletic significance, *Zool. Scr.* **1**:139–155.

Ørvig, T., 1973a, Acanthodian dentition and its bearing on the relationships of the group, *Palaeontogr. Abt. A* **143**:119–150.

Ørvig, T., 1973b, Fossila fisktänder i svepelektronmikroskopet: Gamla frägeställningar i ny belysning, *Fauna Flora* **68**:166–173.

Ørvig, T., 1975, Description, with special reference to the dermal skeleton, of a new radotinid arthrodire from the Gedinnian of Arctic Canada, *Colloq. Int. CNRS* **218**:41–71.

Ørvig, T., 1976a, Palaeohistological notes. 3: The interpretation of pleromin (pleromic hard tissue) in the dermal skeleton of psammosteid heterostracans, *Zool. Scr.* **5:**35–47.

Ørvig, T., 1976b, Palaeohistological notes: The interpretation of osteodentine, with remarks on the dentition in the Devonian dipnoan *Griphognathus, Zool. Scr.* **5:**79–96.

Ørvig, T., 1977, A survey of odontodes ('dermal teeth') from developmental, structural, functional and phyletic points of view, in: *Problems in Vertebrate Evolution* (S. M. Andrews, R. S. Miles, and A. D. Walker, eds.), *Linnean Soc. Symp. Ser.* **4:**53–75.

Ørvig, T., 1978a, Microstructure and growth of the dermal skeleton in fossil actinopterygian fishes: *Birgeria* and *Scanilepis, Zool. Scr.* **7:**33–56.

Ørvig, T., 1978b, Microstructure and growth of the dermal skeleton in fossil actinopterygian fishes: *Boreosomus, Plegmolepis* and *Gyrolepis, Zool. Scr.* **7:**125–144.

Ørvig, T., 1978c, Microstructure and growth of the dermal skeleton in fossil actinopterygian fishes: *Nephrotus* and *Colobodus*: with remarks on the dentition in other forms, *Zool. Scr.* **7:**297–326.

Ørvig, T., 1980a, Histologic studies of ostracoderms, placoderms and fossil elasmobranchs. 3: Structure and growth of the gnathalia of certain arthrodires, *Zool. Scr.* **9:**141–159.

Ørvig, T., 1980b, Histologic studies of ostracoderms, placoderms and fossil elasmobranchs. 4: Ptyctodontid tooth plates and their bearing on holocephalan ancestry: The condition of *Ctenurella* and *Ptyctodus, Zool. Scr.* **9:**219–239.

Osborn, J. W., 1970, New approach to Zahnreihen, *Nature* **225:**343–346.

Osborn, J. W., 1971, The ontogeny of tooth succession in *Lacerta vivipara* Jacquin (1787), *Proc. R. Soc. Lond. B* **179:**261–289.

Osborn, J. W., 1973a, The evolution of dentitions, *Am. Sci.* **61:**548–559.

Osborn, J. W., 1973b, On the biological improbability of Zahnreihen as embryological units, *Evolution* **26:**601–607.

Osborn, J. W., 1974, On the control of tooth replacement in reptiles and its relationship to growth, *J. Theor. Biol.* **46:**509–527.

Osborn, J. W., 1977, The interpretation of patterns in dentitions, *Biol. J. Linn. Soc.* **9:**217–229.

Osborn, J. W., 1978, Morphogenetic gradients: Fields versus clones, in: *Development, Function and Evolution of Teeth* (P. M. Butler and K. A. Joysey, eds.), pp. 171–202, Academic Press, London.

Parker, H. W., and Boeseman, M., 1954, The basking shark *Cetorhinus* in winter, *Proc. Zool. Soc. Lond.* **124:**185–194.

Parrington, F. R., 1958, On the nature of the Anaspida, in: *Studies on Fossil Vertebrates* (T. S. Westoll, ed.), pp. 108–128, University of London, The Athlone Press, London.

Patterson, C., 1965, The phylogeny of the chimaeroids, *Phil. Trans. R. Soc. Lond. B* **249:**101–219.

Patterson, C., 1968, *Menaspis* and the bradyodonts, in: *Current Problems of Lower Vertebrate Phylogeny*, Nobel Symposium 4 (T. Ørvig, ed.), pp. 171–206, Almqvist & Wiksell, Stockholm.

Patterson, C., 1977, Cartilage bones, dermal bones and membrane bones, or the exoskeleton versus the endoskeleton, in: *Problems in Vertebrate Evolution* (S. M. Andrews, R. S. Miles, and A. D. Walker, eds.), *Linn. Soc. Symp. Ser.* **4:**77–121.

Pearson, D. M., 1981, Functional aspects of the integument in polypterid fishes, *Zool. J. Linn. Soc.* **72:**72–93.

Peyer, B., 1919, Die Flossenstacheln der Welse, *Anat. Anz.* **52.**

Peyer, B., 1931, Hartgebilde des Integuments, in: *Handbuch der vergleichenden Anatomie der Wirbeltiere* (L. Bolk, E. Göppert, E. Kallius, and W. Lubosch, eds.), Volume 1, pp. 703–752, Urban & Schwarzenberg, Berlin.

Peyer, B., 1937, Zähne und Gebiss, in: *Handbuch der vergleichenden Anatomie* (L. Bolk, E. Göppert, E. Kallius, and W. Lubosch, eds.), Volume 3, Urban und Schwarzenberg, Berlin.

Peyer, B., 1968, Comparative Odontology, University of Chicago Press, Chicago.

Pflugfelder, O., 1930, Das Gebiß der Gymnodonten. Ein Beitrag zur Histogenese des Dentins, *Z. Anat. Entwicklungsgesch.* **93:**543–566.

Poole, D. F. G., 1956, Fine structure of scales and teeth of *Raja clavata, Q. J. Microsc. Sci.* **97:**99–107.

Poole, D. F. G., 1967, Phylogeny of tooth tissues: Enameloid and enamel in Recent vertebrates, with a note on the history of cementum, in: *Structural and Chemical Organization of Teeth* (A. E. W. Miles, ed.), Volume 1, pp. 111–149, Academic Press, New York.

Poole, D. F. G., 1971, An introduction to the phylogeny of calcified tissues, in: *Dental Morphology and Evolution* (A. A. Dahlberg, ed.), pp. 65–79, University of Chicago Press, Chicago.

Poole, D. F. G., and Shellis, R. P., 1976, Eruptive tooth movements in non-mammalian vertebrates, in: *Eruption and Occlusion of Teeth* (D. F. G. Poole and M. V. Stack, ed.), Colston Papers no. 27, pp. 65–79, Butterworths, London.

Preuschoft, H., Reif, W.-E., and Müller, W.-H., 1974, Funktionsanpassungen in Form und Struktur an Haifischzähnen, *Z. Anat. Entwicklungsgesch.* **143:**315–344.

Rauther, M., 1940, Echte Fische, Integument, in: *Bronns Klassen und Ordnungen des Tierreichs,* 6. Bd., 1. Abt., 2. Buch, Teil 1, pp. 206–335.

Regan, C. T., 1924, Reversible evolution with examples from fishes, *Proc. Zool. Soc. Lond.* **1:**175–176.

Reif, W. E., 1973a, Morphologie und Ultrastruktur des Hai-"Schmelzes," *Zool. Scr.* **2:**231–250.

Reif, W. E., 1973b, Ontogenese des Hautskeletts von *Heterodontus facifer* (Selachii) aus dem Untertithon, *Stuttg. Beitr. Naturkd.* **7:**1–16.

Reif, W. E., 1974, Morphogenese und Musterbildung im Hautzähnchen-Skelett von *Heterodontus, Lethaia* **7:**25–42.

Reif, W. E., 1975, Lenkende und limitierende Faktoren in der Evolution, *Acta Biotheoret.* **24:**136–162.

Reif, W. E., 1976, Morphogenesis, pattern formation and function of the dentition of *Heterodontus* (Selachii), *Zoomorphologie* **83:**1–47.

Reif, W. E., 1977, Tooth enameloid as taxonomic criterion: 1. A new euselachian shark from the Rhaetic-Liassic boundary, *Neues Jahrb. Geol. Palaeontol. Monatsh.* **1977:**565–576.

Reif, W. E., 1978a, Types of morphogenesis of the dermal skeleton in fossil sharks, *Palaeontol. Z.* **52:**110–128.

Reif, W. E., 1978b, Shark dentitions: Morphogenetic processes and evolution, *Neues Jahrb. Geol. Palaeontol. Abh.* **157:**107–115.

Reif, W. E., 1978c, Tooth enameloid as a taxonomic criterion: 2. Is *"Dalatias" barnstonensis* Sykes, 1971 (Triassic, England) a squalomorphic shark? *Neues Jahrb. Geol. Palaeontol. Monatsh.* **1978:**42–58.

Reif, W. E., 1978d, Bending-resistant enameloid in carnivorous teleosts, *Neues Jahrb. Geol. Palaeontol. Abh.* **157:**173–175.

Reif, W. E., 1979a, Morphogenesis and histology of large scales of batoids (Elasmobranchii), *Palaeontol. Z.* **53:**26–37.

Reif, W. E., 1979b, Structural convergences between enameloid of actinopterygian teeth and of shark teeth, *Scanning Electron Microscopy* **1979**(II)**:**546–554.

Reif, W. E., 1980a, A model of morphogenetic processes in the dermal skeleton of elasmobranchs, *Neues Jahrb. Geol. Palaeontol. Abh.* **159**:339–359.

Reif, W. E., 1980b, Development of dentition and dermal skeleton in embryonic *Scyliorhinus canicula, J. Morphol.* **166**:275–288.

Reif, W. E., 1980c, Tooth enameloid as a taxonomic criterion: 3. A new primitive shark family from the lower Keuper, *Neues Jahrb. Geol. Palaeontol. Abh.* **160**:61–72.

Reif, W. E., unpublished, SEM atlas of scales of Recent sharks, to be deposited at the library of the Department of Geology, University of Tübingen.

Reif, W. E., and Goto, M., 1979, Placoid scales from the Permian of Japan, *Neues Jahrb. Geol. Palaeontol. Monatsh.* **1979**:201–207.

Reif, W. E., McGill, D., and Motta, P., 1978, Tooth replacement rates of the sharks *Triakis semifasciata* and *Ginglymostoma cirratum, Zool. Jahrb. Abt. Anat.* **99**:151–156.

Repetski, J. E., 1978, A fish from the Upper Cambrian of North America, *Science* **200**:529–531.

Richey, F. T., 1977, An histochemical basis for the origin and differentiation of dentine and enamel, in: *South. Calif. Acad. Sci. Ann. Meet.* 1977, p. 135.

Ritchie, A., 1968, *Phlebolepis elegans* Pander, an Upper Silurian thelodont from Oesel, with remarks on the morphology of thelodonts, in: *Current Problems of Lower Vertebrate Phylogeny,* Nobel Symposium 4 (T. Ørvig, ed.), pp. 81–88, Almsqvist & Wiksell, Stockholm.

Romer, A. S., 1966, *Vertebrate Paleontology,* 3rd ed., University of Chicago Press, Chicago.

Romer, A. S., 1976, *Vergleichende Anatomie der Wirbeltiere,* 4th ed., Verlag Paul Parey, Hamburg.

Röse, C., 1892, Über Zahnbau und Zahnwechsel der Dipnoer, *Anat. Anz.* **7**:821–839.

Rosen, D. E., Forey, P. L., Gardiner, B. G., and Patterson, C., 1981, Lungfishes, tetrapods, paleontology, and plesiomorphy, *Bull. Am. Mus. Nat. Hist.* **167**:157–276.

Rosén, N., 1913, Studies on the plectognaths. 3. The integument, *Ark. Zool.* **8**(10):1–29.

Rosén, N., 1914, Wie wachsen die Ktenoidschuppen?, *Ark. Zool.* **9**(20):1–6.

Roux, C. H., 1942, The microscopic anatomy of the *Latimeria* scale, *S. Afr. J. Med. Sci. (Biol. Suppl.)* **7**:1–18.

Ryder, J. A., 1892, On the mechanical genesis of the scales of fishes, *Proc. Acad. Nat. Sci. Philadelphia* **1892**:219–224.

Schaeffer, B., 1963, Cretaceous fishes from Bolivia, with comments on pristid evolution, *Am. Mus. Novit.* **2159**:1–20.

Schaeffer, B., 1975, Comments on the origin and basic radiation of the gnathostome fishes with particular reference to the feeding mechanism, *Colloq. Int. CNRS* **218**:101–109.

Schaeffer, B., 1977, The dermal skeleton in fishes, in: *Problems in Vertebrate Evolution* (S. M. Andrews, R. S. Miles, and A. D. Walker, eds.), *Linnean Soc. Symp. Ser.* **4**:25–52.

Schaeffer, B., and Williams, M., 1977, Relationships of fossil and Recent elasmobranchs, *Am. Zool.* **17**:293–302.

Schauinsland, H., 1903, Beiträge zur Entwicklungsgeschichte und Anatomie der Wirbeltiere, Teil 1: *Sphenodon, Callorhynchus, Chamaeleo, Zoologica (Stuttgart)* **39**:1–98.

Schmidt, W. J., 1958, Faserung und Durodentin-Metaplasie bei Fischzähnen, *Anat. Anz.* **105**:349–360.

Schmidt, W. J., and Keil, A., 1971, *Polarizing Microscopy of Dental Tissues,* Pergamon Press, Oxford.

Schnakenbeck, W., 1962, Pisces, in: *Hand. Zool. 6,* 1 Hälfte, 1 Teil, pp. 551–1115, Walter de Gruyter & Co., Berlin.

Schultze, H. P., 1966, Morphologische und histologische Untersuchungen an Schuppen mesozoischer Actinopterygier (Übergang von Ganoid- zu Rundschuppen), *Neues Jahrb. Geol. Palaeontol. Abh.* **126:**232–314.

Schultze, H. P., 1968, Palaeoniscoiden-Schuppen aus dem Unterdevon Australiens und Kanadas und aus dem Mitteldevon Spitzbergens, *Bull. Br. Mus. (Nat. Hist.) Geol.* **16:**341–368.

Schultze, H. P., 1969, Die Faltenzähne der rhipidistiiden Crossopterygier, der Tetrapoden und der Actinopterygier-Gattung *Lepisosteus, Palaeontogr. Ital. N.S.* **35:**63–136.

Schultze, H. P., 1977a, Ausgangsform und Entwicklung der rhombischen Schuppen der Osteichthyes (Pisces), *Palaeontol. Z.* **51:**152–168.

Schultze, H. P., 1977b, The origin of the tetrapod limb within the rhipidistian fishes, in: *Major Patterns in Vertebrate Evolution* (M. K. Hecht, P. C. Goody, and B. M. Hecht, eds.), pp. 542–543, Plenum Press, New York.

Schultze, H. P., 1980, Crossopterygier-Schuppen aus dem obersten Oberdevon Lettlands (Osteichthyes, Pisces), *Neues Jahrb. Geol. Palaeontol. Monatsh.* **1980:**215–228.

Schultze, H. P., 1981, Hennig und der Ursprung der Tetrapoda, *Palaeontol. Z.* **55:**71–86.

Semon, R., 1899, Die Zahnentwicklung von *Ceratodus forsteri,* in: *Zoologische Forschungsreisen in Australien und des Malayischen Archipel* (R. Semon, ed.), Volume 1, part 3, *Denkschr. Med.-Naturwiss. Ges. Jena* **4:**115–135.

Sewertzoff, A. N., 1932, Die Entwicklung der Knochenschuppe von *Polypterus delhesi, Jen. Z. Med. Naturwiss.* **67**(N.F. 60):387–418.

Shellis, R. P., 1975, A histological and histochemical study of the matrices of enameloid and dentine in teleost fishes, *Arch. Oral Biol.* **20:**183–187.

Shellis, R. P., 1978, The role of the inner dental epithelium in the formation of the teeth in fish, in: *Development, Function, and Evolution of Teeth* (P. M. Butler and K. A. Joysey, eds.), pp. 31–42, Academic Press, New York.

Shellis, R. P., 1982, Comparative anatomy of tooth attachment, in: *The Periodontal Ligament in Health and Disease* (B. K. B. Berkovitz, B. Moxham, and H. N. Newman, eds.), pp. 3–24, Pergamon Press, Oxford.

Shellis, R. P., and Berkovitz, B. K. B., 1976, Observations on the dental anatomy of piranhas (Characidae) with special reference to tooth structure, *J. Zool. Lond.* **180:**69–84.

Shellis, R. P., and Berkovitz, B. K. B., 1980, Dentine structure in the rostral teeth of the sawfish *Pristis* (Elasmobranchii), *Arch. Oral Biol.* **25:**339–343.

Shellis, R. P., and Miles, A. E. W., 1974, Autoradiographic study of the formation of enameloid and dentine matrices in teleost fishes using tritiated amino acids, *Proc. R. Soc. Lond. B* **185:**51–72.

Shellis, R. P., and Miles, A. E. W., 1976, Observations with the electron microscope on enameloid formation in the common eel (*Anguilla anguilla;* Teleostei), *Proc. R. Soc. Lond. B* **194:**253–269.

Shellis, R. P., and Poole, D. F. G., 1978, The structure of the dental hard tissues of the coelacanthid fish *Latimeria chalumnae* Smith, *Arch. Oral Biol.* **23:**1105–1113.

Slaughter, B. H., and Springer, St., 1968, Replacement of rostral teeth in sawfishes and sawsharks, *Copeia* **1968**(3):499–506.

Smith. B. G., 1937, The anatomy of the frilled shark, *Chlamydoselachus anguineus,* in: *Archaic Fishes* (E. W. Gudger, ed.), Volume II, pp. 331–520, American Museum of Natural History, New York.

Smith, M. M., 1977, The microstructure of the dentition and dermal ornament of three dipnoans from the Devonian of Western Australia: A contribution towards dipnoan interrelations, and morphogenesis, growth and adaption of the skeletal tissues, *Phil. Trans. R. Soc. Lond. B* **281:**29–72.

Smith, M. M., 1978, Enamel in the oral teeth of *Latimeria chalumnae* (Pisces: Actinistia): A scanning electron microscope study, *J. Zool. Lond.* **185**:355–369.

Smith, M. M., 1979a, Scanning electron microscopy of dontodes in the scales of a coelacanth embryo, *Latimeria chalumnae* Smith, *Arch. Oral Biol.* **24**:179–183.

Smith, M. M., 1979b, Structure and histogenesis of tooth plates in *Sagenodus inaequalis* Owen considered in relation to the phylogeny of post-Devonian dipnoans, *Proc. R. Soc. Lond. B* **204**:15–39.

Smith, M. M., 1979c, SEM of the enamel layer in oral teeth of fossil and extant crossopterygian and dipnoan fishes, *Scanning Electron Microscopy* **1979**(II):483–489.

Smith, M. M., and Miles, A. E. W., 1969, An autoradiographic investigation with the light microscope of proline-H^3 incorporation during tooth development in the crested newt (*Triturus cristatus*), *Arch. Oral Biol.* **14**:479–490.

Smith, M. M., and Miles, A. E. W., 1971, The ultrastructure of odontogenesis in larval and adult urodeles: Differentiation of the dental epithelial cells, *Z. Zellforsch. Mikrosk. Anat.* **121**:470–498.

Smith, M. M., Hobdell, M. H., and Miller, W. A., 1972, The structure of the scales of *Latimeria chalumnae*, *J. Zool. Lond.* **167**:501–509.

Stadtmüller, F., 1936a, Kranium und Visceralskelett der Stegocephalen und Amphibien, in: *Handbuch der vergleichenden Anatomie der Wirbeltiere* (L. Bolk, E. Göppert, E. Kallius, and W. Lubosch, eds.), Volume 4, pp. 501–698, Urban & Schwarzenberg, Berlin.

Stadtmüller, F., 1936b, Kranium und Visceralskelett der Säugetiere, in: *Handbuch der vergleichenden Anatomie der Wirbeltiere* (L. Bolk, E. Göppert, E. Kallius, and W. Lubosch, eds.), Volume 4, pp. 501–698, Urban & Schwarzenberg, Berlin.

Starck, D., 1979, *Vergleichende Anatomie der Wirbeltiere*, Volume 2, Springer, Berlin.

Stensiö, E. A., 1927, *The Downtonian and Devonian Vertebrates of Spitzbergen. Part I: Family Cephalaspidae*, Skrift. Svalbard og Nordishavet, No. 12, Hos Jacob Dybwad, Oslo.

Stensiö, E. A., 1932, *The Cephalaspids of Great Britain*, British Museum (Natural History), London.

Stensiö, E., 1958, Les Cyclostomes fossiles ou Ostracoderms, in: *Traité de Zoologie* (P.-P. Grassé, ed.), Volume 13, part 1, pp. 173–428, Masson, Paris.

Stensiö, E., 1961, Permian vertebrates, in: *Geology of the Arctic* (G. O. Raasch, ed.), Volume 1, pp. 231–247, University of Toronto Press, Toronto.

Stensiö, E. A., 1962, Origine et nature des écailles placoides et des dents, in: *Problèmes actuels de Paléontologie*, pp. 75–85, Centre National de la Recherche Scientifique, Paris.

Stensiö, E., 1964, Les Cyclostomes fossiles ou Ostracodermes, in: *Traite de Paléontologie* (J. Piveteau, ed.), Volume 4, part 1, pp. 96–382, Masson, Paris.

Stensiö, E., 1968, The cyclostomes with special reference to the diphyletic origin of the Petromyzontida and Myxinoidea, in: *Current Problems of Lower Vertebrate Phylogeny*, Nobel Symposium 4 (T. Ørvig, ed.), pp. 13–70, Almqvist & Wiksell, Stockholm.

Studnička, F. K., 1942, Über die Zahnersatzgruppen bei *Myliobatis aquila* L., *Anat. Anz.* **93**:85–95.

Tarlo, L. B., and Tarlo, B. J., 1961, Histological sections of the dermal armour of psammosteid ostracoderms, *Proc. Geol. Soc. Lond.* **1953**:3–4.

Thomasset, J., 1930, Recherches sur les tissus dentaires des poissons fossiles, *Arch. Anat. Histol. Embryol.* **11**:5–153.

Thomson, K. S., 1975, *On the Biology of Cosmine*, Yale Univ. Peabody Mus. Nat. Hist. Bull. 40.

Thomson, K. S., 1977, On the individual history of cosmine and a possible electroreceptive function of the pore-canal system in fossil fishes, in: *Problems in Vertebrate Evolution* (S. M. Andrews, R. S. Miles, and A. D. Walker, eds.), *Linn. Soc. Symp. Ser.* **4**:247–270.

Thomson, K. S., 1980, The ecology of Devonian lobe-finned fishes, in: *The Terrestrial Environment and the Origin of Land Vertebrates* (A. Z. Panchen, ed.), pp. 187–222, Academic Press, London.

Thorsteinsson, R., 1973, Dermal elements of a new lower vertebrate from middle Silurian (upper Wenlockian) rocks of the Canadian Arctic archipelago, *Palaeontogr. Abt. A* **143**:51–57.

Tretjakoff, D., and Chinkus, F., 1927, Das Knochengewebe der Fische, *Z. Ges. Anat. Abt. 1* **83**:363–396.

Turner, S., 1976, *Thelodonti (Agnathi)*, Fossil Catalogues, I: Animalia, pars 122, Junk, s'Gravenhage.

Twai, L. E., 1979, *Pennsylvanian Ichthyoliths from the Shawnee Group of Eastern Kansas*, Univ. Kansas Paleontol. Contrib. 96.

Verluys, J., 1936, Kranium und Visceralskelett der Sauropsiden. Teil 1: Reptilien, in: *Handbuch der vergleichenden Anatomie der Wirbeltiere* (L. Bolk, E. Göppert, E. Kallius, and W. Lubosch, eds.), Volume 4, pp. 699–808, Urban & Schwarzenberg, Berlin.

Vieth, J., 1980, *Thelodontier-, Acanthodier und Elasmobranchier-Schuppen aus dem Unter-Devon der Kanadischen Arktis (Agnatha, Pisces)*, Göttinger Arb. Geol. Paläont. 23.

Wahlert, G. von, 1966a, Biologie und Evolution der Atemwege bei Haien und Rochen, *Veroeff. Inst. Meeresforsch. Bremerhaven, Sonderband* **II**:337–356.

Wahlert, G. von, 1966b, *Atemwege und Schädelbau der Fische*, Stuttgarter Beitr. Naturk. 159.

Wahlert, G. von, 1968, *Latimeria und die Geschichte der Wirbeltiere. Eine evolutionsbiologische Untersuchung*, Gustav Fischer Verlag, Stuttgart.

Wahlert, G. von, 1970, Die Entstehung des Kieferapparates der Gnathostomen, *Verh. Dtsch. Zool. Ges.* **64**:344–347.

Wainwright, S. A., Vosburgh, F., and Hebrank, J. H., 1978, Shark skin: Function in locomotion, *Science* **202**:747–749.

Wängsjö, G., 1944, On the genus *Dartmuthia* Patten with special reference to the minute structure of the exoskeleton, *Bull. Geol. Inst. Uppsala* **31**:349–362.

Watson, D. M. S., 1937, The acanthodian fishes, *Phil. Trans. R. Soc. B* **228**:49–146.

Wells, J. W., 1944, Fish remains from the Middle Devonian bone beds of the Cincinnati Arch region, *Palaeontogr. Am.* **3**:5–46.

Westoll, T. S., 1962, Ptyctodontid fishes and the ancestry of Holocephali, *Nature* **194**:949–952.

Westoll, T. S., 1980, Prologue: Problems of tetrapod origin, in: *The Terrestrial Environment and the Origin of Land Vertebrates* (A. L. Panchen, ed.), pp. 1–10, Academic Press, London.

Westphal, F., 1975, Bauprinzipien im Panzer der Placodonten, (Reptilia triadica), *Palaeontol. Z.* **49**:97–125.

Westphal, F., 1976, The dermal armour of some Triassic placodont reptiles, in: *Morphology and Biology of Reptiles* (A. d'A. Bellairs and C. B. Cox, eds.), *Linn. Soc. Symp. Ser.* **3**:31–41.

White, E. J., 1946, The genus *Phialaspis* and the "*Psammosteus* Limestones," *Q. J. Geol. Soc. Lond.* **101**:207–242.

White, E. J., 1958, Original environments of the craniates, in: *Studies on Fossil Vertebrates* (T. S. Westoll, ed.), pp. 212–234, University of London, The Athlone Press, London.

White, E. J., 1968, Discussion to E. A. Stensiö, in: *Current Problems of Lower Vertebrate*

Phylogeny, Nobel Symposium 4 (T. Ørvig, ed.), pp. 70–71, Almqvist & Wiksell, Stockholm.

White, E. J., 1973, Form and growth of *Belgicaspis* (Heterostraci), *Palaeontogr. Abt. A* **143**:11–24.

Williamson, W. C., 1849, On the microscopic structure of the scales and dermal teeth of some ganoid and placoid fish, *Phil. Trans. R. Soc. Lond.* **139**:435–475.

Williamson, W. S., 1851, Investigation into the structure and development of the scales and bones of fishes, *Phil. Trans. R. Soc. Lond.* **141**:643–702.

Woodward, A. S., and White, E. J., 1938, The dermal tubercles of the Upper Devonian shark, *Cladoselache, Ann. Mag. Nat. Hist.* **11**(2):367–368.

Yakovlev, V. N., 1966, Funktsional'naya evolyutsiya skeleta ryb, *Paleont. Zh.* **1966**(3):3–12 [Functional evolution of the fish skeleton, *Int. Geol. Rev.* **9**:525–532].

Zangerl, R., 1966, A new shark of the family Edestidae, *Ornithoperion hertwigi, Fieldiana Geol.* **16**(1):1–43.

Zangerl, R., 1968, The morphology and the developmental history of the scales of the Paleozoic sharks *Holmesella*? sp. and *Orodus,* in: *Current Problems of Vertebrate Phylogeny,* Nobel Symposium 4 (T. Ørvig, ed.), pp. 399–412, Almqvist & Wiksell, Stockholm.

Zangerl, R., and Case, G. R., 1976, *Cobelodus aculeatus* (Cope), an anacanthous shark from Pennsylvanian black shales of North America, *Palaeontogr. Abt. A* **154**:107–157.

Zangerl, R., and Williams, M. E., 1975, New evidence on the nature of the jaw suspension in Palaeozoic anacanthous sharks, *Palaeontology* **18**:333–341.

8

Evolution of *Chesapecten* (Mollusca: Bivalvia, Miocene–Pliocene) and the Biogenetic Law

JOAN M. MIYAZAKI

Department of Biology
Queens College, City University of New York
Flushing, New York 11367

and

M. F. MICKEVICH

Department of Ichthyology
American Museum of Natural History
New York, New York 10024

INTRODUCTION

The theory of recapitulation associates the developmental sequence with evolutionary change. Haeckel (1866) and von Baer (1866) were prominent in formulating the rules asociating ontogeny and phylogeny. Haeckel's Biogenetic Law ("ontogeny recapitulates phylogeny") emphasizes terminal additions to the developmental sequence during the evolutionary history of a group. Von Baer's Rule emphasizes the primitive or general nature of early developmental events and the specialized nature of features that appear late in development.

In classifying some early pelecypods (bivalves), Jackson (1890) applied the "law of concentration and acceleration of development" (attributed to Hyatt and Cope). According to this law, developmental stages are condensed into earlier growth stages, resulting in more rapid rates of organic differentiation. Jackson's groups were based on observations of ontogenetic similarities and divergence in bivalve shell shape and patterns. Newell's (1937) monograph on Paleozoic Pectinacea included the statement, "If the principle of recapitulation is applicable in pelecypods, and I believe that it is, each fundamental type of ornamentation observed in an adult pectinoid shell represents the culmination of a series of evolutionary modifications." Newell's interpretation was derived from extensive morphologic observations on these pectinids and their stratigraphic position, but in his statement also lies the hesitancy that a test of his suspicion was yet forthcoming.

Hypothesis testing is critical in determining whether a theory such as recapitulation summarizes the pattern of changes through time. A diverse array of related taxa, aligned simply in a sequence with the most juvenile-like forms followed successively by more gerontic forms, would constitute a hypothesis of phylogeny among those taxa according to recapitulation. It would not demonstrate that recapitulation had taken place. It would be encouraging if the fossil sequence duplicated exactly the pattern predicted by the recapitulation hypothesis. However, this would also not be sufficient alone to verify the biogenetic law. The actual genealogy could have been considerably more complex than either sequence. A method for estimating phylogeny independent of ontogenetic information is necessary in order to perform a test of the recapitulation theory.

Complexity can be introduced by accelerations and reversals in a mosaic of character changes which would obscure interpretation of the true evolutionary sequence. This argument was advocated by de Beer (1958) and prevented literal acceptance of the recapitulation theory. De Beer's examples are an important contribution to our understanding of potential effects of developmental rates and the influence their changes have on morphologic differences between taxa. These examples demonstrate the developmental plasticity of morphologic characters in multicellular organisms. He cites this phenomenon, called heterochrony, as a major factor in evolution, and a process tending to obscure the direct association between development and evolution. To de Beer, recapitulating evolutionary patterns were just a special kind of heterochrony which he called hypermorphosis.

However, few investigators would entirely discount developmental considerations in phylogeny construction; many would cite von Baer's Rule. Early developmental stages are generally conservative or undifferentiated relative to later stages. Critical events in embryogenesis or the constancy of an early developmental environment would contribute to the similarities among the early stages of ancestor and descendant. Though different in emphasis, the Biogenetic Law and von Baer's Rule have the same value in evaluating characters in systematics. They predict that characters of early development are primitive relative to those of later development. In light of Haeckel's and von Baer's contributions, Nelson (1973, 1978) suggested that clues to the polarity of character transformations may be determined from the general correlation seen between the appearance of evolutionarily derived character states and the ontogenetic sequence in morphogenesis. Whereas Haeckel had applied the concept of terminal additions to the whole phenotype of related organisms, Nelson restated the Biogenetic Law, restricting its definition to phylogenetic changes that occur within single characters. By formulating evolutionary hypotheses for characters individually, using their ontogenetic character sequence, determination of the presumed primitive character state is simplified. This is particularly advantageous when mosaic evolution has occurred. With mosiac evolution, asynchronous ontogenetic patterns in different characters may admit no simple interpretation of the primitive taxon.

The major purposes of this study were to test whether phylogenetic change occurs along a developmental axis and to test whether developmental modifications are useful in discriminating primitive character states from derived character states. In effect, the Biogenetic Law is treated as a hypothesis. We compared a phylogenetic (cladistic) network of a fossil pectinid lineage of the genus *Chesapecten* to ontogenetic changes in characters which occur in all the included taxa. The cladistic network is obtained independently of developmental information. Correspondence between the order of character state change in ontogeny and phylogeny would support the Biogenetic Law. If heterochrony is more prevalent in the evolution of these pectinids, we would interpret that character evolution is not constrained by ontogeny in the way this simplifying hypothesis would predict.

This study will also be a test of Newell's statement as applied to *Chesapecten*. The fossil character data will provide information about specific morphologic changes and life habit. The sediments containing the *Chesapecten* provide information about the depositional environments. Hence we will not only observe a pattern of change in ontogeny and phylogeny, but ecologic factors may be taken into consideration as well.

THE GENUS *CHESAPECTEN*

The *Chesapecten* fossil record is fairly continuous and restricted in geographic and temporal extent. However, as commonly occurs with paleontologic data, the record of fossil species is interrupted by depositional breaks, which conveniently separate the successional forms into time-unit species. The species seem to be members of a single evolutionary (monophyletic) lineage. The characters that unite *Chesapecten* as a monophyletic group distinct from the other pectinid genera are (1) its reduced or absent cardinal crura, (2) a right valve which is flatter than the left, and (3) scabrous lirae covering the outer disk of the valves (Ward and Blackwelder, 1975). *Chesapecten* is primarily limited to the Miocene and Early Pliocene deposits of the western Atlantic, especially around Chesapeake Bay and southward to North Carolina (Fig. 1), although one Upper Miocene group can be found as far south as Florida. Apparently this lineage was endemic to the study region. These sediments represent deposition over approximately 11 million yr (23–13 my BP); the strata are not complicated by deformation or major unconformities. Hence the vertical sequence of strata and the fossils contained within them probably approximate a record of one lineage within a major depositional embayment.

Ward and Blackwelder (1975) have described seven species of *Chesapecten* found in temporal succession along the stratigraphic sequence

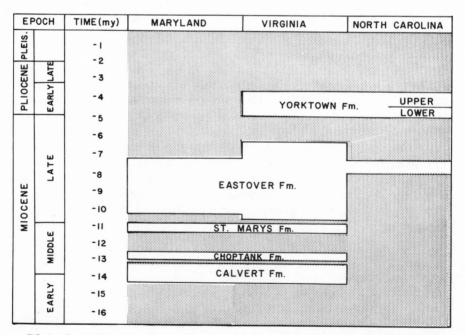

FIG. 1.   Formations of the Chesapeake Group [Adapted from Ward and Blackwelder (1980).]

(Plate I). Within the lineage, species may be distinguished by a number of characters that show change or variation. We assume that much of the variability of the fossil populations is represented by our samples taken from single horizons at single localities (Appendix). Within most collecting localities a broad size range of *Chesapecten* shells may be found, representing many stages in a growth series within samples. Thus, ontogenetic information from a range of organism sizes within locality samples or from growth lines within individuals is easily obtainable. From these, patterns of character differentiation are inferred from individual development. Small to increasingly larger specimens within one locality horizon contain a record of the morphologic changes that accompany ontogeny in a single species. Comparisons of character state differences between species provide information for phylogenetic analysis. Thus the ontogenetic pattern within species may be compared directly with the phylogenetic pattern among species. We feel that it is reasonable to attempt this test with *Chesapecten* because it appears that we have included most or much of the preserved variability seen in this monophyletic group.

GEOLOGIC SETTING AND DEPOSITIONAL ENVIRONMENTS

Chesapecten specimens are abundant and well preserved in generally unlithified Coastal Plain sediments which are particularly well exposed along the west side of Chesapeake Bay. The gently northeast-dipping strata are exposed in Maryland, Virginia, and North Carolina. The formations belong to the Chesapeake Bay Group whose deposits are found from Chesapeake Bay to South Carolina (Gibson, 1970, Gernant *et al*, 1971, Ward and Blackwelder, 1980).

Sediments containing *Chesapecten* were deposited in marine environments. The two lower formations, the Calvert and the Choptank Formations, were lagoonal and shallow neritic deposits (Andrews, 1978; Ward and Blackwelder, 1980). Shorelines shifted southward and the St. Mary's and Eastover Formations were deposited during a series of marine transgressions from Late Middle to Early Late Miocene time (Gibson, 1970). The Yorktown Formation sediments were deposited while the Yorktown Sea covered eastern Virginia and parts of North Carolina. The sediments of the Yorktown Formation are represented by transgressive, maximal transgressive, stable, and regressive stages (Ward and Blackwelder, 1980). A list of the stratigraphic relationships of the formations and the depositional environments is given in Table I.

In this study, 17 samples of *Chesapecten* were obtained from 15

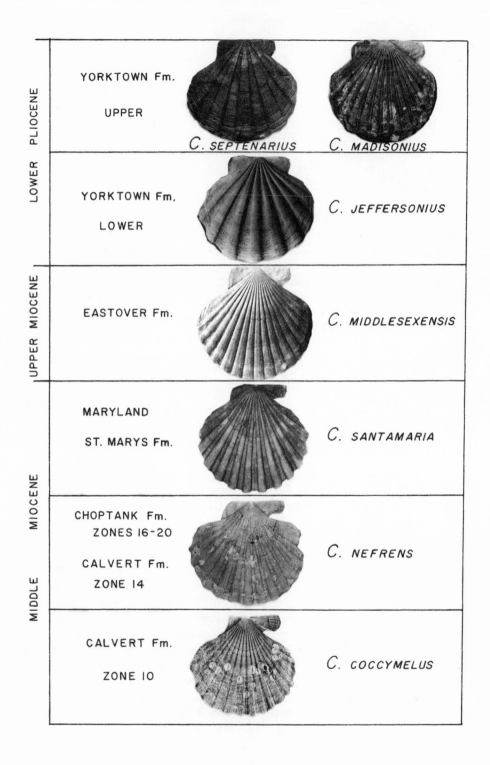

LOWER PLIOCENE	YORKTOWN Fm. UPPER	*C. SEPTENARIUS* *C. MADISONIUS*
	YORKTOWN Fm. LOWER	*C. JEFFERSONIUS*
UPPER MIOCENE	EASTOVER Fm.	*C. MIDDLESEXENSIS*
	MARYLAND ST. MARYS Fm.	*C. SANTAMARIA*
MIDDLE MIOCENE	CHOPTANK Fm. ZONES 16-20 CALVERT Fm. ZONE 14	*C. NEFRENS*
	CALVERT Fm. ZONE 10	*C. COCCYMELUS*

TABLE I. List of Formations and Depositional Environments[a]

Yorktown Formation:
  Upper
  Moore House Section. Contains three members with similar molluscan assemblages; Sample 6', 6", 7 come from this section
  a. Moore House Member. Shallowing regressive sea, higher current, and wave energy condition than previous member, normal salinity
  b. Mogart Beach Member. Normal marine conditions, shallower and quieter than Rushmere Member; clayey beds deposited behind barrier islands
  c. Rushmere Member. Maximum transgressive phase, open marine, shallow shelf, some tropical influence, fine, well-sorted shelly sand
  Lower
  Sunken Meadow Member. Normal saline, transgressive, shallow shelf, sandy, mild-warm temperate, less diverse than Cobham Bay; samples 5' and 5" come from this member.
Eastover Formation
  a. Cobham Bay Member. Shallow open marine embayment, well sorted, warm temperate to subtropical; all samples from the Eastover Formation come from this member, 4A, 4B, 4C
  b. Claremont Manor Member. Transgressive open marine sea, lower temperature, and fine sediment; lower diversity than the St. Mary's Formation samples from this member are rare; sample sizes from this member were too small to analyze and expect meaningful results
St. Mary's Formation. Transgressive, fine sand, well sorted, shelly; samples 2–3, 3', 3" come from this formation
Choptank Formation
  Boston Cliffs Member. Fossiliferous, shallow shelf, lagoonal, series of small embayments; sample 2 comes from this member
Calvert Formation zone 10–12
  Plum Point Marl Member. Shallow shelf lagoonal; samples 1', 1", 1''' come from this member

---

[a] From Andrews (1978) and Ward and Blackwelder (1980).

localities (Appendix). All individuals collected from a single stratigraphic horizon of a given locality comprise one sample and are our basic taxonomic units. They are in the museum collections of the Smithsonian Institution, Department of Paleobiology, and the U.S. Geological Survey at the National Museum of Natural History, Washington, D.C. Most of the samples were collected by Dr. Thomas Waller of the Smithsonian Institution. The samples from Windmill Point and Chancellor Point were collected by Dr. Thomas G. Gibson of the U.S. Geological Survey.

We feel that our samples represent much of the variability that could be found in the stratigraphic sequence. All species as defined by Ward

---

PLATE I. Stratigraphic succession of *Chesapecten* species as presented in Ward and Blackwelder (1975).

and Blackwelder (1975) are respresented by at least one of the 17 samples. Each sample includes representatives from the stage of the larval prodissoconch preserved on the umbonal beak of most valves, to the largest available shell size within a given sample horizon and locality. Continuous within sample variation in rib number indicates the samples are monospecific.

MATERIALS

Characters should reflect differences between homologous components. They are fundamental to cladistic analysis. Right valves were used because they show more species differences between the *Chesapecten* species than the left valves. However, in at least one pectinid species (*Pecten irradians*) the right valves have been shown to be less variable than the left (Davenport, 1900). The right valve also lies below the left in scallops, so it is the valve that interacts with the substrate and generally has features that we may associate with a specific life habit.

Time-specific species differences have been described by Ward and Blackwelder (1975), who defined the *Chesapecten* species as stratigraphic indicators (Plate I). The earliest species have more ribs or plications on the valve disk, a relatively narrow and long anterior auricle, and a deeper byssal notch than later species.

Scallop shells or valves have two major parts, the disk and auricles. Auricles delimit the dorsal portion of the valves. The disk refers to the main part of the shell exclusive of the auricles. Ribs or plications cover the disk (Plate II).

Character-state differences between the growth stages within and between samples of *Chesapecten* are changes in the relative shape and size of homologous components. To express and analyze the shape changes of characters, morphometric observations were made on each right valve as illustrated in Fig. 2. All linear measurements are to the nearest half millimeter; angular measurements are to the nearest degree. Fourteen measurements were used in this study (Fig. 2). Additional measurements were taken, but were either too constant or incomplete over samples.

Measurements

1. Height. The distance between the dorsal-most and ventral-most margins on the valve along the midline perpendicular to the hingeline (*AB*).

ples. The equation is

$$K_{ij} = \frac{u_{ij}}{(\sum_{i=1}^{n} SD_{ij})/n}$$

where i denotes the sample, j the character, u is the mean character value for each sample, n is the number of samples, K is the charcter condition of each sample, and $\sum SD_{ij}/n$ is the mean standard deviation for a character across samples. The resulting data table is shown in Table II.

Because there are ontogenetic changes in most characters with increasing size, a preliminary study was done to examine changes with increasing valve size (Miyazaki, 1978). It was found that in all species a byssal notch was present in specimens up to 30 mm in height. Above 30 mm, the notch was occluded in some of the species, but remained open in others (the earlier ones). We chose this size as a benchmark, and used specimens above 30 mm height in this study (designated as "adults"). Above 30 mm, character values for most characters show small change with increasing size. However, except for the earliest species, *C. coccymelus*, the mean size of "adults" did not differ substantially among

TABLE II. Data Matrix[a]

1	2	3	4	5	6	7	8	9	10	11	Sample	Fossil species
17	8	7	10	19	6	4	5	6	4	4	1'	*C. coccymelus*
17	7	8	10	19	5	3	3	6	2	4	1''	*C. coccymelus*
17	8	9	11	19	6	4	5	7	3	3	1'''	*C. coccymelus*
18	8	8	10	17	6	2	3	5	2	5	2	*C. nefrens*
18	8	7	10	18	6	2	3	5	2	4	2–3	*C. nefrens–santamaria*
18	7	7	11	17	8	1	2	5	3	4	3'	*C. santamaria*
18	7	7	10	17	7	1	2	5	3	3	3''	*C. santamaria*
17	8	8	11	19	6	3	3	5	3	4	4A'	*C. middlesexensis*
18	8	7	10	17	8	2	3	6	3	4	4A''	*C. middlesexensis*
16	9	9	10	18	6	5	4	5	2	4	4B	*C. middlesexensis*
18	9	7	10	17	8	2	3	6	2	4	4C'	*C. middlesexensis*
18	8	8	10	17	7	3	3	5	4	4	4C''	*C. middlesexensis*
18	8	6	11	16	9	1	3	7	4	4	5'	*C. jeffersonius*
18	8	6	11	16	8	2	3	7	4	4	5''	*C. jeffersonius*
17	8	7	11	16	7	2	4	9	8	3	6'	*C. septenarius*
18	8	7	11	16	8	1	3	9	8	4	6''	*C. septenarius*
19	8	7	10	16	8	2	2	7	2	4	7	*C. madisonius*

[a] Characters: 1, length/height; 2, angle; 3, dorsal half-diameter/anterior half-diameter; 4, anterior half-diameter/posterior half-diameter; 5, anterior auricle length/total auricle length; 6, depth of byssal notch/anterior auricle length; 7, rib number; 8, height anterior auricle/anterior auricle length; 9, height of the central rib/height; 10, convexity/height; 11, diameter of the combined smooth and striated muscle scar/height.

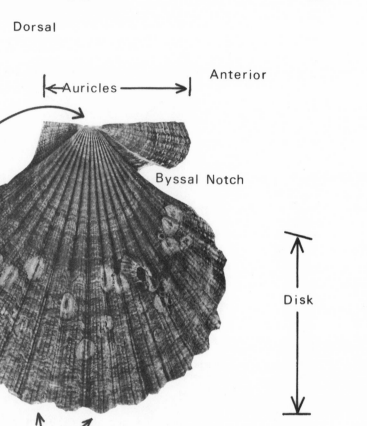

PLATE II.   Terms describing the general shell features of a scallop shell.

2. Length. The distance between the anterior-most and posterior-most points on the disk on a line parallel to the hingeline (*GH*).

3. Height of the anterior auricle. Along a line parallel to the midline, the auricle height is the distance from the most anteriodorsal to the most anterioventral point on the anterior auricle (*DE*).

4. Anterior auricle length. The distance along the outer dorsal ligament between the most anteriodorsal point on the anterior auricle to the beak of the umbo (*AD*).

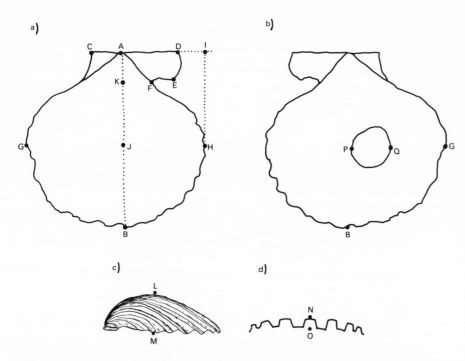

FIG. 2. Measurement points taken on right valves. In this study only specimens above 30 mm in height were used. (a) External view, (b) internal view, (c) side view, (d) ventral edge view.

5. Auricle length. The distance along the outer dorsal ligament between the most anterior and most posterior point of the auricles (*CD*).

6. Anterior half-diameter. The shortest distance between the anterior-most point on the disk and a point *J* along line line *AB* (*HJ*).

7. Dorsal half-diameter. The shortest distance between *H* and a point along the line coincident with the outer ligament (*HI*).

8. Posterior half-diameter. The shortest distance between the posterior-most point on the disk (*G*) to a point along line *AB* (*GJ*).

9. Byssal notch occlusion. As measured from the midline, the shortest distance between the point where the posterioventral edge of the anterior auricle intersects the disk (*F*) to a point (*K*) on line *AB* (*KF*).

10. Rib number. The number of shell plications on the disk margin.

11. Umbonal angle. The angle formed by the umbo along the edges of the disk as measured with a goniometer ($\sphericalangle$).

12. Convexity. The height of the shell from the plane of commissure to the outermost point of the disk exterior (*LM*).

13. Height of the central rib. As measured on the ventral margin of

the disk, the distance from the base to the top of the central plication (*NO*).

14. Diameter of the combined smooth and striated adductor muscle scar (*PQ*).

CHARACTERS

Most measurements were combined into ratios to factor out size variations and to describe the shape of the disk, auricles, and ribs. Umbonal angle and ribs were not transformed. A total of 11 characters were used.

1. Length to height ratio (*GH/AB*). This ratio provides information about the relative elongation of the valve.

2. Umbonal angle ($\sphericalangle$). The umbonal angle gives an absolute measure of the total flare of the disk.

3. Dorsal half-diameter to anterior half-diameter ratio (*HI/JH*). This is a measure of change in anterior flare, which in scallops may change independently of posterior flare.

4. Anterior half-diameter to posterior half-diameter ratio (*GH/GJ*). This is a measure of valve symmetry. Values greater than one commonly occur in scallops that are byssally attached.

5. Convexity to height ratio (*LM/AB*). This is a measure of relative convexity.

6. Diameter of muscle scar to height ratio (*PQ/AB*). This is a measure of proportional changes in the size of the muscle scar.

7. Anterior auricle length to total auricle length ratio (*AD/CD*). This character measures relative symmetry of the auricles.

8. Height of the anterior auricle to anterior auricle length ratio (*DE/AD*). A relatively high value of this ratio is an indication of byssal notch occlusion.

9. Depth of byssal notch to anterior auricle length ratio (*KF/AD*). A larger ratio indicates that the byssal notch is becoming small relative to the anterior auricle.

10. Rib number.

11. Height of the central rib to valve height ratio (*NO/AB*). An increase in this ratio indicates an increase in the sinusoidal disk margin inscribed by the ribs.

The mean value for each character in each sample was divided by the average standard deviation for all samples rather than by individual sample standard deviations to remove bias potentially introduced by the incomplete representation of character variability within some of the sam

samples (see Appendix). Therefore the character state changes we observed among species cannot be due to a simple size–character value relationship that obtains for all species.

METHOD

For the purposes of this study, it was necessary to use a method for constructing phylogeny which does not require assumptions or preconditions about the possible nature of the evolutionary process. Phylogenetic systematic or cladistics is the science of biologic classification which attempts to construct phylogenies objectively without reference to underlying assumptions about the pattern of changes; its purpose is to maximize correspondence between the observed characters and the cladogram, united under the parsimony criterion. Mickevich (1981) has discussed a refinement of the parsimony criterion called "best fit" correspondence. The best fit cladogram is fulfilled by those cladograms that describe the greatest number of characters in the fewest steps, and also that are constructed such that minimal prior information or assumptions about character change or polarity is required in forming the cladogram. Mickevich (1981, 1982) has described a procedure for obtaining a best fit cladogram called transformation series analysis.

Transformation series of character state changes as they are connected on the cladograms are referred to as cladistic transformation series (Mickevich, 1981) or cladistic transformations. An individual character transformation on the best fit cladogram would simply show the sequence of character states as they are linked on the cladogram. According to the best fit criterion in transformation series analysis, the number of *shared* character states should be maximized on the cladogram. The shared state may transform along the cladogram independent of the implied or actual numerical value of the character states. For example, our size-standardized characters may have states of 10, 11, 12, 13, 14, whose numerical order has a sequence which might be considered a transformation ordering. However, transformation series analysis does not use this implied ordering. Therefore it is possible to have a cladistic transformation for such a character of 10-14-12-11-13 that is perfectly compatible with the best fit cladogram. Transformation series analysis thus permits heterochronous character state changes, in accordance with de Beer's model, although best fit ordering of shared states could as easily follow the numerical sequence. This aspect of the flexibility of this technique is important to bear in mind when considering our test of the Biogenetic Law.

A summary of transformation analysis follows;

I. Begin with any hypothesis of phylogenetic relationships for the group, or any set of hypotheses concerning the nature of character evolution, and from this derive a cladogram. All cladograms are most parsimonious trees or Wagner trees (Farris, 1970).

II. From this cladogram determine each transformation series for all characters. The character state transformations may be reordered in this step and need not concur with the previous hypothesis of character state relationships. This procedure is detailed in Mickevich (1982).

III. Calculate a new cladogram from the new set of character transformations.

IV. If in this first pass the new cladogram is different from the previous cladogram, go to II. If both cladograms are identical, stop.

Mickevich (1982) has demonstrated that the solution cladogram resulting from transformation series analysis does not depend on the initial set of character hypotheses. This is an iterative procedure, where identical solutions are obtained despite opposite or near opposite starting hypotheses. Therefore the accuracy of the initial set of character relationships used in step I does not affect the final result. However, it is possible that many cladograms and their corresponding sets of transformations may cycle between steps II and III. Stability cycling of this kind indicates that more than one solution cladogram is possible for the character data. Under such circumstances, one chooses the solution cladogram that explains the most characters according to the "best fit" criterion, minimizing the number of parallelisms and convergences on the cladogram.

RESULTS

Phylogenetic Relationships

The cladogram (Fig. 3) of *Chesapecten* relationships is the result of transition series analysis on the characters. This cladogram was chosen from one of several computed networks by the best fit criterion. When the network is aligned with the stratigraphic fossil sequence, little topographic distortion is required for the network to conform to the fossil sequence (Fig. 4). Cladistic networks may be transformed into cladograms by being "rooted." Every criterion for determining the ancestral group is met by *C. coccymelus* (stratigraphic, ontogenetic, outgroup, smallest, least differentiated).

The cladistic relationships of *Chesapecten* are largely concordant with Ward and Blackwelder's stratigraphic species designations. The re-

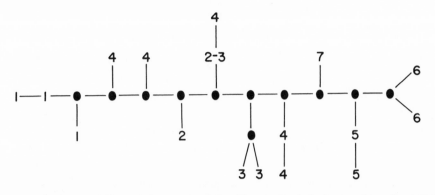

FIG. 3. Cladogram of *Chesapecten* sample relationships constructed using transformation series analysis. Numbers refer to stratigraphic positions of separate samples; however, numbers 6 and 7 occur together.

lationships of *Chesapecten* (Fig. 4) on the cladogram show that the first paleontologic species is linked on the cladogram to the second, third, and so on for most of the tree. Even the sample 2–3 recognized by Ward and Blackwelder (1975) as a transitional form between *C. nefrens* (2) and *C. santamaria* (3) does indeed fall between those groups on the cladogram obtained by transformation series analysis. The cladistic representation of *Chesapecten* evolution is thus strongly corroborated by the fossil record.

The major inconsistency between the fossil and cladistic trees involves the three cladistic lineages of the single stratigraphic species *C. middlesexensis*. The name *C. middlesexensis* has been applied to *Chesapecten* fossils found in the sediments of the Eastover Formation. Ward and Blackwelder's (1975) species description of *C. middlesexensis* is broad enough to cover the range of forms found in that formation. However, they acknowledge that the forms are highly variable (also Dr. Thomas Gibson, personal communication). Samples of *C. middlesexensis* appear in three places on the cladogram; for convenience we will refer to them as 4A, 4B, and 4C. The first branch (4A) lies between the samples of *C. coccymelus* (1) and *C. nefrens* (2), one sample (4B) forms a group with a probable intermediate between *C. nefrens* and *C. santamaria* (2–3), and a third group (4C) of *C. middlesexensis* is derived from *C. santamaria* (3), its immediate stratigraphic predecessor. The point where *C. madisonius* (7) branches from the main lineage might also be interpreted as an inconsistency if one assumed its immediate stratigraphic predecessor (*C. jeffersonius*) was its closest relative. *Chesapecten madisonius* appears on the directed cladogram as a group closest to *C. middlesexensis* (4C), not *C. jeffersonius*.

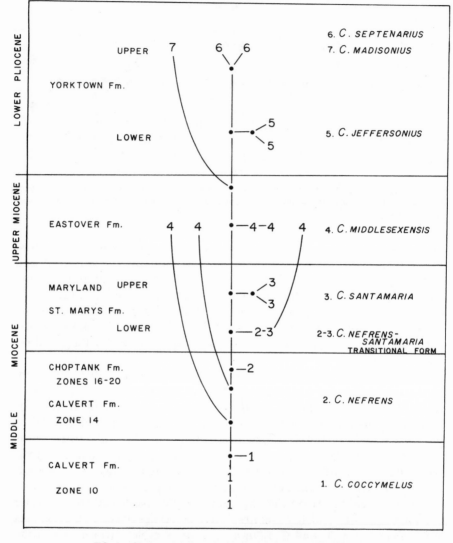

FIG. 4. Cladogram aligned with the stratigraphic sequence.

Phylogenetic and Ontogenetic Character Transformations

The cladogram defines the phylogenetic character state transformations. These character state relationships (A), as they relate on the cladogram, are shown in a side-by-side comparison with the ontogenetic character state transformations (B) that occur in all species (Figs. 5–12 and 14–16; see Appendix for symbols and localities).

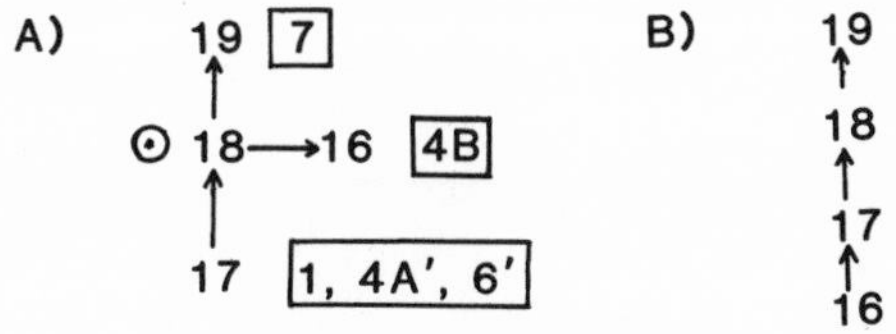

A)   19 ⃞7   B)   19
       ↑
⊙ 18⟶16 ⃞4B   18
       ↑   ↑
                              17
    17 ⃞1, 4A', 6'   16

FIG. 5.   Length/height transformations in on-togeny and phylogeny.

1. Length/height ratio. Length increases relative to height in onto-geny (Plate III). In phylogeny this character is not very variable (Fig. 5). However, the ancestral species (i.e., all *C. coccymelus* and one sample of *C. middlesexensis*, 4A') have a relatively high valve and most derived species are longer. Another highly derived sample of *C. septenarius* (6') is higher valved, representing a reversal in this character.

2. Umbonal angle. Umbonal angle increases during ontogeny (Plate IV). The phylogenetic trends show little variability (Fig. 6). But at least one sample of *C. coccymelus* (1') and both samples of *C. santamaria* are rather narrow, whereas *C. middlesexensis* (4B) is quite broad.

3. Dorsal half-diameter/anterior half-diameter ratio. During onto-geny the dorsal half of the valve decreases relative to the anterior half of the valve. There are multiple reversals in the phylogenetic trend (Fig. 7) in this character, but the earlier species have longer dorsal half-di-ameters, and *C. jeffersonius*, a derived species, has a relatively small dorsal half-diameter.

4. Anterior half-diameter/posterior half-diameter ratio. This char-acter measures the relative symmetry of the valve on either side of the midline. In this study, relative symmetry is not useful in testing the Bio-genetic Law, since there is little ontogenetic or phylogenetic trend in this character (Fig. 8).

5. Anterior auricle length/total auricle length ratio. During individual development this ratio decreases, indicating that the anterior and posterior auricles are becoming equal (Plate V). In phylogeny there is also a tend-ency toward decreasing values of this ratio (Fig. 9 and Plate VI).

6. Depth of the byssal notch/anterior auricle length ratio. The size of the byssal sinus decreases with ontogeny (Plate V). The ontogenetic and phylogenetic trends in this character are identical (Fig. 10 and Plate VI).

7. Anterior auricle height/anterior auricle length ratio. The value of

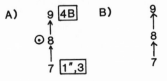

FIG. 6. Umbonal angle transformations in ontogeny and phylogeny.

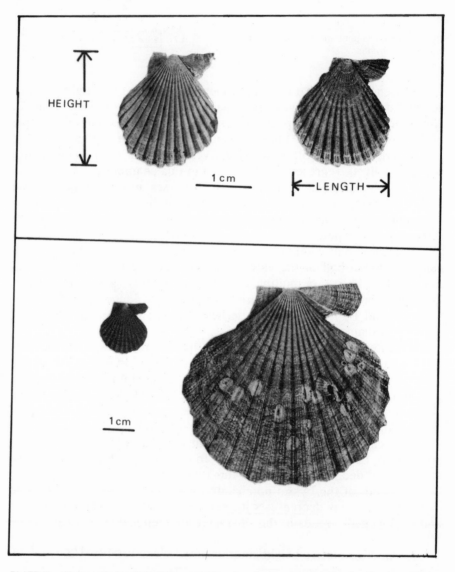

PLATE III.   Changes in length/height proportions in ontogeny, illustrated using *C. coccymelus.*

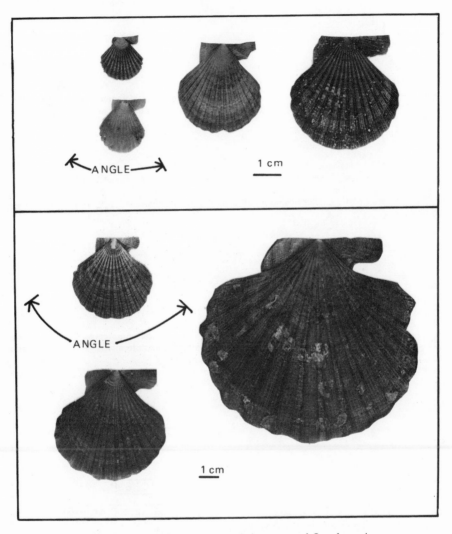

PLATE IV.   Changes in umbonal angle in a range of *C. nefrens* sizes.

this ratio tends to increase during individual development (Plate V). The phylogenetic trend does not coincide with ontogeny (Fig. 11) but the two *C. coccymelus* samples as the ancestral group, which did have rather narrow anterior auricles.

8. Rib number. In later stages of ontogeny, rib number decreases (Fig. 12) by loss at the margins of the disk (Fig. 13). In phylogeny rib number also tends to decrease; however, there is a major reversal in-

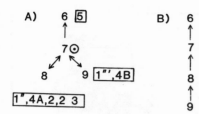

B) 6
    ↑
    7
    ↑
    8
    ↑
    9

FIG. 7. Dorsal half-diameter/anterior half-diameter transformations in ontogeny and phylogeny.

FIG. 8. Anterior half-diameter/posterior half-diameter transformations in phylogeny.

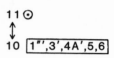

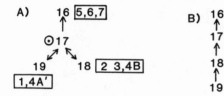

B) 16
    ↑
    17
    ↑
    18
    ↑
    19

FIG. 9. Anterior aurical length/total auricle length transformations in ontogeny and phylogeny.

FIG. 10. Depth of byssal notch/anterior auricle length transformations in ontogeny and phylogeny.

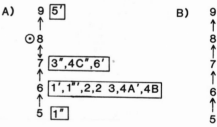

B) 9
    ↑
    8
    ↑
    7
    ↑
    6
    ↑
    5

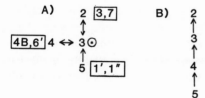

B) 2
    ↑
    3
    ↑
    4
    ↑
    5

FIG. 11. Anterior auricle height/anterior auricle length transformations in ontogeny and phylogeny.

PLATE V. Changes in auricle and byssal notch proportions with ontogeny, shown using *C. jeffersonius.*

volving *C. middlesexensis* (4A). Though rib number decreases with ontogeny, the population variance was higher than the loss with ontogeny, so this character was not size-transformed.

9. Height central rib/height of valve ratio. We discern no trend in ontogeny, but in phylogeny, rib height increases, particularly in later species (Figs. 13 and 14).

PLATE VI. Phylogenetic differences between small and large specimens of early and later species. For comparison purposes, *C. nefrens* is used in place of *C. coccymelus* because adults of this species do not become as large as *C. jeffersonius*.

10. Convexity/valve height ratio. Character state changes in convexity were not informative in constructing this cladogram, so transformation series analysis ignored this character, nor is any ontogenetic sequence recognized (Fig. 15).

11. Diameter muscle scar/valve height ratio. The size of the muscle

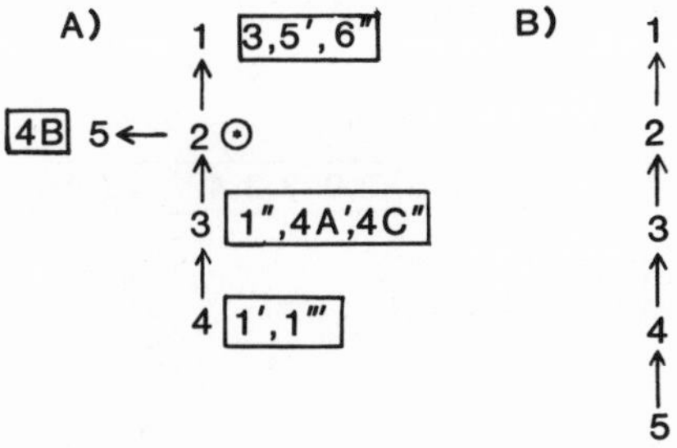

A)   1 $\boxed{3,5',6''}$ B)   1
             ↑   ↑
$\boxed{4B}$ 5← 2 ⊙   2
             ↑   ↑
          3 $\boxed{1'',4A',4C''}$ 3
             ↑   ↑
          4 $\boxed{1',1'''}$ 4
                                        ↑
                                        5

FIG. 12. Rib number transformations in ontogeny and phylogeny.

scar relative to valve height increases with individual development (Fig. 16). In phylogeny this character is of little importance, because this character changes similarly in all taxa (Fig. 17).

Testing the Biogenetic Law

Using transformation series analysis, the computed cladogram and character state transformations may then be compared against the stratigraphic record and Haeckel's Law. A test of association was performed between the directed cladogram and the stratigraphic ordering of species. The number of cladogenic (branching) events between the root (*C. coccymelus*, 1') and each taxon sample and the stratigrahic sequence of formations were compared using Spearman's rank correlation test. The resulting r_s value is 0.79, which is significant at $p < 0.001$. Hence, despite the inconsistences between the fossil sequence and the cladogram, we find that the practice of arranging fossils according to their stratigraphic position is, in the present case, highly correlated with the most parsimonious solution cladogram. Interpretation of phylogenetic relationships

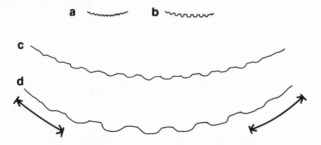

FIG. 13.   Ventral edges of *Chesapecten* right valve. (a, c) *Chespecten nefrens.* (b, d) *Chesapecten jeffersonius.* Both small specimens and both large ones are from individuals of approximately the same height.

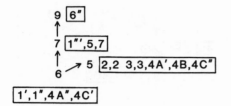

FIG. 14.   Height of the central rib/height transformations in phylogeny.

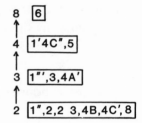

FIG. 15.   Convexity/height transformations in phylogeny.

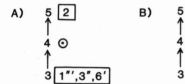

FIG. 16.   Relative muscle scar positions in a range of *C. nefrens* sizes.

FIG. 17. Combined smooth and striated adductor muscle scar/height transformations in ontogeny and phylogeny.

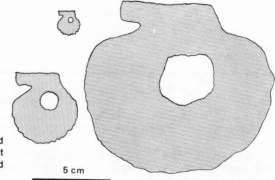

5 cm

from the fossil record, however, is obviously dependent upon the quality and extent of preservation. Sufficient sample density and diversity are required. Appropriate records are probably more prevalent among marine groups, but even in this case, especially for nearshore species, the fossil record is likely to be interrupted during sea regressions. Although the *Chesapecten* fossil record periodically is interrupted, the overall relationships of the taxa on the cladogram and the stratigraphic sequence agree independently. The major uncertainty between the two are the affinities of the *C. middlesexensis* samples.

Once a relatively reliable phylogeny is constructed, tests disclosing the nature of the evolutionary pattern may be performed. The cladistic character state transformations may be used as an independent test of Haeckel's Law. Transformation series show branching patterns of character state change along the cladogram. Character state transformations on the cladogram and during individual development provide comparison data for testing the applicability of the Biogenetic Law to phylogeny construction. We reiterate that transformation series analysis constructs phylogenies independently of the numerical sequences of the character states.

Three aspects of Haeckel's and Nelson's versions of the nature of the relationship between evolution and development were investigated. In the first, we wish to determine whether character state changes in phylogeny are randomly associated or tightly correlated with ontogenetic transformation. This test of association is undirected, since even during reversals the developmental sequence might be useful when forming taxon relationships if character state shifts follow the ontogenetic axis and do so in short or one-step character state changes. For example, in a random model a character that changes with ontogeny from states 1-2-3-4 may change between taxa as one-step (1-2, 2-3, 3-4), two-step (1-3, 2-4), or three-step (1-4) changes. A frequency distribution of these random ex-

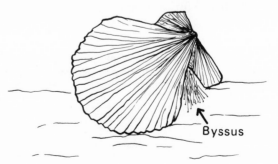

FIG. 18. Illustration of a byssally attached scallop.

Byssus

pectation character associations can be constructed (Table III). From the ontogenetic transformations for each character, the expected number of each pairwise class was noted (Table IV). These expected classes were then compared to the sequence of transformations observed for each character in phylogeny. The expected and observed values for *Chesapecten* are shown in Table IV. The chi-square value is 9.25, which has a probability level of <0.01. The association is nonrandom with respect to the developmental sequence. Therefore we conclude that the study of an ontogenetic sequence for a character within an individual taxon yields information for character evolution for that taxon's monophyletic group.

In the second case, we used the character transformation series to test Nelson's prediction that polarity may be imparted to the cladistic network by using ontogenetic character transitions. Nelson (1973) suggested using character ontogenies to impart actual polarity first to the phylogenetic character state transformations and then to a single taxon whose adult features are most associated with early ontogeny. The samples that possessed the earliest developmental conditions were simply noted and compared (Table V). The combined samples of *C. cocccymelus* possess seven out of eight of the most juvenile-like states for characters that have a recognizable ontogenetic transformation series. Only the length/height ratio character does not indicate *C. coccymelus* as the ancestral taxon. A binomial test under the frequency distribution where $p = q = 0.5$ for 7 out of 8 positive associations is significant at the 0.035 level for a one-tailed test. Therefore we conclude that Nelson's general restatement for individual characters is correct for this example. The most juvenile-like character states are most likely the most plesiomorphic character states.

Finally, Haeckel's theory of recapitulation stated that total adult morphology is transformed by terminal additions onto the developmental sequence. A simple test of Haecke's Law as applied to *Chesapecten* was performed by using ontogenetic character transformations in comparison with the character data for each sample. In order to do this the ontogenetic sequence for each character was ranked so that the earliest developmental

TABLE III. Transformation Series Observed If
Phylogenetic Transformations Were Random with
Respect to the Ontogenetic Axis (Undirected)

Transformation classes	Number	Frequencies
One step (1–2, 2–3, 3–4)	3	0.500
Two step (1–3, 2–4)	2	0.333
Three step (1–4)	1	0.167

TABLE IV. Least Restrictive Model, No Polarity of Character Transformation Is Assumed

Steps in character change	Frequency under random expectation	Expected	Observed
One step differences	0.48	10.08	17
Two step differences	0.32	6.72	3
Three step differences	0.20	4.2	1

$\chi^2 = 9.25$, d.f. $= 2$, $p < 0.01$

states were coded with a value of 1 and subsequent states were coded with successively higher values. For example, for the character length/height, the implied ontogenetic transformation is $16 \rightarrow 17 \rightarrow 18 \rightarrow 19$ and recoded these transitions become $1 \rightarrow 2 \rightarrow 3 \rightarrow 4$. All the characters that undergo recognizable ontogenetic changes were recoded in this fashion and the sample data (Table II) values were reassigned these ontogenetic codes (Table VI). Following a strict interpretation of Haeckel's Law to

TABLE V. Samples Whose Adult Character States Are Most Juvenile-Like[a,b]

Sample	1	2	3	4	5	6	7	8
1'				§		§	§	
1"		§		§	§			
1'''			§	§		§	§	§
2								
2-3								
3'		§						
3"		§						§
4A'				§				
4A"								
4B	§		§					
4C'								
4C"								
5'								
5"								
6'								§
6"								
7								

[a] Denoted by §.
[b] Characters: 1, length/height; 2, angle; 3, dorsal half-diameter/anterior half-diameter; 4, anterior auricle length/total auricle length; 5, byssal notch depth/anterior auricle length; 6, anterior auricle height/anterior auricle length; 7, Rib number; 8, Muscle scar diameter/height.

TABLE VI. Recoded Data Matrix with Respect to Degree of Ontogenetic Derivation for Individual Characters[a]

Sample	1	2	3	4	5	6	7	8	Sum[b]
1'	2	2	3	1	2	1	2	2	15
1"	2	1	2	1	1	3	3	2	15
1'''	2	2	1	1	2	1	2	1	14
2	3	2	2	3	2	3	4	3	22
2–3	3	2	3	2	2	3	4	2	21
3'	3	1	3	3	4	4	5	2	24
3"	3	1	3	3	3	4	5	1	23
4A'	2	2	2	1	2	3	3	2	17
4A"	3	2	3	3	4	3	4	2	24
4B	1	3	1	2	2	2	1	2	14
4C'	3	3	3	3	4	3	4	2	25
4C"	3	2	2	3	3	3	3	2	21
5'	3	2	4	4	5	3	5	2	28
5"	3	2	4	4	4	3	4	2	26
6'	2	2	3	4	3	2	4	1	21
6"	3	2	3	4	4	3	5	2	26
7	4	2	3	4	4	4	4	2	27
Total number of states for each character	4	3	4	4	5	4	5	3	

[a] See Table V, Footnote b for key to characters.
[b] Sum across all characters for each sample. Highest possible total is 32.

determine relative developmental condition across all characters, ontogenetic character codes were summed for each sample and the total values were compared (Table VI). A sample with relatively adult characteristics (i.e., most character states with values of 3 or 4) will have a higher sum total value than a more juvenile-like form (i.e., whose character state values are primarily 1's and 2's). The amount of phylogenetic taxon derivation was determined by counting the number of internodes (branch points) between each taxon and the root of the tree (*C. coccymelus*, 1'). A test of association using Spearman's rank correlation was performed by comparing the cladistic network of taxon branching (cladogenic branching) events and the sum-ranked measure of total ontogenetic derivation for each taxon.

The Spearman rank correlation is 0.679 (df = 15). The probability that this occurred by chance alone is significant at less than the 0.01 level. Therefore we can conclude that the strict interpretation of Haeckel's Biogenetic Law applies to cladogenesis as it proceeds through the accumulation of terminal additions of new ontogenetic stages in *Chesapec-*

ten. This trend is probabilistic. There are some taxa that have evolved more juvenile-like morphologies than their ancestors. This would classically be called neoteny, yet these occurrences are not so abundant as to overwhelm the predictive power of Haeckel's Law. The overall morphology of later *Chesapecten* taxa is more adult in form than the ancestral taxa.

DISCUSSION

Stratigraphic Relationships versus Cladograms

Classically, stratigraphy and comparative study of homologous organs aid our determination of evolutionary patterns and relationships. Our results show that the stratigraphic sequence of fossil taxa and cladistic transformation series analysis on homologous characters are significantly congruent. These are independent ways for determining sequences and both are independent of the ontogeny criterion. We have tried to determine the best possible phylogeny in order to examine the pattern of character evolution and its possible association with the Biogenetic Law.

Is the parsimony, or best fit criterion, truly reflecting the sequence of events in *Chesapecten* evolution? Ward and Blackwelder (1975) designated species for their value as stratigraphic indicators. A range of forms, even though possibly polyphyletic, which can generally be distinguished from earlier and later fossil species may still be useful as age indicators. However, in an evolutionary sense the stratigraphic sequence may also be correct. *Chesapecten middlesexensis* may have diversified in the process of reversals and accelerations over the relatively long duration (4 my) of Eastover Formation sediment deposition. More work remains to be done on the stratigraphic relationships of the various morphotypes of *C. middlesexensis*. This species has a broad geographic range, extending as far south as Florida. The assumption that reversals and accelerations within the period of Eastover deposition were the cause of the variable forms of *C. middlesexensis* would cause us to use a cladogram that would not represent the most parsimonious solution for the character data. A conflict is thus raised between the parsimony criterion and the stratigraphic sequence.

Under ideal conditions of preservation through time, it would seem that fossil evidence should be the most useful guide to the phylogenetic relationships between taxa. However, the quality of fossil data may be poor because their depositional environments may not have been constant. As a result, it is often difficult or unreliable to infer the geneology of taxa by observing their stratigraphic sequence. Conceivably, as well, two stratigraphic sequences at even moderately separated localities may present differing patterns of taxon relationships, particularly if the taxa in question are rare. One of the attractive features of the *Chesapecten* lineage is that it is abundant and well preserved in the fossil record. However, some of the formations were ecologically complex, which in turn makes it more difficult to read phylogeny directly from the fossil sequence. For the present study, we did not do detailed stratigraphic analysis and have had to rely on published accounts and discussions with experts on the area. However, even if we knew the precise stratigraphic location of each sample in relation to each other, discrepancies or even complete concordance with the cladogram would not necessarily mean that the stratigraphic superpositioning of sample relationships is incontrovertible. Taxon relationships constructed according to the parsimony criterion, on the other hand, make use of a single interpretation (the shortest network), based strictly on the morphologic data of the individual taxa using their shared derived similarities. The cladogram is a representation of the simplest explanation of all the morphologic data at hand. The simplicity and lack of ambiguity in understanding the factors or criteria used in constructing taxon relationships by cladistic techniques make the evolutionary interpretations derived by this method more useful in our analysis.

Acceptance of the cladistic interpretation, however, requires that we assume that there are "gaps" in the fossil sequence, particularly in the lineage leading to *C. middlesexensis* (4A) (see Fig. 4). That lineage apparently emerged from an ancestor between the first and second stratigraphic species *C. coccymelus* and *C. nefrens*. Ancestral forms of *C. middlesexensis* (4A) do not appear in the fossil collections we used from the Choptank Formation, where *C. nefrens* is found, nor from the St. Mary's Formation, where *C. santamaria* is found. The *C. middlesexensis* (4B) branching from the transitional form of *C. nefrens-santamaria* (2-3) may also indicate a sampling gap in the St. Mary's Formation. *Chesapecten*-rich St. Mary's localities are so rare that gaps in lineage representatives from this formation are neither surprising nor disturbing to us. There are many more Choptank Formation localities where abundant *C. nefrens* are found and no other forms are recognized, but this does not preclude the possibility of a gap in ancestral representation of *C. middlesexensis* (4A). Gibson's (1970) study of the depositional basins of the Chesapeake Group sediments showed that the Choptank and St. Mary's Formations are considerably more restricted in geographic extent than the earlier and later formations. The size of their basins was determined not only from the extent of exposure, but also the depth of sediment deposition. Although no firm conclusions can be drawn from this fact, it is still of interest to mention that deposition was restricted during the times when the proposed gaps would have occurred.

The variability between paleontologic *C. middlesexensis* samples arose either in time throughout the Early Middle Miocene or in space, within the Eastover Formation. The cladogram indicates that time might be the more important factor, since some *C. middlesexensis* (4A) are more closely related morphologically to a shape intermediate between the first and second stratigraphic species. Group B is linked to a transitional shape between the second and third stratigraphic species. A third group (C) morphologically succeeds its immediate stratigraphic ancestor. In contrast, the presence of all three in the Eastover Formation indicates that the variability arose in space. Since we cannot prove the relative validity of one over the other, the choice must be one of preference, and for the reasons stated above we opt for the most parsimonious solution. Further details might only be elucidated by new finds.

Chesapecten madisonius from the Lower Pliocene is more closely related to *C. middlesexensis* (4C) than its immediate stratigraphic predecessor *C. jeffersonius* of the lowermost Pliocene. A sampling gap in the Late Miocene is a plausible explanation for this. The "gap" in fact may no longer exist. Dr. Thomas G. Gibson of the U.S. Geological Survey has detailed samples from this interval which show apparent intermediate forms between *C. middlesexensis* and *C. madisonius* (personal communication).

We find that the results of parsimony analysis when compared to the fossil record have enhanced our understanding of the complexity of the *C. middlesexensis* forms and may be supported with regard to the relationships of *C. madisonius*. Thus we feel *Chesapecten* evolution as represented in the cladogram (Fig. 4) is reasonably correct and is strongly supported by the fossil record.

Ontogeny versus Phylogeny

From what we consider to be a reasonably well established phylogeny, the individual character state transformations could be investigated more closely. The test results of the usefulness of the ontogenetic axis, Haeckel's Law, and Nelson's version of his law to phylogeny construction were all significant.

The positive association between character state transformations and the ontogenetic axis showed that when changes in character state occurred, they generally did not undergo as many larger step transformations as would occur if the transformations followed a random pattern. The taxa showed more small differences in character state change than would occur in a random manner. Thus stepwise ontogenetic character transformations indicated the degree of relatedness in phylogenetic character state transformation. The diverse heterochrony patterns which de Beer (1958) had suggested would tend to obscure recapitulating patterns or hypermorphosis did not dominate evolution in this group.

As Nelson (1973) suggested, juvenile character states were useful in imparting polarity to the cladogram. Most characters that are characteristic of early ontogeny indicate *C. coccymelus* as the ancestral taxon. This concurs with its stratigraphic position, size, and degree of morphologic derivation for assigning it as the root of the cladogram.

The test of Haeckel's prediction of evolution through terminal addition was also significant even though reversals did occur, particularly in samples from the St. Mary's and Eastover Formations. The prevalence of small-step character state transitions and the general pattern of terminal additions causes intermediate taxa to appear intermediate in morphology. This pattern of transformation led Miyazaki (1978) to conclude that evolution in this group was gradual. However, depending upon how one defines the term gradual or even how one defines a species, this conclusion may be open to debate. We do not perform a test of association against time in this study, nor attempt to redefine the species, recognizable as fossil taxa, because we feel that more refined collections especially of *C. middlesexensis* and further analysis is required before we will enter into this debate (e.g., Levinton and Simon 1980). If the cladogram more truly represents *Chesapecten* evolution, the early branch of *C. middlesexensis* (A) might be considered a long-lived species. Smooth morphological transformations (i.e., small changes between species along the developmental axis) are recognized from *C. coccymelus* to *C. middlesexensis* (C) and from *C. middlesexensis* (C) to *C. septenarius*. Further information may be obtained from more detailed investigation of the *C. middlesexensis–C. madisonius* transition. Whether the species in this lineage arose quickly or slowly, however, is irrelevant to the fact that the individual character morphologies between our samples have usually changed from one developmental stage to the next along the ontogenetic axis such that intermediate morphologic forms appear successively in phylogeny. Jackson (1890) and Newell's (1937) observations of recapitulating patterns in pectinids is supported by this study of a single monophyletic pectinid lineage.

Phyletic Size Increase

Newell (1949) has suggested that phyletic size increase results in the tendency to observe recapitulation or hypermorphosis in fossil invertebrates. Phyletic size increase, also known as Cope's Rule (Cope, 1885, 1896), has been recognized in many groups, including bivalves (Jackson, 1890; Newell, 1937, 1942), and the *Chesapecten* lineage is no exception to this rule. The earlier *Chesapecten* species were light and small; later species are large and heavy, although *C. septenarius* shows a slight reversal in this trend. We must emphasize that the species differences are not simple allometric differences, but include changes in the timing of the appearance of new character states; generalized increased shell size and weight accompany these timing changes. Characters, for example, byssal notch occlusion, appear at successively earlier stages in development (see Plate IV). Also, with the exception of the *C. coccymelus* samples, sample means for height are overlapping; the *maximum* values for successive taxa on the cladogram undergo a generalized increase (see Appendix).

Phyletic size increase occurs in such diverse groups that general explanations have been sought to account for this phenomenon. Newell (1949) and Stanley (1973) proposed that selection for increased size or changes in timing of reproductive age probably results in increased evolutionary size. The oyster stock *Cubitostrea sellaeformis* (Stenzel, 1949) is similar to *Chesapecten* in that it also undergoes progressive size increase. The early paleontologic species are small and juvenile in appearance, while later species are larger and attain a more "mature" appearance with development. Stenzel (1949) believed that phyletic size increase progressed in "enormously large interbreeding populations." He felt that intraspecific competition was high in large populations and thus the advantage of rapid growth to large sizes was transmitted to successive generations. In a contrasting paper, covering a number of groups, Boucot (1976) has argued that low-density interbreeding populations may increase rates of phyletic evolution which in turn are positively correlated with phyletic size increase.

We cannot offer much information about the selective pressures which might have been imposed on *Chesapecten*. However, information on the depositional environments and life habit of these fossil species can be inferred; possibly these factors could be of interest in observations of the evolutionary pattern of the lineage.

The succession of depositional environments containing this lineage were not constant over time (Table I). The earliest environments were shallow and nearshore sequences, whereas the later fossiliferous basins were deeper. This general shift from shallow to deeper sequences in time may have some bearing on evolution in this group. In order to present this argument clearly, it is now useful to briefly discuss changes in life habit and functional morphology which occur during scallop development.

Life Habit

Scallops may undergo morphologic changes during growth, with corresponding changes in life habit. All known scallop larvae and juveniles attach to the substratum by thin threads called the byssus (Merrill, 1961; Stanley, 1970; Kauffman, 1969) (Fig. 18). The byssal notch is the shell feature that marks the position of the byssus (Plate II). Some scallop species retain the byssus throughout their lives, although other species, after reaching a certain size, abandon the byssus and lie on their right valves upon the sediment surface.

Very large scallop shells, too large to be supported by the byssal threads, generally lose the byssal notch with growth. A prominent notch was noted in all *Chesapecten* shells up to approximately 30 mm in height, when, depending on the species, the notch may either persist or become lost. Since this height marks an important transformation toward adult morphologies, only specimens above 30 mm in height were used in the cladistic part of this study.

At larger sizes, scallops are apparently heavy enough to remain stable under normal current conditions. By utilizing a natural cleansing reaction of valve clapping, many scallop species became "swimmers" by regulating the points at which water will be ejected from between the valves (Yonge, 1936; Fairbridge, 1953; Stanley, 1970; Gould, 1971). Although many shell characteristics are believed to influence swimming efficiency (Waller, 1969; Thayer, 1972), the presence of a relatively broad, flat right valve seems useful as a hydrofoil (Gould, 1971). With even further growth, swimming efficiency tends to be lost due to the sheer weight of the growing shell and the individuals become sedentary (Waller, 1969; Gould 1971). Morton (1980) has documented this for the *Amusium* group, which are among the best swimmers in the Pectinidae.

In phylogeny, a fairly clear trend of species shell morphologies parallels these changes in life habit. The adults of the early species, particularly *C. coccymelus*, have shell characteristics typical of byssally attached species. *Chesapecten nefrens* and *C. santamaria*, although they retain the byssal notch as adults, grow to be much larger than their *C. coccymelus* ancestor. They also have the light, broad, and flat right valve that Gould (1971) has suggested would make an efficient hydrofoil for swimming scallops. It appears that these species, although retaining some ability to attach to the substratum, were also capable of swimming.

The *C. middlesexensis* morphs had heavier shells than their ancestral species. It is likely that these forms could swim, particularly at intermediate sizes, but with further growth became sedentary.

Still later, *C. madisonius* and *C. jeffersonius* developed massive right and left valves with continued growth. It is likely that these species became mainly sedentary forms at smaller sizes than their ancestors. *Chesapecten septenarius* is similar to *C. jeffersonius* but has a narrower shell and does not tend to be as large or massive as its ancestor.

Arguments concerning the selective advantages of large size or increased swimming ability in the evolution of *Chesapecten* cannot be effectively addressed or substantiated here. Although these may be ways of "escaping" predators, thus imparting selective advantage, it is interesting to note that observations on living scallops not only show that swimming is used in flight reactions (Morton, 1980) but may also be important in scallop migrations (Baird, 1966; Morton, 1980, p. 401).

Life Habit and Environment

The depositional environments of *C. coccymelus* and *C. nefrens* were predominantly shallow and lagoonal. Later depositional basins were generally deeper water embayments, although a number of transgressive–regressive sequences also occur over these time periods. The early species of *Chesapecten* were byssate forms; the later species lost the byssus earlier in development and presumably became swimmers. Still later forms became sedentary as adults.

Kauffman (1969) has associated byssate pectinid species with near-shore, shallow environments where attachment is beneficial to forms exposed to high current energy. Kauffman suggests that epifaunal unattached forms are mainly found in quieter, deeper water. This correlation occurs in the *Chesapecten* species, although on the basis of sediment characteristics the later species probably lived under occasional periods of high-energy conditions. Periodic high-energy conditions surrounding the larger *Chesapecten* might have contributed to the massiveness of their shells, in order to increase stability.

If Kauffman's generalization of correlated environments and functional types applies to *Chesapecten* evolution, the tendency to observe a correlation between ontogeny and phylogeny could be related to changes in the characteristics of the depositional basin which contain the ancestors and descendants of this lineage.

Developmental Timing

Among-species differences in homologous shell characters may be at least partially attributed to the timing of appearance of shell features. Features relevant to this discussion involve the radial ribs, byssal notch, and shell thickness. Ribs appear on the postlarval shell as plications of the mantle tissue which forms them. With growth, the advancing mantle edge is folded into waves; the crests and troughs form the radial ribs. With growth, ribs on the anterior and posterior margins of the disk are lost as the amplitudes of the plications flatten (Fig. 13). Increased shell breadth is a partial consequence of the loss of ribs in ontogeny. The plications at the center of the valve do not flatten as much as do the marginal ones. It would seem that there is a physiologic difference between these shell regions. Perhaps this is an aging phenomenon which occurs more quickly at the anterior and posterior mantle edges.

In phylogeny ribs are also lost, and this is an important character in phylogenetic transformation. Whatever causes rib loss in ontogeny may not be the same as the factor(s) that cause(s) differing and decreasing initial numbers of mantle-edge plications in the postlarval shell of successional species. If it is not, the recapitulation pattern is a spurious correlation which has nothing to do with condensation of timing events and terminal addition (in this case loss). However, the timing process that regulates postlarval shell rib number might also be operative in influencing the initiation of mantle plications on the postlarval shell. Related living species with a similar pattern of rib development could be used to study whether the pattern and process that would unify development with observed evolutionary patterns are truly justified.

Byssal notch occlusion undergoes strong recapitulation in *Chesapecten*. Unlike rib loss, where the source of the change is not yet certain, the byssal notch is rather clearly associated with the byssus and the larval foot in scallops. These protrude from the byssal opening and can physically push the mantle in the notch region back. In the absence of a foot or byssus, the mantle expands and occludes the notch. The degree to which the mantle affects this character alone or in combination with other organs is not clear. In any event, the later *Chesapecten* species lose the byssal notch earlier in development than the earlier species. In this case, timing of byssal notch occlusion is clearly accelerated. Development of increased shell thickness may also be characterized in this light, but we have not investigated this character in sufficient detail.

An increased role of physiologic and developmental effects in morphologic character changes would help to clarify ambiguities in the interpretation of character changes. Garstang's (1922) restatement of the recapitulation law (ontogeny proceeds through successive grades of differentiation, which result in successive grades of evolution) encourages increased understanding of differentiation in development rather than a literal acceptance of Haeckel. This level of understanding is an ideal that has been or will always be elusive or difficult to obtain for many kinds of evolutionary data. Still, there are living representatives of fossil species which may be studied in light of their differentiation during maturation which in turn may explain, or be explained by, their evolutionary history.

Knowledge of the details of character development, at perhaps a cellular or tissue level, would contribute much more to our ability to evaluate a theory such as recapitulation or heterochrony. Character states thought to be qualitatively or numerically homologous may be unrelated and our conclusions concerning evolution will therefore be meaningless. If recapitulation is a factor in the evolution of pectinids, then the process(es) that control(s) the timing of the appearance of shell features would most likely aid our understanding of the pattern behind their evolution.

CONCLUSIONS

1. The evolution of *Chesapecten* as shown in Fig. 4 is independently supported by the fossil record and cladistic transition series analysis. The numerical cladistic methodology of transition series analysis, which forms taxon associations without reliance on prior information concerning polarity or character transitions, makes the resulting cladogram a useful reference in testing other evolutionary hypotheses.
2. Evolutionary changes generally occur along an ontogenetic axis in *Chesapecten*; therefore this axis is useful in determining degree of relationship.
3. Nelson's prediction that early developmental stages are useful in determining cladistic network polarity concurs with other criteria of stratigraphic superpositioning, size, and complexity in this group.
4. The overall measure of developmental transformation shows that juvenile forms occur early in phylogeny, and gerontic forms are more derived. This observation supports Jackson's (1890) and Newell's (1937) suggestion that the principle of recapitulation is probably applicable in pectinid evolution.
5. Morphologic differences between taxa are associated with differ-

ences in functional morphology, hence with differences in their respective life habits. Among the possible explanations for this change in life habit, following Kauffman (1969), is the correlation between functional pectinid species types and the environments in which they lived. Selection for increased size may also have contributed to this shift, but we have no independent way of demonstrating this.

6. The change in developmental patterns between species may be most strongly linked to timing of appearance of shell features, particularly at the anterior region of the disk and auricles and to the posterior region of the disk.

ACKNOWLEDGMENTS

We are grateful to a number of people for their help and support in the course of this work. Measurements were taken on collections of Thomas R. Waller and Thomas G. Gibson and both provided hours of helpful discussion. Lauck W. Ward was helpful in providing sample locality information. Versions of the manuscript during preparation were critically read by Stefan Bengtson, Arthur J. Boucot, Sara S. Bretsky, Niles Eldredge, Leslie F. Marcus, Nancy A. Neff, Normal D. Newell, and Thomas R. Waller. Warren Blow and Lars-Eric Jonsson provided technical assistance. The Paleobiology Department of Uppsala University in Sweden provided a snug harbor for J.M. in the final stages of manuscript preparation. Finally, we wish to thank James S. Farris and Jeffrey S. Levinton for their continued support and encouragement through all phases of this work. Portions of this study were supported by NSF Grant DEB-7824647.

APPENDIX. Key to Symbols for Species and Localities[a]

Species	Sample no.[b]	Mean height (mm)	Maximum sample height (mm)
C. coccymelus	1'	38.67	48
	1"	55.67	77
	1‴	39.04	69
C. nefrens	2	73.20	127
C. santamaria	3'	92.50	151
	3"	95.82	140
C. nefrens-santamaria	2-3	85.90	140

C. middlesexensis	4A′	66.10	109
	4A″	83.7	151
	4B	39.25	56
	4C′	83.00	139
	4C″	54.00	112
C. jeffersonius	5′	77.45	154
	5″	68.33	112
C. septenarius	6′	42.5	52
	6″	66.86	104
C. madisonius	7	86.43	145

[a] All numbers refer to species as they occur stratigraphically, except *C. septenarius* (7) and *C. madisonius* (8), which occur together. The three *C. middlesexensis* lineages are labeled A, B, and C, as in the text. To distinguish samples of the same taxon, primes have been added. Species numbers without primes indicate that all samples have that character state. To simplify Figs. 5–17, only some of the character states are labeled as to sample. The symbol ⊛ indicates that all remaining samples have the associated character state.

[b] Localities and samples:

1′   Camp Roosevelt. U.S.G.S. 25295, Zone 10, Shattuck, west shore of Chesapeake Bay, Calvert Formation.

1″   Chesapeake Heights. North of Davis Beach, Zone 10(A&B), Calvert Formation.

1‴   Chesapeake Heights. Same locality as 1″, Zone 10C, Calvert Formation.

2   Boston Cliffs. Choptank River from river bed to 10 ft, Zone 19, Calvert Formation.

2-3   Langley's Bluff.U.S.G.S. 25303, west shore of Chesapeake Bay, 2 miles southeast of Hermanville, Maryland, St. Mary's Formation.

3′   Windmill Point. U.S.G.S. 24304, right bank of St. Mary's River, at mouth of St. Inigoes Creek, Maryland, St. Mary's Formation.

3″   Chancellor Point. Left bank of St. Mary's River, St. Mary's Formation.

4A′   Meherrin River Boat Camp. U.S.G.S. 25321, near Marfreesboro, right bank up from bridge, off Highway 258, Eastover Formation.

4A″   Corbin Creek. U.S.G.S. 25313, at Highway 14, bridge in stream cut, Eastover Formation.

4B   Rappahannock River above Punch Bowl Point beach, at end of County Maintenance Road 648, Eastover Formation.

4C′   Yorktown Cliffs. East of bridge of Colonial Park. Collections from cliffs as far as Coast Guard fuel wharf, layer of fossils at top of cross bedding about three levels above beach, Eastover Formation.

4C″   Rappahannock River. Near Bowler's Wharf, U.S.G.S. 25308, Eastover Formation.

5′   Haynes Pond Dam. On Carter Creek, on left bank about 3 ft from foundation of mill house, Lower Yorktown Formation.

5″   Lanexa, Virginia. West side of railroad cut, float, Lower Yorktown Formation.

6′   Rice Road Fill Pit. U.S.G.S. 25330, 29 Harris Creek Road, Hampton, Virginia, Upper Yorktown Formation.

6″   Lee Creek Mine. U.S.G.S. 25334, Texas Gulf Sulfur phosphate mine, north of Aurora, North Carolina, on right bank of the Pamlico River, Upper Yorktown Formation.

7   Rice Road Fill Pit. Same as 6′, Upper Yorktown Formation.

REFERENCES

Andrews, G. T., 1978, Marine diatom sequence in the Miocene strata of the Chesapeake Bay region, Maryland, *Micropaleontology* **24**:371–406.

Baird, R. H., 1966, Notes on an escallop (*Pecten maximus*) population in Holyhead harbour, *J. Mar. Biol. Assoc. UK* **46**:33–47.

Boucot, A. J., 1976, Rates of size increase and of phyletic evolution, *Nature* **261**:594–696.

Cope, E. D., 1885, On the evolution of Vertebrata, *Am. Nat.* **19**:140–148, 234, 247, 341–353.

Cope, E. D., 1896, *Primary Factors of Organic Evolution*, University of Chicago, Chicago.

Davenport, C. B., 1900, On the variation of the shell of *Pecten irradians* Lamark from Long Island, *Am. Nat.* **34**:863–877.

de Beer, G. R., 1958, *Embryos and Ancestors*, Clarendon Press, Oxford.

Fairbridge, W. S., 1953, A population study of the Tasmanian "commercial" scallop, *Novotola meridionalis* (Tate) (Lamellibranchiata, Pectinidae), *Austr. J. Mar. Fr. Res.* **4**(1):1–40.

Farris, J. S., 1970, Methods for computing Wagner trees, *Syst. Zool.* **19**:83–92.

Garstang, W., 1922, The theory of recapitulation: A critical restatement of the Biogenetic Law, *Linnean Soc. J.* **35**(232):81–101.

Gernant, R. E., Gibson, T. G., and Whitmore, Jr., F. C., 1976, Environmental history of the Maryland Miocene, in: *Field Trip, Geological Society of America Annual Meeting, Maryland Geological Survey Guidebook 3*, pp. 1–15.

Gibson, T. G., 1970, Late Mesozoic–Cenozoic tectonic aspects of the Atlantic Coastal Plain margin, *Geol. Soc. Am. Bull.* **81**:1813–1822.

Gould, S. J., 1971, Muscular mechanics and the ontogeny of swimming in scallops, *J. Paleontol.* **14**:61–94.

Haeckel, E., 1866, *Generelle Morphologie der Organismen, Allgemeine Grundzüge der Organischen formen–Wissenschaft, mechanisch begründet durch die von Charles Darwin reformerte Decendenz-Theorie*, Vols. 1 and 2.

Hennig, W., 1966, *Phylogenetic Systematics*, University of Illinois Press, Urbana, Illinois.

Jackson, R. T., 1890, Phylogeny of the Pelecypoda. The Aviculidae and their allies, *Boston Soc. Nat. Hist. Mem.* **4**:277–400.

Kauffman, E. G., 1969, Form, function, and evolution: Functional morphology of the bivalve shell, in: *Treatise on Invertebrate Paleontology (N) Mollusca (!)* (R. C. Moore, ed.), pp. N140–N193, Geological Society of America, University of Kansas.

Levinton, J. S., and Simon, C., 1980, A critique of the punctuated equilibria model and implications for the detection of speciation in the fossil record, *Syst. Zool.* **29**:130–142.

Merrill, A. S., 1961, Shell morphology in the larval and post-larval stages of the sea scallop, *Placopecten magellanicus* (Gmelin), *Bull. Mus. Comp. Zool. Harv.* **125**:1–20.

Mickevich, M. F., 1981, Quantitative phyletic biogeography, in: *Advances in Cladistics*, Proceedings of the First Meeting of the Willi Hennig Society (V. A. Funk and D. R. Brooks, eds.), pp. 209–222, New York Botanical Garden, Bronx, N.Y.

Mickevich, M. F., 1982, Transformation series analysis, *Syst. Zool.* **31** (in press).

Mickevich, M. F., and Mitter, C., 1981, Treating polymorphic characters in systematics: A phylogenetic treatment, in: *Advances in Cladistics*, Proceedings of the First Meeting of the Willi Hennig Society (V. A. Funk and D. R. Brooks, eds.), pp. 45–58, New York Botanical Garden, Bronx, NY.

Miyazaki, J. M., 1978, Evolution of a Tertiary Scallop and Haeckel's Law, M.S. Thesis, State University of New York at Stony Brook.

Morton, B., 1980, Swimming in *Amusium pleuronectes* (Bivalvia: Pectinidae), *J. Zool. Lond.* **190**:375–404.

Nelson, G. J., 1973, The higher level phylogeny of the vertebrates, *Syst. Zool.* **22**:87–91.

Nelson, G. J., 1978, Ontogeny, phylogeny, paleontology, and the biogenetic law, *Syst. Zool.* **27**:324–345.

Newell, N. D., 1937, *Late Paleozoic Pelecypods: Pectinacea*, Kansas Geological Survey Bulletin 10(1).

Newell, N. D., 1942, *Late Paleozoic Pelecypods: Mytilacea*, Kansas Geological Survey Bulletin 10(2).

Newell, N. D., 1949, Phyletic size increase, an important trend illustrated by fossil invertebrates, *Evolution* **3**:103–124.

Stanley, S. M., 1970, *Relation of Shell Form to Life Habits of the Bivalvia (Mollusca)*, Geological Society of America, Boulder, Colorado.

Stanley, S. M., 1973, An Explanation of Cope's Rule, *Evolution* **27**:1–26.

Stenzel, H. B., 1949, Successive speciation in palenotology: The case of oysters of the *sellaeformis* stock, *Evolution* **3**:34–50.

Thayer, C. W., 1972, Adaptive features of swimming monomyarian bivales (Mollusca), *Forma Functio* **5**:1–31.

Verrill, A. E., 1897, A study of the Family Pectinidae, with a revision of the genera and sub-genera, *Conn. Acad. Arts Sci. Trans.* **10**:41–95.

Von Baer, K. E., 1866, Über Prof. Nic. Wagners Entdeckung von Larven, die sich fortplanzen, Herrn Garrens Verwandte und ergänzende Beobachtung und über die Pädogenesis überhaupt, *Bull. Acad. Imp. des Sciences, St. Petersbourg* **9**:63–137.

Waller, T. R., 1969, The evolution of the *Argopecten gibbus* stock (Mollusca: Bivalvia), with emphasis on the Tertiary and Quaternary species of eastern North America, Paleontol. Soc. Mem. 3, *J. Paleontol.* **43**(5), Suppl.

Ward, L. W., and B. W. Blackwelder, 1975, *Chesapecten, A New Genus of Pectinidae (Mollusca: Bivalvia) from the Miocene and Pliocene of Eastern North America*, U. S. Geological Survey Prof. Paper 861.

Ward, L. B., and Blackwelder, B. W., 1980, *Stratigraphic Revision of Upper Miocene and Lower Pliocene Beds of Chesapeake Group, Middle Atlantic Coastal Plain*, U.S. Geological Survey Bull. 1482-D.

Yonge, C. M., 1936, The evolution of the swimming habit in the Lamellibranchia, *Mus. Roy. Hist. Nat. Belg. Mem. Ser. 2* **1936**(3):77–100.

9

Punctuated versus Gradual Mode of Evolution
A Reconsideration

ANTONI HOFFMAN*

Geological Institute
University of Tübingen
Tübingen, West Germany

INTRODUCTION

The concept of punctuated equilibria (Eldredge and Gould, 1972; Gould and Eldredge, 1977) continues to attract the attention of evolutionary biologists and paleontologists (Bock, 1979; Stanley, 1979; Levinton and Simon, 1980; Hoffman, 1981; Mahé and Devillers, 1981; Schopf, 1981a). This is so not only because of its heterodox appearance, contradicting the paleontologic folk wisdom that views species as transforming gradually into new species. The attractiveness of punctuationalism is first of all due to its far-reaching implications for the study of macroevolution, since dismissal of the old gradualistic concept forces us to invoke some new biologic principles and/or mechanisms to account for macroevolutionary change. Most commonly, epigenetics and development are referred to in this context (Gould, 1980, 1982; Alberch, 1982; Maderson *et al.*, 1982; Rachootin and Thomson, 1981). To explain large-scale evolutionary trends the theory of species selection has been put forth. It claims that some major features of the biosphere (e.g., numerical prevalence of sexual over asexual species, competitive exclusion of major taxa in geologic time, etc.) are not to be explained by the evolution of populations through natural selection, but rather by the differential extinction and

* Permanent address: Wiejska 14 m. 8, PL-OO-490 Warszawa, Poland.

speciation rates in various lineages (Stanley, 1975; 1979; Gould and Eldredge, 1977; Vrba, 1980; Gilinsky, 1981).

Futhermore, the concept of punctuated equilibria lies at the core of computer models applied to investigate the evolution of organic diversity (Raup et al., 1973; Gould et al., 1977), even though one might also apply the latter approach while rejecting the constraints of the punctuationist view of species evolution. This is also the case with the continuing debate on the methodology of phylogenetic reconstruction (Cracraft and Eldredge, 1979; Eldredge and Cracraft, 1980), because the Hennigian method of phylogenetic systematics assumes the association of evolutionary change with speciation only for the sake of convenience (Hennig, 1966), while it does not expect those punctuations to be real. Finally, the concept of punctuated equilibria may be of some significance for biostratigraphers, too, because it implies that the gaps in the fossil record of evolutionary lineages are real (Eldredge and Gould, 1977; see also Krassilov, 1978).

With all these far-reaching consequences of the concept of punctuated equilibria in mind, there can be little doubt that it is worth discussing. As a matter of fact, there has been much controversy on the validity and generality of the punctuationist view of species evolution. This controversy, being concerned with the empirical support for this concept as well as with its theoretical background, is partly due to the fact that the terminology is far from unequivocal. However, the focus has been mainly on real issues concerning the relative frequency and the biologic controls of the punctuated versus the gradual mode of species evolution. The aim of this paper is to discuss these two major issues; but to this end, some terminologic confusion first must be clarified.

The theory of species selection is not considered because it has been critically evaluated in a recent paper by Maynard Smith (1982).

PUNCTUATED EQUILIBRIA AND PHYLETIC GRADUALISM

The concept of punctuated equilibria was first presented (Eldredge, 1971; Eldredge and Gould, 1972, 1977) as an empirical generalization of what can be considered as a common experience of biostratigraphers; namely, that the fossil record is full of long-ranging morphospecies and missing links in between. Later, however, it has been conceived of in terms of an evolutionary theory (Gould and Eldredge, 1977; Gould, 1980), or even a paradigmatic view of evolution (Eldredge, 1979). No wonder that the testability and empirical content of the idea have been discussed at great length but without any satisfactory conclusion. To this end, punctuated equilibria must be considered a model, and should also be clearly defined.

Definitions

The model of punctuated equilibria claims that the large majority of morphologic changes in evolution take place very rapidly, during so-called speciation events, in populations that are so small that their discovery in the fossil record is virtually impossible; whereas, changes are negligible along those segments of a lineage which, by usual taxonomic standards, are to be recognized as single chronospecies.

The opposite model, of phyletic gradualism, claims that morphologic change is accomplished by a gradual shift of the range of intraspecific variation; this shift does not necessarily have to be unidirectional or to proceed at a constant rate, but it must eventually exceed the limits set by the usual taxonomic standards to a single chronospecies.

These characteristics are modified from the original descriptions of the two modes of evolution (Eldredge and Gould, 1972; Gould and Eldredge, 1977; see also Levinton and Simon, 1980). The term speciation event, as meant here, may refer either to the splitting of a lineage or to quantum evolution in Simpson's (1944) sense. Originally, the concept of punctuated equilibria was intimately related to the theory of allopatric speciation through a small subpopulation of an ancestral species confined to an isolated area at the periphery of the total geographic range (Mayr, 1963). Later, however, Gould and Eldredge (1977; see also Gould, 1980) did accept the possibility of sympatric speciation in populations subdivided into genetically isolated subpopulations (Bush, 1975; White, 1978). They referred also to the supposedly punctuationist "flush and crash" model of Carson (1975). Still other genetic models of speciation might be cited in this context (Lande, 1980a; Templeton, 1980), because the only requirement to fit a genetic model with the model of punctuated equilibria is that it allows for very rapid morphologic change in very small populations.

The original claim that the bulk of evolutionary change did happen in association with speciation events in the genetic sense of the term must be weakened. As a matter of fact, there is no way to prove or disprove it, because a speciation event can be inferred in the fossil record only from the associated change in morphology, which inference may or may not be correct (Levinton and Simon, 1980; Schopf, 1981a,b; Raup and Crick, 1981). Consequently, the model of punctuated equilibria, as meant here, claims only that apart from very short periods of very rapid, but not necessarily discontinuous, evolution in very small populations, the rate of morphologic evolution is close to zero.

On the other hand, morphologic evolution need not be unidirectional or at a constant rate within the framework of the model of phyletic gradualism. This model does not claim that orthoselection is acting. Even if the random-walk explanation for the evolutionary record of a lineage cannot be refuted (Raup, 1977; Raup and Crick, 1981), this does not contradict the model of phyletic gradualism as long as the net total change is really significant by usual taxonomic standards. In fact, the gradualistic model simply does not address the question of whether the change is effected by directional selection or by genetic drift. After all, to claim that the rate of phyletic evolution is roughly constant for each particular taxon would seem absurd to anyone who appreciates Simpson's (1944, 1953) evidence for variation in evolutionary tempo both within and between lineages, as well as to anyone aware of the dependence of evolutionary rates upon population size and structure, mutation rate, selection coefficients, etc. Therefore, the only tenet of the model of phyletic gradualism, as meant here, is that morphologic change is gradual and large enough to be recognized as transspecific.

Recognizability

The two models are not mutually exclusive in the sense that a continuous spectrum of modes of evolution could not occur even within a single evolutionary lineage. Morphologic gaps between the segments of a lineage may be either nonexisting, or smaller, or larger, or much larger than the net change along the continuous segments (or chronospecies). The former case would be entirely consistent with the model of phyletic gradualism, the latter one with the model of punctuated equilibria, but intermediate cases cannot be ruled out *a priori*.

Consequently, if a significant morphologic change in a vertical sequence of populations with overlapping ranges of variation is recurrent over a large geographic area, that is indicative of gradual evolution. One might of course argue that such a sequence of populations reflects a directed punctuated rather than a gradual evolution, with the punctuations just between the sampling levels, but this would certainly be a nonfalsifiable and counterproductive claim leading inevitably to *regressus ad infinitum*.

This is why I cannot accept that part of Gould and Eldredge's (1977) criticism against the gradualistic interpretation (Kellogg, 1975) of the fossil record of the Late Cenozoic radiolarian *Pseudocubus vema* which refers to the supposedly punctuated pattern of morphologic change. Certainly, the rate of morphologic change in *P. vema* was highly variable and permits the recognition of three successive morphologic stages, but the occurrence of intermediate samples (omitted by Gould and Eldredge in their reanalysis of the data) assures us that the transitions took place in large populations, even though at a rather rapid rate. Nevertheless, Gould and Eldredge (1977) are right: Kellogg's (1975; see also Kellogg and Hays, 1975) study cannot be accepted as a case of phyletic gradualism, because there is as yet no geographic control to make sure that this is not the record of a shift in an ecologically controlled cline. The importance of this objection is well illustrated by Schankler (1981), who documented in an Eocene sedimentary basin the record of local extinction and reentry of mammals representative of lineages evolving elsewhere.

In order to fit the model of phyletic gradualism, a lineage must not only change gradually, but also over a considerable geographic area. There is no reason to expect that such data cannot be obtained from the fossil record. Deep-Sea Drilling Project cores, for instance, provide an excellent opportunity for case studies, and some gradualistic stories have indeed been produced on this basis (Prothero and Lazarus, 1980; Malmgren and Kennett, 1981).

It is much more difficult, if not impossible, to provide positive evidence for punctuated evolution, if only because one can hardly prove that punctuation events observed in one area reflect more than immigrations from another, perhaps unknown or inaccessible or even nonpreserved, area where the lineage did evolve gradually.

This difficulty is well exemplified by the study of dactylioceratid ammonites from the Lower Jurassic Grey Shales of Great Britain (Howarth 1973), which has been cited by Sylvester-Bradley (1977) as a case of punctuated equilibria. There is in fact a vertical sequence of distinctive dactylioceratid populations (*Dactylioceras crosbeyi, D. clevelandicum, D. tenuicostatum*, and *D. semicelatum*) with no intermediates in between, which seems to be recurrent throughout Europe and perhaps even beyond. However, the record of this evolutionary lineage includes time gaps between the particular populations, or chronospecies, and there is no way to determine whether the intermediates are lacking because of their extreme rarity (as is claimed by the model of punctuated equilibria), or whether each of the British dactylioceratids simply immigrated from some other area (as would be expected by the gradualists).

The same problem is involved in the evaluation of detailed studies on Mediterranean Neogene foraminifers. Gradstein (1974), for instance, investigated biometrically the Pliocene assemblages of the planktic foraminifer *Globorotalia* and recognized some distinct and stratigraphically separated groups within that genus (*G. margaritae* and *G. puncticulata*, *G. bononiensis*, *G. inflata*). He was able to document considerable and often statistically significant morphologic oscillations within the groups, but no morphologic transitions between them. This might be considered a case of punctuated equilibria, but there seem to be stratigraphic gaps between the ranges of the particular groups and gradual evolution could have taken place somewhere else. In turn, E. Thomas (1980) documented the absence of any significant net change in the benthic foraminifer *Uvigerina cylindrica* in the Mio-Pliocene of Crete, which refutes the traditional picture of gradualistic evolutionary trends in *Uvigerina* in the Mediterranean Neogene (Meulenkamp, 1969). Whether or not this new interpretation fits the model of punctuated equilibria, depends, however, on the nature of the origin of this lineage. Unfortunately, the pattern of its derivation from the supposed ancestor *U. bononiensis* remains unknown.

No wonder that where the fossil record is spottier and the samples are smaller, discussions become really endless. This is the case with some Eocene mammals of North America (Gingerich, 1976, 1977; Gould and Eldredge, 1977; West, 1979; Schankler, 1981) and also with the hominids (Eldredge and Tattersall, 1975; Cronin *et al.*, 1981; Rightmire, 1981; Delson, 1981).

RELATIVE FREQUENCY

Having defined the models of punctuated equilibria and phyletic gradualism, the relative frequency of the two modes of species evolution might seem to be ascertainable. This is indeed what has been considered as crucial for the further evaluation of the implications of this supposed revolution in evolutionary paleontology (Eldredge and Gould, 1972; Gould and Eldredge, 1977; Stanley, 1979; see also Lande, 1980b). The debate on this issue, however, has already lasted almost a decade and I must confess that I do not see any possibility to settle it through empirical arguments, because, with all the notorious deficiencies of the fossil record taken into account, each position can be defended endlessly with the use of more or less plausible *ad hoc* hypotheses.

Punctuated Equilibria

One may agree with Gould and Eldredge (1977), Hallam (1978), and Stanley (1979) that the commonness of morphologic stasis in evolutionary time can hardly be overestimated. (I emphasize the term morphologic stasis in contrast to species stasis because we know nothing about the underlying genetic changes.) For instance, recent studies on bivalve species durations point out that the figures given by the authors cited above may underestimate rather than overestimate the mean duration of an invertebrate, morphologically defined chronospecies, which may considerably exceed a dozen million years (Jablonski, 1980; Stanley *et al.*, Chinzei, 1980; Hoffman and Szubzda-Studencka, 1982; but see Koch, 1980). These data seem to indicate that the rates of morphologic evolution are indeed close to bimodal in distribution. When taken in conjunction with the well-known rarity of unquestionable ancestor–descendant relationships among fossil species (Hallam, 1978; Hoffman and Szubzda-Studencka, 1982), they provide support, though in a purely qualitative sense, for the model of punctuated equilibria. The criticism raised by Harper (1975, 1976) loses its validity in this context because the model of punctuated equilibria has been defined less radically here than by Stanley (1975, 1979), allowing for rapid evolution events other than allopatric speciation.

However, the other tests proposed by Stanley (1975, 1979) are irrelevant to the problem, because a positive correlation between species-richness and morphologic diversity of a group does not say anything about the mode of evolution and the timing of morphologic change in that group. The pontian cockles or the cichlid fishes, for instance, could well have achieved their diversity in accordance with the model of punctuated equilibria. On the other hand, they could also have split and evolved gradually but at much faster rates than did their relatives in other areas (Greenwood, 1974, 1979). What is at issue here is not the rate of net overall change, but the pattern of variation in that rate within particular lineages. This pattern, however, is, and probably will remain, unknown for any reasonably large sample of evolutionary lineages.

Phyletic Gradualism

The commonness of morphologic stasis corroborates the model of punctuated equilibria. On the other hand, quite a large number of examples can be cited in support of the model of phyletic gradualism. In addition to the lineage of the Permian foraminifer *Lepidolina* (Ozawa, 1975), which was the only case to withstand the critical scrutiny of Gould and Eldredge (1977), one may also refer to the Triassic conodontophorid *Gondolella mombergensis* lineage (Dzik and Trammer, 1980) and the Cenozoic foraminifer *Globorotalia conoidea–G. inflata* lineage (Malmgren and Kennett, 1981). The latter two cases seem to satisfy all the requirements set up even by the radical version of the model of punctuated equilibria. Furthermore, they concern some real morphologic changes and not only an increase in size, which has been commonly accepted as an exception to the model of punctuated equilibria (Hallam, 1978; Stanley, 1979).

Bettenstaedt (1962) was ridiculed by Gould and Eldredge (1977) for his explicit expression of, and outright adherence to, gradualistic prejudices. However, his case studies have not been critically evaluated. Actually, these are biometric investigations on benthic foraminifers and ostracodes from several Lower to Upper Cretaceous sections in northwest Germany (but they often extend far beyond that area), correlated with each other by a number of biostratigraphic and other geologic time markers; the morphologic changes involved do, again, concern not as much size increase as shell form (Hiltermann and Koch, 1950; Albers, 1952; Bettenstaedt, 1952, 1958; Grabert, 1959; Zedler, 1960, 1961; Bartenstein and Bettenstaedt, 1962; Michael, 1966). Similar, even though perhaps a little weaker, cases for the model of phyletic gradualism are provided by the Late Cretaceous *Helicorbitoides–Lepidorbitoides* lineage (Van Gorsel 1975) and by the Tertiary large foraminifer *Lepidocyclina* from southeast Asia (Van Vessem, 1978).

One might of course argue, as did Gould and Eldredge (1977), that protists are the most likely to undergo gradual morphologic evolution, due to their genetic simplicity and the lack of morphologic integration. However, the ammonites *Otoscaphites puerculus* from the Cretaceous of Japan (Tanabe 1975, 1977) and *Texanites soutoni* and *Submortoniceras* from the Cretaceous of South Africa (Klinger and Kennedy 1980), the graptolites *Isographtus victoriae* from the Ordovician of Australasia (Cooper 1973), and the Neogene freshwater snails *Melanopsis*, *Rhodopyrgula*, and *Viviparus* from the Greek islands (Willmann 1981) also fit the model of phyletic gradualism. Moreover, the study of A. M. Ziegler (1966) of the brachiopod *Eocoelia* lineage from the Silurian of the Welsh Borderland was all too easily rejected by Gould and Eldredge (1977). Their arguments focused on the wide within-sample variation in *Eocoelia* but not on the evident shift of the variation range, as if the wide intrapopulation variation itself would make impossible any evolutionary change; they also pointed out that not all measured characters of *Eocoelia* did change, as if it were established that evolution always affects all the characteristics of an organism in a concerted way.

There is also the Brinkmann's (1929) classic study of the ammonites *Kosmoceras* from the Oxford Clay, which was commonly, but contrastingly, cited in the context of the discussion of the punctuated versus gradual mode of evolution (Hewitt and Hurst, 1977; Kennedy, 1977; Sylvester-Bradley, 1977; Dzik and Trammer, 1980). The original data have been recently reanalyzed by Raup and Crick (1981), who concluded that the gradual mode of morphologic change within the investigated time interval is undisputable. Still, one may argue whether this is a case of real evolution, or just of ecophenotypic oscillations within the limits of morphologic stasis. Nevertheless, taking Brinkmann's experience with Jurassic ammonites into account, his taxonomic decisions should be appreciated, and hence the *Kosmoceras* lineage should be considered to fit the model of phyletic gradualism, as defined here.

Certainly, even the largest set of counterexamples to the model of punctuated equilibria cannot disprove it, especially in view of the impressive data on long-term morphologic stasis. I suggest, therefore, to accept the conservative proposition that both modes of morphologic evolution do occur in nature, as well as a series of intermediate modes (Cisne *et al.*, 1980).

CONTROL MECHANISMS: CRITIQUE

With that conclusion in mind, the most intriguing question is why some species-level lineages undergo gradual morphologic evolution, while others undergo punctuations.

The differential mode of morphologic evolution might be attributed to differential characteristics of habitat or ecospace development through time. This would, however, imply a very strict adherence either to the billiard-ball model of living beings, or to the adaptationist program in its extreme form of the Panglossian paradigm (Gould and Lewontin, 1979; Reif, 1982). The point being made here is that if we consider the evolutionary behavior of species to be controlled entirely by extrinsic, environmental factors, the organisms appear as puppets driven with perfect efficiency from outside; or else they appear as optimal machines able to rearrange their form and function immediately in order to adapt to new environmental challenges. In both cases there would be virtually no causes for extinction other than the instantaneous disappearance of the whole habitat. This is indeed what has been claimed, for instance, by Williams (1975).

In my opinion, however, the fossil record demonstrates clearly that quite many species became extinct because the tempo of some environmental, biotic or abiotic, changes exceeded their adaptive potential, rather than due to an instantaneous and total destruction of their habitat. This

indicates that even though organic evolution is stimulated, or even required, by environmental challenges, its extrinsic controls are nonetheless
mediated by some biologic mechanisms. Consequently, organisms may be
justifiably regarded as striving continuously for optimality within the constraints of their phylogenetic heritage and constructional possibilities, but
not as being optimal under any given environmental regime (Lewontin,
1978).

Incidentally, this is also implied by an analysis of the optimal strategy
of species evolution. As pointed out by Slobodkin and Rapoport (1974),
the evolution of each particular species can be treated in terms of the
Gambler's Ruin game, with the only payoff to the species being its survival
or its ability to continue playing. Under such conditions, the optimal
strategy for the player is to minimize the stake. Therefore, adaptation to
environmental changes can be expected to involve first of all tactical
adjustments (behavior, population dynamics, physiology), to be followed
by genetic evolution only when the former means turn out to be insufficient. There should always be a time lag in the evolutionary response
of species to changes in their habitat. This may hold true especially for
sexual organisms, because it has been shown by Thompson (1976) that
the presence of sex reduces the ability of populations to follow environmental vagaries promptly.

Now, if the evolutionary game may indeed be considered as a kind
of Gambler's Ruin game, the evolutionary behavior of species can be
expected to vary from one species to another, depending upon their biologic and ecologic characteristics. This is expressed in terms of the punctuated versus the gradual mode of morphologic evolution. An explanation
of this variation has been attempted in terms of the developmental homeostasis of individuals, the genetic homeostasis of populations, and the
ecologic homeostasis of communities.

Developmental Homeostasis of Individuals

To point out the biologic mechanism responsible for the commonness
of the punctuated evolution of species, Eldredge and Gould (1972) referred
to the developmental, or morphogenetic, homeostasis of individual organisms, as advocated by Lerner (1954). They claimed that morphologic
stability of a species in evolutionary time may arise from the generally
higher fitness of heterozygous versus homozygous genotypes, as the former can follow the normal developmental pathway more easily. This
normal developmental pathway permits the achievement of a highly integrated adult form involving several adaptive complexes closely inter-

related with each other. Virtually each deviation from that "normal" form, or phenotype, is expected to be less fit. Therefore, a genetic revolution by homozygosity, or by release of the developmental homeostasis, is a prerequisite to morphologic evolution.

This concept, however, appears unconvincing because it excludes phyletic gradualism entirely, so that any evidence for the latter process provides a critical test. On the other hand, Lerner's (1954) idea of heterozygosity as the main cause for developmental stability of organisms has been seriously challenged by Soulé (1979). Finally, the enormous evidence for clinal variation (Rensch, 1959; Mayr, 1963; Endler, 1977), quite often exceeding in range the morphologic gaps between distinct species, shows clearly that a deviation from the "normal" form of a species may not be counterselected, and also that such a deviation may be produced by minute genetic and/or purely phenotypic changes rather than by a true genetic revolution.

Genetic Homeostasis of Populations

In their original paper, Eldredge and Gould (1972) also emphasized the genetic homeostasis of organic populations, as proposed by Lerner (1954). This concept arose from the observation that populations affected by strong directional selection may partly revert to their original phenotype after this selection pressure has been released. This phenomenon represents a kind of evolutionary inertia of populations. Eldredge and Gould (1972) claimed that very rapid allopatric speciation provides the main mode of evolution because the genetic homeostasis can be overcome only in small, isolated populations where the evolutionary inertia of great numbers of individuals is considerably reduced.

However, once the genetic homeostasis of populations is invoked as an explanation for evolutionary punctuations through allopatric speciation, both phyletic gradualism and sympatric speciation turn out to be inexplicable violations to the theory because a disturbance of homeostasis has been argued to be possible only during very short events in small peripheral isolates. Yet several cases of phyletic gradualsim are cited above, and it has also been clearly shown by Bush (1975) and Rosenzweig (1978; see also Maynard Smith, 1966; Thoday, 1972; White, 1978, Kirkpatrick and Selander, 1979; Rosenzweig and Taylor, 1980) that sympatric speciation may occur under a wide variety of conditions. The later reconciliation of the model of punctuated equilibria with sympatric speciation (Gould and Eldredge, 1977; Gould, 1980) does not help very much, because it was demonstrated by Maynard Smith (1976a) that the tempo of genetic evolution does not necessarily increase with a decrease in population size (see also Lande, 1980a; Templeton, 1980), while Anderson and Evensen (1978) presented data suggesting that speciation may not be confined to small populations. There is also Endler's (1977) model of gradual morphologic divergence and speciation in a cline, which can hardly be reconciled with the simplistic view of allopatric speciation as advocated by the proponents of the model of punctuated equilibria.

Ecologic Homeostasis of Communities

Boucot (1978) rejected the model of punctuated equilibria because the ecologic integration of communities, which he strongly advocated, could not permit allopatrically originated species to enter established communities. This is certainly an extremely radical and hardly acceptable interpretation of the community paradigm of modern systems ecology (Hoffman, 1979). In its conservative form, however, that paradigm, or the assignment of great significance to the ecologic homeostasis of communities, may also be applied to explain the control mechanisms of the punctuated versus the gradual mode of evolution. In fact, I proposed a few years ago (Hoffman 1978) that, instead of the genetic and developmental homeostases, some other homeostatic mechanisms may be involved in maintaining the long-term morphologic stability of species through evolutionary time, while underdevelopment of those mechanisms may account for phyletic gradualism. I claimed that stasis may result from a centripetal selection for maintenance of the ecologic role of a species in a stable community structure. Community structure was meant there as the sum of all biologic interactions among community members, and of their autecologic relations to the abiotic environment. Indeed, community structures often seem to be highly complex and integrated; some communities are also very stable in ecologic time and exhibit long-term persistence in evolutionary time. With these community properties taken into account, the system-analytic, or cybernetic, approach might appear appropriate to analyze community behavior in time (Hoffman, 1980). Following from this assumption and some simple cybernetic models analyzed by Gecow (1975), I claimed that the structural homeostasis of complex communities accounts for the commonness of stasis and punctuations in the evolution of species, whereas the virtual absence of such mechanisms from simple communities allows for phyletic gradualism.

As a matter of fact, I have found data to support this hypothesis

(Hoffman, 1979). Shallow-water marine benthic mollusk species seem to arise in large groups or clusters, supposedly during major ecologic reorganizations, while planktic foraminifer and coccolithophorid species derived from extremely simple communities seem to evolve independently from one another. These data, however, are inconclusive because the apparent difference in evolutionary behavior between the sublittoral benthos and the pelagic plankton may be effected by a difference in environmental inertia between the sublittoral and the pelagic realms, rather than by a difference in complexity of the respective communities.

Furthermore, the systems approach to ecologic communities may be inadequate because communities should be considered as an epiphenomenon of species evolution rather than as real and discrete units (Hoffman, 1979). This obviously refutes any hypothesis that builds on considerable structural homeostasis of ecologic communities. A reconciliation of the system-ecologic concepts with the modern evolutionary theory has been recently attempted (Maynard Smith, 1976b; Stenseth, 1979; Wilson, 1980), but one may still remain skeptical because some important lines of argument against the significance of ecologic homeostasis of communities have been left untouched by those new theoretical developments. This holds true for the commonness of physical disturbances which usually prevent competing species from coevolving and coadapting to any significant extent (Connell, 1978), as well as for the extreme rarity of actual cases of coevolution among competitors (Connell, 1980).

CONTROL MECHANISMS: HYPOTHESIS AND EVIDENCE

The attempts to explain the variation in the punctuated to the gradual mode of evolution have thus failed. All those explanations, however, referred to the process of speciation and its biologic controls, whereas what one really has evidenced in the fossil record as being punctuated or gradual are only patterns in morphologic change. The mutual identity of the patterns of morphologic and genetic change has been merely assumed by the radical version of the model of punctuated equilibria, but it seems very unlikely that it exists (Schopf 1981a,b). There is no one-to-one, unequivocal relationship between the genotype and its resultant phenotype. This is why I am proposing to refocus the models of punctuated equilibria and phyletic gradualism with emphasis on the patterns of morphologic change through evolutionary time, instead of speaking in terms of species multiplication and transformation.

Developmental Canalization and Plasticity

It has been emphasized by Schmalhausen (1949) and Waddington (1957, 1975) that the ability of organisms to develop into the "normal" adult form under a wide variety of more or less unusual environmental conditions is of considerable selective advantage. Such a developmental resilience and equifinality of organisms is indeed entirely consistent with the optimum strategy of minimizing the stake in the evolutionary game, because the adaptations required by a new environmental regime can then be achieved without any genetic and hence irreversible changes (Slobodkin and Rapoport, 1974). The ability of organisms to follow the normal developmental pathway, or chreod, but also their inability to enter it if the environmental conditions are too far off the norm, has been termed developmental canalization, or homeorhesis (Waddington 1957, 1975). This fundamental property of individual development is brought about by the action of the epigenetic system, which is responsible that only certain genetic potentialities of the organism are revealed during ontogeny. It permits genetic variation to accumulate in a population over time, without being perceived by natural selection. Under environmental stress, the canalization may break down. The epigenetic system reacts by exposing the phenotypic variation and consequently provides natural selection with more material to operate upon. As a result, a new chreod may be entered by the morphogenetic system, provided it offers an adaptive solution to the new environmental regime. The opposite to developmental canalization is the well-known phenomenon of developmental plasticity, whereby a single genotype can produce highly variable phenotypes under variable environmental regimes. What is of major importance in the present context is that species are widely variable in the stabilizing power of their homeorhetic mechanisms, which ranges from strong canalization to considerable plasticity (Waddington, 1975; Stearns, 1982, and references therein).

One may then expect that species displaying very powerful homeorhetic mechanisms should undergo evolutionary punctuations rather than gradual changes, while the opposite should hold true for species with very large developmental plasticity (Stearns, 1980; 1982; Hoffman, 1981). In fact, the more a morphogenetic system is able to absorb environmental and genetic changes and to produce nonetheless the "normal" phenotype, the more stable should be the species morphology in evolutionary time. However, once the limits of homeorhesis are trespassed, the morphogenetic system can be expected to fall into another domain of attraction and hence to immediately enter a new chreod (Ho and Saunders, 1979). If the morphogenetic system of a species is, in turn, very sensitive in its response to environmental and genetic changes, the phenotypes can be expected to show a considerable range of variability from one microhabitat to another; in the absence of a well-defined chreod the morphogenetic system may shift gradually from one domain of attraction to another.

Briefly, the hypothesis can be proposed that punctuations in morphologic evolution occur in species with considerable amounts of developmental canalization, while gradual evolution is confined to species with relatively high developmental plasticity; morphologic stasis, however, may be brought about by canalization as well as by plasticity. This is to say that any difference in genetic and/or ecologic mechanisms may not be responsible for the apparently differential modes of morphologic evolution. Both the punctuated and the gradual modes of evolution might be accounted for by the accumulation of micromutations as well as by large-effect mutations in regulatory genes. Both of them might also occur in species with a very great or very small potential for tactical adjustments to new environmental challenges. The patterns of morphologic change through evolutionary time seem to present a continuous spectrum rather than a binary opposition. What is contended here is that this variation in pattern may be brought about by a variation in the action of the epigenetic system and in its homeorhetic properties that mask the underlying genetic processes, rather than by the latter themselves.

Morphologic Variation and the Mode of Morphologic Evolution

Were there no independent measure for the power of a homeorhetic mechanism, this hypothesis would be nothing but a mere, useless tautology, because powerful homeorhesis could be recognized only from the occurrence of evolutionary punctuations and its absence only from the occurrence of gradual changes through time. Fortunately, the pattern and range of intraspecific variability in the developmental pathway can be expected to correlate with homeorhetic properties of the morphogenetic system. Thus, the variability may provide a criterion, if only an approximate one, to assess the stabilizing effects of homeorhesis. Developmental plasticity should bring about very large and continuous morphologic variation in space. Developmental canalization, in turn, should be correlated with much smaller and discontinuous variation. To test the hypothesis put forth above, one must therefore analyze the relationship between the pattern of developmental variation and the mode of morphologic evolution. This relationship, however, may be highly variable among taxa and it may even vary among different morphologic characters in a single lineage. It is therefore crucial to look at single lineages that undergo both modes of morphologic evolution, or at least exhibit stasis as well as punctuations.

The best test case is certainly provided by the Cenozoic freshwater mollusks from East Africa investigated recently by Williamson (1981). The considered species are highly variable in autecology and reproductive strategy but nevertheless show a very similar evolutionary pattern. There are no gradual morphologic changes whatsoever, but only some intervals of morphologic stasis, punctuated by very rapid phenotypic transformations happening coevally in various lineages. Those rapid transformations are correlated with environmental events and occur in relatively large populations that exhibit a considerable, statistically significant increase in phenotypic variation. This indicates that a breakdown induced by environmental stress in the developmental canalization achieved by the species during the preceding periods of stasis was of paramount importance for the punctuations, rather than genetic drift or a founder effect.

Another case of this kind is provided by the Lower Oxfordian to Lower Kimmeridgian ammonite lineage of *Creniceras renggeri, C. lophotum*, and *C. dentatum* investigated by B. Ziegler (1957, 1959). This lineage was originally studied in south Germany, but the succession of forms can be traced much beyond that area, over the whole of Central Europe (Malinowska 1963). This assures us that, regardless of its disputable sexual dimorphic relationships to other oppeliid ammonites (Makowski, 1962; Palframan, 1966; B. Zieger, 1974), *Creniceras* can be considered to represent a homogeneous lineage. This lineage displays a slow and more or less gradual increase in shell size (from 15 mm on the average in the Lower Oxfordian *C. renggeri* to 25 mm on the average in the Lower Kimmeridgian *C. dentatum*) and a parallel gradual decrease in conspicuousness of the ventral denticles (from 20% of the whorl height in *C. renggeri* to 8% in *C. dentatum*). These changes are associated with a gradual loss of phenotypic variability, as demonstrated by the drop in coefficient of variation from 20 to 7 for the shell size, and from 42 to 28 for the height of the ventral denticles [this drop is statistically significant at the 0.05 level by Levene's test; cf. Van Valen (1978)]. Toward the end of its stratigraphic range, however, *C. dentatum* underwent a jerky increase in shell size, well above 30 mm on the average. Simultaneously, the coefficient of variation increased, again significantly (from 7 to 13). As indicated by the geologic and paleontogic evidence, the environmental conditions remained largely constant. This suggests that the increase in variation does not reflect increased environmental heterogeneity. It may rather show that the previously established developmental canalization with respect to shell size had to be broken down in order to permit the achievement of much larger size.

Some additional support for the correlation of morphologic punctuations with low phenotypic variability, as is claimed here, can be found
in the original prime examples for the model of punctuated equilibria. As
assessed by Gould (1969), the conservative stock of Bermudan land snails
Poecilozonites bermudensis zonatus, which underwent iterative punctua-
tional events leading to paedomorphic offshoots, exhibits strikingly low
morphologic variation when compared to its closely related congeners.
The phacopid trilobites investigated by Eldredge (1972) show generally
very little intrapopulation variation in number of lens files in the eye,
which is the diagnostic character. A notable exception, however, is provided by two highly variable samples considered to mark the onset of
new, punctuationally originated lineages. In addition to the ancestral and
descendant morphs, these two samples also include a number of other
morphologic varieties with a lens file reduced to a variable degree.

That the stasis may well be promoted also by developmental plasticity
and its correlate, wide morphologic variation, is clearly documented by
two demosponges, *Cylindrophyma milleporata* and *Cnemidiastrum stellatum*. They co-occur in the whole Upper Jurassic sponge megafacies of
Europe, even though *C. stellatum* displays enormous developmental plasticity which is apparently lacking in *C. milleporata* (Trammer, 1981).
Other examples of this kind include, for instance, the Cretaceous ammonite *Neogastroplites* from the North American Western Interior (Reeside and Cobban, 1960), and some long-ranging Cretaceous ammonites
from South Africa (*Plesiotexanites stangeri, Paratexanites pseudotricar-
inatum*) which considerably exceed in morphologic variation the gradually
evolving lineage of *Texanites soutoni* and *Submortoniceras* (Klinger and
Kennedy, 1980).

On the other hand, the best-documented examples of gradual morphologic evolution all show very large amounts of morphologic variation.
The coefficients of variation range from 15 up to 40, and even more. This
holds true for the foraminifers *Lepidolina, Lepidocyclina*, and *Globorotalia* investigated by Ozawa (1975), Van Vessem (1978), and Malmgren
and Kennett (1981), respectively; and also for the conodontophorids *Gondolella* (Dzik and Trammer, 1980), brachiopods *Eocoelia* (A. M. Ziegler,
1966), graptolites *Isograptus* (Cooper, 1973), and ammonites *Kosmoceras*
[Brinkmann's data reanalyzed by Raup and Crick (1981)].

Parallel with a gradual shift in morphology, the range of variation
very commonly undergoes a considerable decrease. This is well documented quantitatively, for instance, in the Cretaceous foraminifers *Bolivinoides* and *Conorotalites* (Bettenstaedt, 1962). A statistically significant decrease in variation, but apparently independent of any environmental
changes, has been found also in successive populations of a Devonian

spiriferid brachiopod (Flessa and Bray, 1977). A purely qualitative but nevertheless convincing example is provided by the Central European Tithonian or Volgian pseudovirgatitine and virgatitine ammonites (Kutek and Zeiss, 1974). This is certainly a monospecific but highly variable lineage rather than a group of parallel-evolving species as was originally described. In the *pseudoscythica* Zone the lineage is represented by *Pseudovirgatites* with all the morphologic varieties that have been thus far distinguished at the specific or subspecific level. What is important is that *Pseudovirgatites*, having somehow evolved from *Ilowaiskya* with which it partly co-occurs, displays very wide variation in density of the biplicate ribs on the inner whorls and in mode of ribbing on the middle whorls (biplicate, polygyrate, fasciculate, virgatotome). At the base of the overlying *scythicus* Zone the lineage is represented by *Zaraiskites* with its two morphologic varieties *Z. quenstedti* and *Z. scythicus*. Their individuals still vary in density of biplicate ribbing on the inner whorls, while the variation in mode of ribbing on the middle whorls has disappeared (there are only virgatotome and very few polygyrate ribs on each specimen). Higher up in the section the variety *quenstedti* disappears, leading to reduced variability also in density of biplicate ribbing on the inner whorls. The mode of derivation of this lineage from *Ilowaiskya* is unknown; nevertheless, the wide range of morphologic variation at the onset of the lineage and its subsequent gradual reduction suggest the role of developmental canalization in evolution.

MACROEVOLUTIONARY IMPLICATIONS

If punctuated equilibria were the only, or even a predominant, mode of evolution, macroevolutionary theory demanded by Gould (1980) would really be indispensable. Stasis and punctuations might be attributed to the very nature of the epigenetic system, because cybernetic analysis of the latter would predict its discontinuous behavior, i.e., its rapid transitions from one domain of attraction to another (Ho and Saunders, 1979; Alberch, 1982). Punctuated patterns of morphologic evolution might then be considered to reflect a jerkiness in macroevolutionary processes producing evolutionary novelties. Consequently, macroevolution would be irreducible to microevolutionary processes operating within the limits of a single domain of attraction available to each specific epigenetic system. Macroevolutionary patterns would appear explicable solely in morphologic or morphogenetic, but not in genetic terms.

However, this may not be the case. There are punctuated as well as gradual patterns of morphologic evolution. These differential modes of evolution may be attributed to a variation in action of the epigenetic system, as expressed by developmental canalization and plasticity. Canalization and plasticity, however, may well be considered adaptive and genetically determined (Waddington, 1957, 1975). The punctuated versus the gradual mode of morphologic evolution appears then as a mere epiphenomenon of the underlying genetic processes which are controlled by selection and constrained by a variety of historical, physical, and mechanical factors (Seilacher, 1970; R. D. K. Thomas, 1979). Macroevolution (= origin of evolutionary novelty) turns out to be beyond the reach of paleontologic analysis, but not inexplicable in genetic terms. In this context, the ecologic approach of Stearns (1982) and the experimental approach outlined by Levins (1968) might be promising.

Certainly, evolutionary novelties may arise very rapidly, due to single genetic or developmental alterations, if a functional threshold is being crossed (Jaanusson, 1981). This is well exemplified by the jaw of bolyerine snakes (Frazzetta, 1970), the uniserial rhabdosome of monograptid graptolites (Jaanusson, 1973), and the features characteristic of the plethodontid salamander *Aneides* (Larson et al., 1981), which have apparently arisen in single steps, without any adaptive intermediate stages. All the associated morphologic modifications, however, have occurred gradually. This mechanism resembles very closely those envisaged previously for cichlid fishes (Liem, 1974) and birds (Bock, 1979). The point being made here is that this mechanism is entirely consistent with the concept of developmental canalization of characters which is adaptively advantageous under a certain environmental regime, but is broken down under another regime which permits achievement of a key innovation.

Macroevolution then becomes reducible to microevolutionary processes. To this end, however, developmental biology and evolutionary ecology must be reconciled with each other and applied to explain how the observed key innovations did appear and why other conceivable innovations did not happen.

SUMMARY

1. The model of punctuated equilibria claims that a large majority of morphologic changes in evolution take place very rapidly in populations that are so small that their discovery in the fossil record is virtually impossible; whereas changes are negligible along those segments of a lineage which, by usual taxonomic standards, are recognized as single

chronospecies. The opposite model of phyletic gradualism claims that morphologic change is accomplished by a gradual shift of the range of intraspecific variation which must eventually exceed the limits set by usual taxonomic standards for a single chronospecies.

2. There is no way to decide which of the two ideal models is more commonly fitted by real evolutionary lineages. Available data on long-term morphologic stasis corroborate the commonness of punctuated equilibria, while a large set of examples can also be cited in support of phyletic gradualism. Therefore, the conservative proposition that both modes of morphologic evolution do occur is accepted.

3. The question of why some lineages undergo gradual morphologic evolution while others display punctuations has thus far been answered in terms of developmental homeostasis of individuals, genetic homeostasis of populations, and ecologic homeostasis of communities. These explanations, however, have failed.

4. The variation in patterns of morphologic evolution can be explained in terms of developmental canalization versus developmental plasticity. Punctuations are confined to lineages with much canalization, while gradual evolution occurs in lineages with much plasticity; stasis, however, may be promoted by canalization as well as by plasticity. There are in fact paleontologic data to corroborate this hypothesis.

5. Developmental canalization and plasticity may be adaptive and genetically determined. Punctuated equilibria cannot, therefore, serve as an argument for macroevolution being decoupled from microevolution.

Acknowledgments

Many of the ideas presented in this paper stem from discussion and/or correspondence with Stephen Gould, Adam Tomnicki, John Maynard Smith, David Raup, Wolf Reif, Adolf Seilacher, Steven Stearns, Roger Thomas, and Jerzy Trammer, whose interest and criticism are here gratefully acknowledged. Thanks are due to the Alexander von Humboldt-Stiftung, Bonn, for financial support.

REFERENCES

Alberch, P., 1982, Developmental constraints in evolutionary processes, in: *Evolution and Development* (J. T. Bonner, ed.), pp. 313–332, Springer, Heidelberg.

Albers, J., 1952, Taxonomie und Entwicklung einiger Arten von *Vaginulina* d'Orb. aus dem Barrême bei Hannover (Foram.), *Mitt. Geol. Staatsints. Hamburg* **21:**75–112.

Anderson, S., and Evensen, M. K., 1978, Randomness in allopatric speciation, *Syst. Zool.* **27:**421–430.

Bartenstein, H., and Bettenstaedt, F., 1962, Marine Unterkreide (Boreal und Tethys), in: *Leitfossilien der Mikropaläontologie* (Arbeitskreis deutsch. Mikropal., ed.), pp. 225–297, Berlin.

Bettenstaedt, F., 1952, Stratigraphisch wichtige Foraminiferen-Arten aus dem Barrême vorwiegend Nordwest-Deutschlands, *Senckenbergiana* **33:**263–295.

Bettenstaedt, F., 1958, Phylogenetische Beobachtungen in der Mikropaläontologie, *Palaeontol. Z.* **32:**115–140.

Bettenstaedt, F., 1962, Evolutionsvorgänge bei fossilen Foraminiferen, *Mitt. Geol. Staatsinst. Hamburg* **31:**385–460.

Bock, W. J., 1979, The synthetic explanation of macroevolutionary change—A reductionist approach, *Bull. Carnegie Mus. Nat. Hist.* **13:**20–69.

Boucot, A. J., 1978, Community evolution and rates of cladogenesis, *Evol. Biol.* **11:**545–655.

Brinkmann, R., 1929, Statistisch-biostratigraphische Untersuchungen an mitteljurassischen Ammoniten über Artbegriff und Stammesentwicklung, *Abh. Ges. Wiss. Göttingen, Math.-Phys. Kl. N.F.* **13:**1–249.

Bush, G. L., 1975, Modes of animal speciation, *Annu. Rev. Ecol. Syst.* **6:**339–364.

Carson, H. L., 1975, The genetics of speciation at the diploid level, *Am. Nat.* **109:**83–92.

Cisne, J. L., Chandlee, G. O., Rabe, B. D., and Cohen, J. A., 1980, Geographic variation and episodic evolution in an Ordovician trilobite, *Science* **209:**925–927.

Connell, J. H., 1978, Diversity in tropical rain forests and coral reefs, *Science* **199:**1302–1310.

Connell, J. H., 1980, Diversity and the coevolution of competitors, or the ghost of competition past, *Oikos* **35:**131–138.

Cooper, R. A., 1973, Taxonomy and evolution of *Isograptus* Moberg in Australasia, *Palaeontology* **16:**45–115.

Cracraft, J., and Eldredge, N. (eds.), 1979, *Phylogenetic Analysis and Paleontology*, Columbia University Press, New York.

Cronin, J. E., Boaz, N. T., Stringer, C. B., and Rak, Y., 1981, Tempo and mode in hominid evolution, *Nature* **292:**113–122.

Delson, E., 1981, Paleoanthropology: Pliocene and Pleistocene human evolution, *Paleobiology* **7:**298–305.

Dzik, J., and Trammer, J., 1980, Gradual evolution of conodontophorids in the Polish Triassic, *Acta Palaeont. Pol.* **25:**55–89.

Eldredge, N., 1971, The allopatric model and phylogeny in Paleozoic invertebrates, *Evolution* **25:**156–167.

Eldredge, N., 1972, Systematics and evolution of *Phacops rana* (Green, 1832) and *Phacops iowensis* Delo, 1935 (Trilobita) in the Middle Devonian of North America, *Bull. Am. Mus. Nat. Hist.* **47:**45–114.

Eldredge, N., 1979, Alternative approaches to evolutionary theory, *Bull. Carnegie Mus. Nat. Hist.* **13:**7–19.

Eldredge, N., and Cracraft, J., 1980, *Phylogeneric Patterns and the Evolutionary Process*, Columbia University Press, New York.

Eldredge, N., and Gould, S. J., 1972, Punctuated equilibria: An alternative to phyletic gradualism, in: *Models in Paleobiology* (T. J. M. Schopf, ed.), pp. 82–115, Freeman, San Francisco.

Eldredge, N., and Gould, S. J., 1977, Evolutionary models and biostratigraphic strategies, in: *Concepts and Methods of Biostratigraphy* (E. G. Kauffman and J. E. Hazel, eds.), pp. 25–40, Dowde, Hutchinson & Ross, Stroudsburg.

Eldredge, N., and Tattersall, I., 1975, Evolutionary models, phylogenetic reconstruction, and another look at hominid phylogeny, *Contrib. Primatol.* **5:**218–242.

Endler, J. A., 1977, *Geographic Variation, Speciation, and Clines*, Princeton University Press, Princeton.

Flessa, K. W., and Bray, R. G., 1977, On the measurement of size-independent morphological variability: An example using successive populations of a Devonian spiriferid brachiopod, *Paleobiology* **3:**350–359.

Frazzetta, T., 1970, From hopeful monsters to bolyerine snakes, *Am. Nat.* **104:**55–72.

Gecow, A., 1975, A cybernetical model of improving and its application to the evolution and ontogenesis description, Presented at 5th Int. Bio.-Math. Congress, Paris, September 1975.

Gilinsky, N. L., 1981, Stabilizing species selection in the Archaeogastropoda, *Paleobiology* **7:**316–331.

Gingerich, P. D., 1976, Paleontology and phylogeny: Patterns of evolution at the species level in early Tertiary mammals, *Am. J. Sci.* **276:**1–28.

Gingerich, P. D., 1977, Patterns of evolution in the mammalian fossil record, in: *Patterns of Evolution* (A. Hallam, ed.), pp. 469–500, Elsevier, Amsterdam.

Gould, S. J., 1969, An evolutionary microcosm: Pleistocene and Recent history of the land snail *P.* (*Poecilozonites*) in Bermuda, *Bull. Mus. Comp. Zool. Harv. Univ.* **138:**407–532.

Gould, S. J., 1980, Is a new and general theory of evolution emerging?, *Paleobiology* **6:**119–130.

Gould, S. J., 1982, Change in developmental timing as a mechanism of macroevolution, in: *Evolution and Development* (J. T. Bonner, ed.), pp. 333–346, Springer, Heidelberg.

Gould, S. J., and Eldredge, N., 1977, Punctuated equilibria: The tempo and mode of evolution reconsidered, *Paleobiology* **3:**115–151.

Gould, S. J., and Lewontin, R. C., 1979, The spandrels of San Marco and the Panglossian paradigm: A critique of the adaptationist program, *Proc. R. Soc. Lond. B* **205:**581–598.

Gould, S. J., Raup, D. M., Sepkoski, J. J., Schopf, T. J. M., and Simberloff, D. S., 1977, The shape of evolution: A comparison of real and random clades, *Paleobiology* **3:**23–40.

Grabert, B., 1959, Phylogenetische Untersuchungen an *Gaudryina* und *Spiroplectinata* (Foram.) besonders aus dem nordwestdeutschen Apt und Alb, *Abh. Senckenb. Naturforsch. Ges.* **498:**1–71.

Gradstein, F. M., 1974, Mediterranean Pliocene *Globorotalia*—A biometrical approach, *Utrecht Micropaleontal. Bull.* **7:**1–128.

Greenwood, P. H., 1974, The cichlid fishes of Lake Victoria, East Africa: The biology and evolution of a species flock, *Bull. Br. Mus.* (*Nat. Hist.*) *Zool.* (*Suppl.*) **6:**1–134.

Greenwood, P. H., 1979, Macroevolution—Myth or reality?, *Biol. J. Linn. Soc.* **12:**293–304.

Hallam, A., 1978, How rare is phyletic gradualism and what is its evolutionary significance? Evidence from Jurassic bivalves, *Paleobiology* **4:**16–25.

Harper, C. W., 1975, Origin of species in geologic time: Alternatives to the Eldredge–Gould model, *Science* **190:**47–48.

Harper, C. W., 1976, Stability of species in geologic time, *Science* **192:**269.

Hennig, W., 1966, *Phylogenetic Systematics*, University of Illinois Press, Urbana.

Hewitt, R. A., and Hurst, J. M., 1977, Size changes in Jurassic liparoceratid ammonites and their stratigraphical and ecological significance, *Lethaia* **10:**287–301.

Hiltermann, H., and Koch, W., 1950, Taxonomie und Vertikalverbreitung von *Bolivinoides*-Arten im Senon Nordwestdeutschlands, *Geol. Jahrb.* **64:**595–632.

Ho, M.-W., and Saunders, P. T., 1979, Beyond neo-Darwinism—An epigenetic approach to evolution, *J. Theor. Biol.* **78**:573–591.

Hoffman, A., 1978, Punctuated-equilibria evolutionary model and paleoecology, *Ann. Soc. Géol. Pologne* **48**:327–331.

Hoffman, A., 1979, Community paleoecology as an epiphenomenal science, *Paleobiology* **5**:357–379.

Hoffman, A., 1980, System-analytic conceptual framework for community paleoecology, *Ann. Soc. Geol. Pol.* **50**:161–172.

Hoffman, A., 1981, Biological controls of the punctuated versus gradual mode of species evolution, in: *International Symposium "Concept and Method in Paleontology,* (J. Martinell, ed.), pp. 57–63, Universidad de Barcelona, Barcelona.

Hoffman, A., and Szubzda-Studencka, B., 1982, Bivalve species duration and ecologic characteristics in the Badenian (Miocene) marine sandy facies of Poland, *Neues Jahrb. Geol. Palaeontol. Abh.* **163**:122–135.

Howarth, M. K., 1973, The stratigraphy and ammonite fauna of the Upper Liassic Grey Shales of the Yorkshire Coast, *Bull. Br. Mus. (Nat. Hist.) Geol.* **24**:237–277.

Jaanusson, V., 1973, Morphological discontinuities in the evolution of graptolite colonies, in: *Animal Colonies* (R. S. Boardman, A. H. Cheetham, and W. A. Oliver, eds.), pp. 515–521, Dowden, Hutchinson & Ross, Stroudsburg.

Jaanusson, V., 1981, Functional thresholds in evolutionary progress, *Lethaia* **14**:251–260.

Jablonski, D., 1980, Apparent versus real biotic effects of transgressions and regressions, *Paleobiology* **6**:397–407.

Kellogg, D. E., 1975, The rate of phyletic change in the evolution of *Pseudocubus vema* (Radiolaria), *Paleobiology* **1**:359–370.

Kellogg, D. E., and Hays, J. D., 1975, Microevolutionary patterns in Late Cenozoic Radiolaria, *Paleobiology* **1**:150–160.

Kennedy, W. J., 1977, Ammonite evolution, in: *Patterns of Evolution* (A. Hallam, ed.), pp. 251–304, Elsevier, Amsterdam.

Kirkpatrick, M., and Selander, R. K., 1979, Genetics of speciation in lake whitefishes in the Allegash Basin, *Evolution* **33**:478–485.

Klinger, H. C., and Kennedy, W. J., 1980, Cretaceous faunas from Zululand and Natal, South Africa. The ammonite subfamily Texanitinae Collignon, 1948, *Ann. S. Afr. Mus.* **80**:1–357.

Koch, C. F., 1980, Bivalve species duration, areal extent and population size in a Cretaceous sea, *Paleobiology* **6**:184–192.

Krassilov, V. A., 1978, Organic evolution and natural stratigraphic classification, *Lethaia* **11**:93–104.

Kutek, J., and Zeiss, A., 1974, Tithonian-Volgian ammonites from Brzostówka near Tomaszów Mazowiecki, *Acta Geol. Pol.* **24**:505–542.

Lande, R., 1980a, Genetic variation and genotypic evolution during allopatric speciation, *Am. Nat.* **116**:463–479.

Lande, R., 1980b, *Macroevolution*, by Steven M. Stanley (review), *Paleobiology* **6**:233–238.

Larson, A., Wake, D. B., Maxson, L. R., and Highton, R., 1981, A molecular phylogenetic perspective on the origin of morphological novelties in the salamanders of the tribe Plethodontini (Amphibia, Plethodontidae), *Evolution* **35**:405–422.

Lerner, I. M., 1954, *Genetic Homeostasis*, Wiley, New York.

Levins, R., 1968, *Evolution in Changing Environments*, Princeton University Press, Princeton.

Levinton, J. S., and Simon, C. M., 1980, A critique of the punctuated equilibria and implications for the detection of speciation in the fossil record, *Syst. Zool.* **29**:130–142.

Lewontin, R. C., 1978, Adaptation, *Sci. Am.* **239**:213–230.

Liem, K. F., 1974, Evolutionary strategies and morphological innovations: Cichlid pharyngeal jaws, *Syst. Zool.* **22**:425–441.

Maderson, P. F. A., Alberch, P., Gould, S. J., Goodwin, B. C., Hoffman, A., Murray, J. D., Raup, D. M., Ricqlès, A. de, Seilacher, A., Wagner, G. P., and Wake, D. B., 1982, Role of development in macroevolutionary change, in: *Evolution and Development* (J. T. Bonner, ed.), pp. 279–312, Springer, Heidelberg.

Mahé, J., and Devillers, C., 1981, Stabilité de l'éspèce et évolution: La théorie de l'équilibre intermittent ("punctuated equilibrium"), *Géobios* **14**:477–491.

Makowski, H., 1962, Problem of sexual dimorphism in ammonites, *Palaeontol. Pol.* **12**:1–92.

Malinowska, L., 1963, Stratygrafia oksfordu Jury Czestochowskiej na podstawie amonitów, *Pr. Inst. Geol.* **36**:1–165.

Malmgren, B. A., and Kennett, J. P., 1981, Phyletic gradualism in a Late Cenozoic planktonic foraminiferal lineage; DSDP Site 284, southwest Pacific, *Paleobiology* **7**:230–240.

Maynard Smith, J., 1966, Sympatric speciation, *Am. Nat.* **100**:637–650.

Maynard Smith, J., 1976a, What determines the rate of evolution, *Am. Nat.* **110**:331–338.

Maynard Smith, J., 1976b, A comment on the Red Queen, *Am. Nat.* **110**:325–330.

Maynard Smith, J., 1982, Current controversies in evolutionary theory, in: *Dimensions of Darwinism: Themes and Counterthemes in Twentieth Century Evolutionary Theory* (M. Grene, ed.), Cambridge University Press, Cambridge, in press.

Mayr, E., 1963, *Animal Species and Evolution*, Harvard University Press, Cambridge, Mass.

Meulenkamp, J. E., 1969, Stratigraphy of Neogene deposits in the Rethymnon Province, Crete, with special reference to the phylogeny of uniserial *Uvigerina* from the Mediterranean region, *Utrecht Micropaleontol. Bull.* **2**:1–168.

Michael, E., 1966, Die Evolution der Gavelinelliden (Foram.) in der nordwestdeutschen Unterkreide, *Senckenbergiana Lethaea* **47**:411–459.

Ozawa, T., 1975, Evolution of *Lepidolina multiseptata* (Permian foraminifer) in East Asia, *Mem. Fac. Sci., Kyushu Univ. D (Geol.)* **23**:117–164.

Palframan, D. F. B., 1966, Variation and ontogeny of some Oxfordian ammonites: *Taramelliceras richei* (de Loriol) and *Creniceras renggeri* (Oppel) from Woodham, Buckinghamshire, *Palaeontology* **9**:290–311.

Prothero, D. R., and Lazarus, D. B., 1980, Planktonic microfossils and the recognition of ancestors, *Syst. Zool.* **29**:119–129.

Rachootin, S., and Thompson, K. S., 1981, Epigenetics, paleontology, and evolution, in: *Evolution Today* (G. Scudder and J. J. Reveal, eds.), pp. 181–194, Hunt Institute for Botanical Documentation, Pittsburgh.

Raup, D. M., 1977, Stochastic models in evolutionary paleontology, in: *Patterns of Evolution* (A. Hallam, ed.), pp. 59–78, Elsevier, Amsterdam.

Raup, D. M., and Crick, R. E., 1981, Evolution of single characters in the Jurassic ammonite *Kosmoceras, Paleobiology* **7**:200–215.

Raup, D. M., Gould, S. J., Schopf, T. J. M., and Simberloff, D. S., 1973, Stochastic models of phylogeny and the evolution of diversity, *J. Geol.* **81**:525–542.

Reeside, J. B., and Cobban, W. A., 1960, *Studies of the Mowry Shale (Cretaceous) and Contemporary Formations in the United States and Canada*, U.S. Geological Survey Prof. Paper 151.

Reif, W.-E., 1982, Functional morphology on the Procrustean bed of the neutralism–selectionism debate. Notes on the Constructional Morphology approach, *Neues Jahrb. Geol. Palaeontol. Abh.* (in press).

Rensch, B., 1959, *Evolution Above the Species Level*, Columbia University Press, New York.

Rightmire, G. D., 1981, Patterns in the evolution of *Homo erectus, Paleobiology* 7:241–246.

Rosenzweig, M. L., 1978, Competitive speciation, *Biol. J. Linn. Soc.* 10:275–289.

Rosenzweig, M. L., and Taylor, J. A., 1980, Speciation and diversity in Ordovician invertebrates: Filling niches quickly and carefully, *Oikos* 35:236–243.

Schankler, D. M., 1981, Local extinction and ecological re-entry of early Eocene mammals, *Nature* 293:135–138.

Schmalhausen, I. I., 1949, *Factors of Evolution: The Theory of Stabilizing Selection*, Blakiston, Philadelphia.

Schopf, T. J. M., 1981a, Punctuated equilibrium, *Paleobiology* 7:156–166.

Schopf, T. J. M., 1981b, Evidence from findings of molecular biology with regard to the rapidity of genomic change: Implications for species durations, in: *Paleobotany, Paleoecology, and Evolution. Festschrift for Harlan P. Banks* (K. J. Niklas, ed.), Volume 1, pp. 135–192, Praeger, New York.

Seilacher, A., 1970, Arbeitskonzept zur Konstruktionsmorphologie, *Lethaia* 3:393–396.

Simpson, G. G., 1944, *Tempo and Mode in Evolution*, Columbia University Press, New York.

Simpson, G. G., 1953, *The Major Features of Evolution*, Columbia University Press, New York.

Slobodkin, L. B., and Rapoport, A., 1974, An optimal strategy of evolution, *Q. Rev. Biol.* 49:181–200.

Soulé, M. E., 1979, Heterozygosity and developmental stability. Another look, *Evolution* 33:396–401.

Stanley, S. M., 1975, A theory of evolution above the species level, *Proc. Natl. Acad. Sci. USA* 72:646–650.

Stanley, S. M., 1979, *Macroevolution—Pattern and Process*, Freeman, San Francisco.

Stanley, S. M., Addicott, W. O., and Chinzei, K., 1980, Lyellian curves in paleontology: Possibilities and limitations, *Geology* 8:422–426.

Stearns, S. C., 1980, A new view of life-history evolution, *Oikos* 35:266–281.

Stearns, S. C., 1982, The role of development in the evolution of life histories, in: *Evolution and Development* (J. T. Bonner, ed.), pp. 237–258, Springer, Heidelberg.

Stenseth, N. C., 1979, Where have all the species gone? On the nature of extinction and the Red Queen hypothesis, *Oikos* 33:196–227.

Sylvester-Bradley, P. C., 1977, Biostratigraphical tests of evolutionary theory, in: *Concepts and Methods of Biostratigraphy* (E. G. Kauffman and J. E. Hazel, eds.), pp. 41–64, Dowden, Hutchinson & Ross, Stroudsburg.

Tanabe, K., 1975, Functional morphology of *Otoscaphites puerculus* (Jimbo), and Upper Cretaceous ammonite, *Trans. Proc. Palaeontol. Soc. Jpn N.S.* 99:109–132.

Tanabe, K., 1977, Functional evolution of *Otoscaphites puerculus* (Jimbo) and *Scaphites planus* (Yabe), Upper Cretaceous ammonites, *Mem. Fac. Sci. Kyushu Univ. D (Geol.)* 23:367–407.

Templeton, A. R., 1980, The theory of speciation via the founder principle, *Genetics* 94:1011–1038.

Thoday, J. M., 1972, Disruptive selection, *Proc. Roy. Soc. Lond. B* 182:109–143.

Thomas, E., 1980, Details of *Uvigerina* development in the Cretan Mio-Pliocene, *Utrecht Micropaleontol. Bull.* 23:1–167.

Thomas, R. D. K., 1979, Morphology, constructional, in: *Encyclopedia of Paleontology* (R. W. Fairbridge and D. Jablonski, eds.), pp. 482–487, Dowden, Hutchinson & Ross, Stroudsburg.

Thompson, V., 1976, Does sex accelerate evolution?, *Evol. Theory* 1:131–156.

Trammer, J., 1981, Morphological variation and relative growth in two Jurassic demosponges, *Neues Jahrb. Geol. Palaeontol.* Monatsh. 1981:54–64.

Van Gorsel, J. T., 1975, Evolutionary trends and stratigraphic significance of the Late Cretaceous *Helicorbitoides–Lepidorbitoides* lineage, *Utrecht Micropaleontol. Bull.* **12**:1–100.

Van Valen, L., 1978, The statistics of variation, *Evol. Theory* **4**:33–43.

Van Vessem, E. J., 1978, Study of Lepidocyclinidae from South-East Asia, particularly from Java and Borneo, *Utrecht Micropaleontol. Bull.* **19**:1–163.

Vrba, E. S., 1980, Evolution, species, and fossils: How does life evolve?, *S. Afr. J. Sci.* **76**:61–84.

Waddington, C. H., 1957, *The Strategy of the Genes*, Allen & Unwin, London.

Waddington, C. H., 1975, *The Evolution of an Evolutionist*, Cornell University Press, Ithaca.

West, R. M., 1979, Apparent prolonged evolutionary stasis in the primitive Eocene hoofed mammal *Hyopsodus*, *Paleobiology* **5**:252–260.

White, M. J. D., 1978, *Modes of Speciation*, Freeman, San Francisco.

Williams, G. C., 1975, *Sex and Evolution*, Princeton University Press, Princeton.

Williamson, P. G., 1981, Palaeontological documentation of speciation in Cenozoic molluscs from Turkana Basin, *Nature* **293**:437–443.

Willmann, R., 1981, Evolution, Systematik und stratigraphische Bedeutung der neogenen Süsswassergastropoden von Rhodos und Kos/Ägäis, *Palaeontogr. Am. A* **174**:10–235.

Wilson, D. S., 1980, *The Natural Selection of Populations and Communities*, Benjamin/Cummings, Menlo Park.

Zedler, B., 1960, Mikropaläontologische Untersuchungen in den Unterkreide-Aufschlüssen Moorberg und Stöcken bei Hannover, *Ber. Naturhist. Ges. Hannover* **104**:25–45.

Zedler, B., 1961, Stratigraphische Verbreitung und Phylogenie von Foraminiferen des nordwestdeutschen Oberhauterive, *Paläontol. Z.* **35**:28–61.

Ziegler, A. M., 1966, The Silurian brachiopod *Eocoelia hemisphaerica* (J. de C. Sowerby) and related species, *Palaeontology* **9**:523–543.

Ziegler, B., 1957, *Creniceras dentatum* (Ammonitacea) im Mittel-Malm Südwestdeutschlands, *Neues Jahrb. Geol. Palaeontol. Monatsh.* **1956**:553–575.

Ziegler, B., 1959, Evolution in Upper Jurassic ammonites, *Evolution* **13**:229–235.

Ziegler, B., 1974, Über Dimorphismus und Verwandtschaftsbeziehungen bei "Oppelien" des oberen Juras (Ammonoidea: Haplocerataceae), *Stuttg. Beitr. Naturk. B (Geol. Palaeontol.)* **11**:1–42.

Index